QUANTITATIVE ETHOLOGY

QUANTITATIVE ETHOLOGY

Edited by

PATRICK W. COLGAN
Queen's University
Kingston, Ontario

A WILEY-INTERSCIENCE PUBLICATION

JOHN WILEY & SONS, New York • Chichester • Brisbane • Toronto

Library of Congress Cataloging in Publication Data

Main entry under title:
Quantitative ethology.

"A Wiley-Interscience publication."
Bibliography: p.
Includes indexes.
1. Animals, Habits and behavior of—Mathematical models. I. Colgan, Patrick W.
QL751.65.M3Q36 591.5 78-999
ISBN 0-471-02236-5

Printed in the United States of America

10 9 8 7 6 5 4 3 2 1

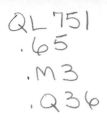

Contributors

PATRICK W. COLGAN. Department of Biology, Queen's University, Kingston, Ontario

VICTOR J. DE GHETT. Department of Psychology, State University College, Potsdam, New York

ROBERT M. FAGEN. Department of Ecology, Ethology, and Evolution, University of Illinois, Urbana, Illinois

DENNIS F. FREY. Biological Sciences Department, California State Polytechnic University, San Luis Obispo, California

GEORGE S. LOSEY JR. Department of Zoology, University of Hawaii, Honolulu, Hawaii

RICHARD A. PIMENTEL. Biological Sciences Department, California State Polytechnic University, San Luis Obispo, California

PETER J. B. SLATER. School of Biological Sciences, University of Sussex, Brighton, England

J. TERRY SMITH. Department of Mathematics, Queen's University, Kingston, Ontario

IAN SPENCE. Department of Psychology, University of Western Ontario, London, Ontario

B. DENNIS SUSTARE. Department of Biology, Clarkson College, Potsdam, New York

DONALD Y. YOUNG. Department of Ecology, Ethology, and Evolution, University of Illinois, Urbana, Illinois

Foreword

Ethology is a science based on observation, and can be fairly said to have begun with the biologist's curiosity about what animals do. It was a step of monumental importance to apply the accepted methods of observation and objective description, so long used in anatomy and physiology, to the seemingly ephemeral world of animal behavior. To consider the behavior of animals to be a proper study in biology, and behavior to be subject to the same forces of natural selection that shaped the anatomy and physiology of animals, was to say that the traditional tools of biology could also be turned with profit to the study of behavior. This approach was practiced by Charles Darwin himself in *The Expression of the Emotions in Man and Animals* and later reintroduced and promoted in neo-Darwinian terms by Konrad Lorenz. It has had an increasing impact on the way in which we view the animal world and ourselves up to the present day, when finally the relationship between animal and human behavior is no longer a matter of supposition.

Like other branches of biology, ethology began with a natural history stage in which biologists were fascinated with a suddenly discovered world of marvelous behavioral adaptations, differences, and similarities. Laws and principles emerged and experimental methods were introduced, leading to a growing sophistication of behavioral inquiry in the past two decades. Ethology is, however, still only on the threshold of utilization of quantitative methods in behavior analysis. This delay has resulted partly from a lack of familiarity with appropriate statistics among ethologists and partly from a lack of appropriate statistical methods for handling data of the type that observational behavior studies usually generate.

The relative frequency of papers employing quantitative methods in data analysis is increasing rapidly in the behavioral sciences, and it is essential that the ethologist keep abreast of advances in quantitative methodology appropriate to his or her discipline. This can only be accomplished by interaction with quantitative ethologists who can present methods in a form comprehensible to, and applicable by, scientists with a less quantitative background. We are today beginning a phase of growth in the depth of our understanding of behavioral phenomena. By applying, as a normal rather than exceptional practice, more sophisticated statistical analyses to behavioral data, including multivariate methods that allow a holistic approach to complex phenomena, our progress can be greatly enhanced. It is just at this time that it is so essential for those ethologists gifted in quantitative manipulation to communicate to ethologists at large the methods that they are developing. The horizons of behavioral inquiry may thus expand and all behavioral scientists may profit from the new possibilities that such methods suggest for probing into the deep structure of animal and human behavior.

It was for these reasons that I organized a symposium on "Quantitative Methods in Ethology" at the 1975 annual meeting of the Animal Behavior Society, from which grew the idea for the present book, under the able editorship of Patrick Colgan. *Quantitative Ethology* is only a start in this direction, but a very important start, and one which will capture the attention of all ethologists. It marks the attainment of an adolescent stage of development in our field, one which is welcomed and promises healthy growth to maturity. Like adolescence, however, it will have its special problems. Overzealous practitioners may find themselves seeking change of method for its own sake. When analysis becomes an end in itself, it is all too easy to forget that it should be related to biologically meaningful questions. Nonetheless, advances in ethology will surely depend on growth and dissemination of quantitative methodology and thinking, and the present volume is an important step in this most welcome direction.

D. W. DUNHAM
Department of Zoology
University of Toronto

January 1978

Acknowledgments

I appreciate the constructive criticism of colleagues and students, especially in the Department of Mathematics of Queen's University, and the support of the National Research Council of Canada (grant A6678).

Victor De Ghett thanks A. T. Yates, now at the Educational Testing Service in Princeton University, for first turning him in the direction of hierarchical cluster analysis several years ago when both were graduate students. The computer programs that produced the dendograms in Chapter 5 were written by him at that time.

Robert Fagen is grateful to E. O. Wilson for suggesting development and application of the methods reported in Chapter 2, to R. N. Goldman for assistance, including advice on negative binomial curve fitting, and to Drs. Wilson, M. L. Corn, and M. Bekoff, who kindly permitted analysis of their unpublished data, and to Balliere Tindall, publishers of *Animal Behaviour*, for their kind permission to use material in Chapter 2 from an article (Fagen and Goldman 1977) which originally appeared in that journal. Robert Fagen and Donald Young thank M. Morse, P. J. B. Slater, W. van der Kloot, and Brill Publishers P. R. Wiepkema, for permission to use their data in the temporal pattern analyses of Section 2 of Chapter 4, and J. R. Baylis for permission to reproduce Table 3.

Peter Slater acknowledges a number of colleagues for beneficial discussions and, in particular, thanks Professor R. J. Andrew, Dr. D. Clayton, Dr. P. Clifton, Dr. T. H. Clutton-Brock, and Dr. J. H. Mackintosh for their comments.

J. Terry Smith and I thank R. Haven Wiley of the University of North Carolina for data used as an example in Chapter 6, and Eric Moore of Queen's University for critical comments on the manuscript. The computations for that chapter were supported by National Research Council grant A8787 to J. T. Smith.

Ian Spence is grateful to E. H. Miller of Dalhousie University for initially proposing that he write Chapter 7 and also for providing him with many relevant ethological references and data sets; to Christine Robertson of the University of Toronto for kindly sending a prepublication copy of her INDSCAL study of Siamese fighting fish as an example; to Richard Harshman for many valuable comments and suggestions; and to the National Research Council of Canada for their continued support (grant A8351).

Dennis Sustare thanks Cheryle Hughes for the drawing of all figures, and Edward H. Burtt Jr. and Jack P. Hailman for their suggestions and critical readings of Chapter 10.

All contributors are grateful to David Dunham, who organized a symposium on quantitative methods in ethology at the 1975 meeting of the Animal Behavior Society out of which this volume has grown, and to the patience of the publisher and the forbearance of their families.

PATRICK W. COLGAN

Kingston, Ontario
January 1978

Contents

QUANTITATIVE ETHOLOGY

PROLOGUE

The rapid growth of the application of advanced quantitative techniques to ethological problems reflects an increasing realization among researchers that sophisticated approaches to the study of behavior are required to achieve a full understanding of the phenomena involved. These techniques, and others yet to be developed, will doubtless become tools of major importance to practicing ethologists. Unfortunately, descriptions of the derivations, methods of applications, and uses and limitations of these techniques lie scattered throughout an extremely diverse literature. It is the purpose of this book to draw together between two covers a comprehensive review of this information. This volume is intended to serve as a reference book enabling the working ethologist to apply the various techniques in an appropriate and enlightened manner to the problems of concern.

The increasing application of mathematics of ethology is only one portion of the broad quantification of organismic biology. The appearance of mathematically oriented volumes in the neighboring disciplines of population genetics (Crow and Kimura 1970), population and community ecology (May 1973, Pielou 1969, Poole 1974), biogeography (MacArthur and Wilson 1967), and systematics (Sneath and Sokal 1973) testifies to the breadth of the trend. And the formal equivalences discovered to be inherent in phenomena in organismic biology and the social sciences, two traditional disciplines that have unnecessarily and unfortunately evolved in relative isolation, assist the extension of Huxley's (1942) "modern synthesis" to the entire life sciences (cf. Wilson 1975). The progress of this development is occasionally and temporarily delayed by certain facts, individuals, or events. Mathematical notation is often incomprehensible to the uninitiated. Perhaps partly because of this, some scientists (often older than most) show a cataleptic abhorrence of anything quantitative. The inappropriate use of a particular technique or interpretation of its results may bring disrepute to the technique and obscure it merits when properly employed. But the growing appreciation of the necessity for, and benefits derived from, more objective and quantitative approaches provide momentum for a fuller understanding of these approaches and hence the problems under study.

Historically, beyond arithmetic, univariate statistics is the mathematics that has been most used in behavioral biology, often in the hope that tortured data will indeed confess anything. The collection of quantitative data, the extraction of statistics from these data (where by "statistic" is meant "any function of the data alone"), and the inferential use of these statistics to make general statements about a population or populations is now well established throughout whole-animal biology. The four major kinds of statements made are:

1: It is likely that these data could be a sample from a such-and-such distribution. For example, students of orientation behavior are interested in random distributions around a circle.

2: It is likely that the value of this population parameter lies within a certain distance of some specified value, and the most likely single value is such-and-such.

3: It is likely that the values of this parameter in this set of populations are not all equal.

4: It is likely that there exist correlations, including auto- and cross-correlations, among the variables of this population (and hence the value of one or more variables can be predicted by knowledge of the values of other variables). (The study of response dependencies and sequences (e.g. Delius 1969, Nelson 1964, Slater 1973) and that of behavioral rhythms involve this statement.)

The statistics involved may be measures of information, central tendency, dispersion, and/or other measures of distribution.

For each of these statements, the problems of the data to be collected, the scale of measurement to be used, and the assumptions of the inference procedure to be made must be faced. With regard to the problem of valid assumptions, one must frequently make a choice. One may choose a more powerful technique derived from the theory of normal populations, whose strict assumptions, such as homoscedasticity, or equal variance of populations, may not be met, and hence one must rely on a certain robustness of the technique (i.e. ability to produce valid conclusions in the face of invalid initial assumptions). Or one may choose a less powerful, generally nonparametric, technique where one exists (and often one does not). These problems are still often neglected and statistics are still often used, like the lamppost by the drunk, more for support than illumination. Nevertheless, as noisier and more complex systems are studied, the inclusion of this process of collecting quantitative data, extracting statistics, and inferring population features as an essential aspect of acknowledged experimental protocol is a heartening development over the prior condition of qualitative, and often anecdotal, data, and conclusions based on intuition, good or bad.

The present book is intended to further the application of quantitative techniques to ethological problems by making more accessible the major recent and advanced mathematical methods of relevance. The scope of this book makes the assembling of the requisite information by any one individual a formidable, if not impossible, task. For this reason, the book consists of contributions by several individuals, it is hoped the standard format of the chapters and the scrutiny of the editor will serve to create sufficient eveness of style that the volume can indeed function as a reference work. In keeping with the aim of preparing a book for the practicing researcher, the contributors are themselves all working ethologists with particular familiarity with the techniques with which they are dealing.

The user of this book is assumed to be familiar with the mathematical topics encountered in a contemporary undergraduate program in the sciences: basic calculus, linear and matrix algebra, and univariate probability theory, and statistics. Review of these topics can be achieved by reading Thomas (1968), Painter and Yantis (1971), Hodges and Lehmann (1970), and Sokal and Rohlf (1969). It should also be noted that the important general problems of experimental design and of data management, including storage and retrieval, are not considered here. Sources of information for these problems are Borko (1962) and Cochran and Cox (1957).

Most chapters of the present book include four numbered sections following an introduction where the general function and history of the technique is described. Under Assumptions the assumptions for the appropriate use of the techniques are outlined. Under Method the mechanics of the computations are given. Under Examples several examples are discussed to illustrate use of the technique. And, finally, under Interpretation and Limitations problems associated with the interpretation of the results, the robustness of the technique, and any general limitations of the technique are examined. References to important works dealing with the theory and/or application of the technique and any existing relevant computer program packages are made in the appropriate places.

The first four chapters of this volume deal with techniques for analyzing four fundamental and different aspects of the structure of behavior. Chapter 1 considers the basic issue of data collection. Chapter 2 treats the topic of estimating the total number of response types in, and the overall diversity of, the behavioral repertory of an animal. In Chapter 3 the measurement of the amount of information transferred in communicative acts is discussed. And in Chapter 4 the problems associated with the quantitative study of durations, intervals, latencies, and sequences of stimuli and responses are examined.

Chapters 5 to 9 deal with the applications of multivariate techniques to ethology. Multivariate data are data that encode more than one dimension of a stimulus or response. For instance, an ethologist may record the direction, duration, and intensity of a limb movement. The many responses of a behavioral system, or the many dimensions of a given stimulus or response, pose essentially multivariate problems. Behaviorists have long realized this. For instance, in his textbook on motivation Bindra (1959) pointed out that goal direction is a multidimensional concept. Paradoxically, multivariate techniques have until very recently been only rarely employed in ethology, perhaps because of lack of accessible information on their use. Yet the need for these techniques is ever clearer. For this reason multivariate techniques are prominently featured in this book.

Chapter 5 examines the use of cluster analysis to estimate the similarities of response types and so to group these types objectively. Chapter 6 describes techniques for analyzing multidimensional contingency tables, a generalization of common two-dimensional tables, using chi-square and likelihood-ratio test statistics. In Chapter 7 multidimensional scaling, a very recent and generalized method for searching for structure in data, is outlined.

Chapters 8 and 9 deal with major aspects of multivariate analysis based on multinormal populations. The structure of a single such population is considered in Chapter 8 through principal components analysis and factor analysis, two allied techniques for reducing the dimensionality of multivariate data to a few important dimensions. Differences among populations are examined in Chapter 9 using multivariate analysis of variance, a generalization of familiar univariate variance analysis, and discriminant analysis, a method of discovering which variables are most important for discriminating among the populations.

The remainder of the book examines more general aspects of the quantitative study of behavior. Chapter 10 presents the types of diagrams available to ethologists and the relations among these types. In Chapter 11 the role of formal simulation models as aids in quantitative studies and hypothesis testing is discussed. Finally, in the Epilogue the impact on ethological theory of the use of the techniques outlined in the preceeding chapters is discussed.

CHAPTER

1

Data Collection

PETER J. B. SLATER
University of Sussex

One of the most notable features of ethological work is its stress on the idea that reasonable hypotheses can only be formulated after a period of purely observational work. This chapter is concerned with preliminaries even more basic: problems that must be considered and choices that must be made even before a piece of ethological research, whether observational or experimental, gets underway. Although the questions to be dealt with are fundamental ones, they have not, until recently, attracted much attention in the ethological literature. Perhaps as a result of this ethologists, in their haste to come to grips with their animals or with the questions that interest them, have often tended to consider only cursorily the issues I discuss. These issues fall under three headings.

First comes the question, raised in the Prologue, of whether or not quantification is necessary and how it may be achieved. Given the great difficulty of measuring behavior, is it worth the trouble? Should one not just settle, as many ethologists have done in the past, for qualitative descriptions and impressions? Second, I discuss the problem of splitting up behavior into categories. This is an essential precursor to any study and one that must be carried out with care and thoroughness, for only a well-defined classification that splits up behavior in a way similar to that in which it is parcelled by the animal is likely to yield interesting and useful results. Though ethologists have always realized the need for classification, they have not always recognized its difficulty. The third problem concerns the form in which data are to be collected, and here the relative merits of a number of possibilities must be considered. The choice is an important one, for some methods will be more appropriate, practicable, and illuminating than others depending on the particular topic under study.

Research in ethology varies along a number of continua: from protozoa to primates, from individual to social, from causation to function, from laboratory to field, from descriptive to experimental, from qualitative to quantitative. Where, on each of these dimensions, a particular research project should lie depends to a large extent on the interests of the ethologist and the problem that has caught his attention. But for some of them more general rules, which apply almost regardless of the specific topic, may be outlined. The most relevant of these to our present purpose is the qualitative-quantitative continuum and I discuss this, and its relationship to some of these other dimensions, in Section 1.

1. WHY QUANTIFY?

It would not be necessary to provide a justification for quantitative treatment in many other branches of science. However, ethology differs from most other scientific fields in two respects. First, it is a young science where much of the work involves a description of the phenomena that must be explained rather than experimentation to test hypotheses. Although it is argued here that quantification is as useful at this stage as it is later on, it is perhaps less imperative.

A second way in which ethology is different is that the ways of quantifying behavior are by no means obvious. A botanist comparing two species of plant may start straight away to measure height, leaf area, and so forth. but the ethologist, and particularly the field-worker, may be confronted by animals that he cannot identify individually, that he may not always be able to find, that have a wide variety of behavior patterns many of which merge into one another, and that may behave differently depending on a host of environmental, experiential, and motivational factors. He has therefore an excuse for offering only a qualitative description of what he sees. After all, he may argue, any figures that he collected would be imprecise and any statistical treatment that he attempted would probably, to the purist, be invalid. Why then should he bother?

It may be reasonable to argue that qualitative description is sufficient at the outset to define the different behavior patterns that the animal under study may show. For many species an adequate ethogram consists simply of a list of possible behaviors and a description of their form. Even here, however, quantification may be required in order to define a realistic borderline between behavior patterns, unless the types of act shown are obviously discrete. Quantitative treatment is necessary here because one is trying to decide whether there is a difference between two rather similar things, and it becomes increasingly so at later stages in research because such comparisons are the essence of the scientific approach to problems. The research worker repeatedly asks such questions as: "Do the controls behave differently from the experimentals?" "Do males fight more than females?" "Does the sequence of acts differ from a random sequence?" Without quantitative data one may be able to answer some of these questions, but one can only do so with assurance where the differences are obvious. It is clearly preferable to collect quantitative data, however difficult this may be, to present them in figures and tables, and to test them statistically. One can then look with confidence at marginal effects and present the results in a way that all can judge.

Perhaps the strongest reason for quantifying behavior is, paradoxically, one that makes its quantification most difficult. This is the problem of individual differences. While environmental factors during development do affect physiological and anatomical features (e.g. better nourished people grow taller), these effects are slight compared with the profound effects of experience on behavior. Furthermore, differences in motivational state at the time of testing may cause individuals to behave differently from each other. So, even with a "homogeneous population" and a "standardized test," a wide spread of readings is likely to be obtained. When one wants to ask if one group is different from another, there is likely to be a considerable overlap between the two sets of figures. Few behavioral results will therefore come into the "obvious" category mentioned earlier.

Statistical tests would never have been developed if all attempts to measure the same variable yielded the same result; they would simply be unnecessary. As variability in behavior tends to be great, statistical treatment here is even more important than in many other areas of biology. Variability may even be so great as to hide a true effect in any but a very large sample of animals. Another approach that may therefore be useful is to try to cut down on individual differences to the greatest possible extent. The laboratory worker often attempts to do this by using inbred strains (to minimize genetic variation), and by ensuring that both the rearing and testing conditions of his animals are as similar as possible. He can thereby hope to obtain results that are as consistent between individuals as possible so that, when he compares two groups, he has a better chance of detecting differences. This approach to behavior does, however, have its disadvantages. One is that he has removed the animal from the environment to which evolution has adapted it, and thus the relevance of laboratory findings may be open to question. Another is that the extent to which he can generalize from his results is limited: perhaps they would have been different had he used another strain or different rearing conditions. It is safer to generalize from such results as can be obtained from a more heterogeneous population. To some extent, therefore, consistency has been bought at the cost of generalizability.

Problems for the field worker come at the opposite end of the spectrum. As his animals are in their natural environment, he might perhaps expect some consistency because their development will have taken place under the rearing conditions to which selection has adapted them. Unfortunately this possible advantage is likely to be offset by the difficulty of controlling variables during data collection. Because of this, both individual differences and differences between watches are likely to be great. One animal may be older than another; it may be raining during one watch and not during another; some observation periods may occur when the animals have just woken up, others when they have recently seen a predator. Thus here consistency between observations will tend to be low, making it hard to reach conclusions. On the other hand, generalizability of such results as can be obtained will be high. The field worker can also take comfort from the fact that, while the presence of the observer is unnatural, there is much less danger of studying artifacts of the procedure he has employed in this situation than there is in the laboratory.

There are thus advantages and disadvantages to both laboratory and field studies, and these should be considered by anyone planning a behavioral project. The most fruitful approach will depend on the particular aspect of behavior under study: it would clearly be of limited use to study the adaptiveness of behavior in the laboratory, but equally there are difficulties in attempting detailed experiments on the causation of behavior in the field.

Having decided upon these general aspects of procedure and, it is hoped, having committed himself to collecting quantititive data as far as it is possible, the ethologist is now all set to begin. The chances are that he will spend a few minutes designing a check sheet for recording his data and a few more deciding on which behavior patterns to include. It is only when he has collected a large amount of data that he will begin to regret that he did not spend longer on these apparently trivial tasks. Decisions taken at this early stage can make or mar the usefulness of the data that are gathered, and yet these decisions are often taken lightly for the simple reason that the person carrying out the research did not realize that he had a choice of options open to him. In the rest of this chapter I consider the two main problems involved: first, that of splitting behavior into categories and, second, that of choosing a useful way to record it.

2. CATEGORIES OF BEHAVIOR

Whether or not the aim of an ethological study is to achieve an ethogram, all such studies must start by splitting behavior into categories, even if only to enable the observer to decide which he should include and which ignore. If all behavior patterns were species-typical and invariant fixed action patterns, this task would be an easy one. But, particularly in higher animals, many behaviors vary considerably in form and intensity, making classification more difficult. Several recent studies have shown fixed action patterns to be not quite so fixed as was previously supposed. It is worth considering some of this work briefly as a basis for discussing the problems of classifying behavior.

Variable Action Patterns

One reason why ethologists have often tended to stress the fixity of the behavior patterns they studied is that much early work concentrated on signal movements, especially in birds and fishes. While some such signals are graded in intensity, many of them are discrete, all or nothing, events of striking stereotypy. Two reasons have been suggested for this. Morris (1957) argued that their "typical intensity" makes them less ambiguous, while Maynard Smith (1972) has suggested that signals may become fixed because it is advantageous to the individual not to betray his motivational state. Be this as it may, the signal movements of lower vertebrates are something of a special case: the ethologist who studies monkeys, or whose interest is in feeding or grooming behavior, will not find it so easy to define his units.

Unlike many ethological concepts, that of the fixed action pattern has been subjected to rather little criticism. Just as the assumption that behavior patterns were inherited led few people to study their ontogeny, so little attempt was made to measure the variability of fixed action patterns when their invariance was taken as the rule. Some recent papers have, however, adopted a more questioning attitude (e.g. Barlow 1968; Schleidt 1974), and the variability of a number of fixed action patterns has now been examined. Unfortunately almost all these studies deal with displays and analyze variation in timing rather than in any other possible parameter. There is certainly a need for further work in this area, especially given that variation is the raw material on which natural selection acts.

The most convenient measure of variability is the coefficient of variation (CV) which is, quite simply, the standard deviation of a measure expressed as a percentage of its mean. Some displays are clearly very fixed when analyzed in this way, at least within a single individual. Wiley (1973) measured the time between the first swish and the first snap in 45 successive repetitions of the strut display of a male sage grouse. The average interval was 1.55 seconds and the standard deviation was 0.011 seconds giving a coefficient of variation of 0.7%. Measurement of the durations of elements in the signature bob display of individual *Anolis* lizards yielded CVs ranging from 2.5 to 9.5%, much of the small amount of variation present probably being attributable to the equipment used (Stamps and Barlow 1973). The intervals between successive notes in two song phrases of a single male chaffinch, measured on over 50 repetitions, yielded CVs of 1.5% and 3.2%, the variation again probably due largely to measurement error (Slater unpublished). On the other hand more sizeable CVs were obtained when the number of notes within a phrase was analyzed using five different phrases from the same individual. These CVs ranged from 9.6% to 14.4%.

Such within-individual measurements of coefficients of variation can be obtained because behavior patterns, unlike morphological features, are repeated. Variation between individuals tends to be greater, though Wiley (1973) found that it was still very small in the sage grouse strut display: he could not put a figure on it because this would have involved comparison between reels of film the speed of which was not precisely calibrated. As Barlow (1968) points out, the displays of the goldeneye studied by Dane, Walcott, and Drury (1959), and often quoted as essentially invariant, mostly show CVs of 10 to 20%, though some are less and some more. Figures taken from Dane and van der Kloot (1964) for the highly stereotyped head-throw display yield a CV of 7.2%. Three measures of duration taken from the signature bob of *Anolis* by Stamps and Barlow (1973) gave CVs ranging from 11.5 to 39% while those for three morphological features were 10 to 12%. Thus it seems that the variability of displays itself varies considerably among them, being sometimes greater, sometimes less, than would be expected from other biological measures. Coefficients of variation provide a useful guide to stereotypy and may, when applied to signals, suggest the information that is being transmitted and the parameters in which it is encoded. In *Anolis*, for example, the signature bob has all the features necessary for individual recognition: stereotypy within an individual and variation between them.

Fixed action patterns can vary in other respects as well as in timing. The study by Stamps and Barlow (1973), for example, showed that different component parts of the fanbob display in *Anolis* varied independently in intensity and that some might even be left out. While the signature bob component was stereotyped, introductory movements might or might not be present depending on the distance away of the intruder stimulating the display. Height, side-flattening, and dewlap extension developed to varying degrees depending, in the last two cases, on the aggressiveness of the animal. Similar findings resulted from a study of the song-spread display of carib grackles by Wiley (1975). Here beak elevation and wing elevation varied independently, the former being stimulated particularly by nearby males, the latter by females.

It is likely that close analysis of other fixed action patterns would reveal similar variations and there thus seems little point in casting behaviors such as these out of the category though they fail to fit in with many of the characteristics that fixed action patterns were originally supposed to possess (listed by Schleidt, 1974). Barlow (1968) is in favor of using the term modal action pattern instead, though few have followed his example. This expression is meant to embody the idea that, while not strictly fixed, these behavior patterns can be identified in a reliable statistical way.

Barlow's point is an important one when one's main interest is in choosing categories for data collection rather than with a particular theoretical framework. One is here more concerned with achieving an operational definition than a definition that is only complied with when precise conditions of ontogeny or causation are met. Whether or not one wants to call them fixed action patterns, the frequent occurrence of behavior patterns having the following characteristics is certainly helpful from the practical point of view:

1. *They are species-typical.* Some allowance must be made here for differences in age or sex. Only infants may cry or only males may sing but, within these classes of animal, such behavior patterns are typically found in all individuals.
2. *The component movements that make them up occur together, simultaneously or sequentially, with a high degree of predictability.* The component parts may vary in intensity independently of each other, one may occasionally be missing or their order may sometimes change, but these variations are not so great as to make it useful to subdivide the behavior further.
3. *They are repeatedly recognizable.* When an animal performs a particular behavior pattern there is no doubt that it belongs to one category rather than another.

Behavior patterns that fulfill these conditions are comparatively easy to categorize, but by no means all of behavior slots into such a scheme. At an extreme a continuum of intermediates may exist between two rather dissimilar patterns making categorization particularly difficult. Again, a given pattern, while distinguishable from others, may vary so much that it is unreasonable to regard all occurrences of it as equivalent point events. Here it may be necessary to use a scale (of, for example, duration, frequency, or intensity) on which individual events are scored. For many studies, however, this would involve going into a greater degree of detail than is necessary to achieve the end in mind. The level of analysis required will also determine the fineness of the categorization called for: what amounts to "a high degree of predictability" (point 2 above) will differ from one study to another. For example, a detailed analysis of grooming may require a category called "scratching right side of head" with scales on which the speed and duration of individual occurrences can be rated. But in a study of how an animal allots its time between different activities at different times of day a grosser level of categorization, yielding a single category called "grooming," may be more practicable and illuminating.

The Choice of Categories

Given that behavior patterns do vary, in some cases considerably, and that studies differ in the level of analysis required, are there any general principles that should be adopted in categorizing behavior?

The position of the ethologist attempting to split up behavior is like that of the nineteenth century taxonomist confronted by a selection of skins on his museum bench and trying to decide which ones to allocate to which species. The taxonomist had to base his judgment on their form, so deciding whether there were enough points of similarity between two specimens to assign them the same name. With the advent of the biological species concept came a less arbitrary rule, based on whether or not two different forms interbreed, but prior to that there was (and still is in some cases) no better definition of a species than "that which a competent taxonomist would call such." This definition might well also apply to the ethologist assigning his behavior patterns to "species" and, as his most useful guide is usually the form of the actions which he observes, there is little substitute for painstaking description before he chooses the categories which he will use. In some studies categorization according to consequences is a possible alternative to that according to form (Hinde, 1970, p. 604) but, particularly at the outset, the ethologist must beware of massing behaviors because of presumed similarities of causation or function. The following suggestions are therefore primarily aimed at those intending to split up behavior according to its form.

1. *Behavior categories must be discrete.* Basically this means that the acts within a category must have clear points of similarity between them that they do not share with any acts outside that group. Where this is the case, it will be possible to assign a particular occurrence to the appropriate category more easily and objectively.

2. *Behavior categories must be homogeneous.* All the acts in a particular category should be similar in form so that there is as little danger as possible of having massed two rather different behaviors in the same category. Once two behaviors have been put in the same category, the observer has committed himself to regarding them as equivalent to each

other. If they are actually causally or functionally different from each other, his analysis will reveal a rather complicated picture of the circumstances in which they occur, which could have been avoided had he kept them separate in the first place.

3. *It is better to split than to lump.* If two behaviors are very similar, but objective criteria exist for separating them, it is better to put them in separate categories. This is simply because they may turn out to be more different from each other than the initial description of their form suggested. Behaviors that are lumped during data collection cannot be split during analysis; those that are split can always be lumped.

4. *Names of categories should avoid causal or functional implications.* Calling a greeting "friendly," a noise a "hunger call," or a facial expression a "smile," carries with it in each case certain implications that the observer may not wish to suggest or that may be found to be unsound by subsequent research. The use of names that simply describe the form of a movement or that, in the case of sounds, are onomatopoeic, will avoid this to some extent, though it may still require some self-discipline to avoid the tacit assumption that the "silent bared-teeth display" one has described is really a smile.

5. *The number of categories used must be manageable.* The earlier suggestion that acts should be split rather than lumped, may have left primate watchers picturing themselves seated in the bush with a keyboard that would do justice to a Chinese typewriter. There are dangers in using too many categories as well as too few. With too few, the observer is not making the best possible use of his time and his attention is liable to drift. With too many, his reaction time will be long and his recording accuracy reduced. This may to some extent be overcome by practice and by the choice of a suitable recording method (see next section), but for many studies it may be necessary either to ignore certain behavior patterns or to mass those that are most similar, or of peripheral interest to the study.

These remarks have largely been concerned with defining behavior patterns on the basis of their form but, as mentioned earlier, intensity is another feature that may vary. If this is not central to his interest, the investigator may wish to ignore it, especially if he already has to record a large number of different types of behavior. If, on the other hand, he is making a more specific study and has few categories to cope with, he will lose a lot of information if he simply records that a behavior occurred without any measure of its intensity.

As with distinguishing between behavior patterns of different form, it is essential that the intensity levels employed should be as discrete and objectively defined as possible. An act of behavior can be regarded as consisting of a combination of different muscular contractions. When performed at low intensity certain of these features may be poorly developed or absent. A possible way of assigning a particular occurrence to a level on an intensity scale is to rate it according to the presence or absence of these. Failing this, in a displaying mammal, for example, such features as the angle of the tail to the horizontal, whether or not the front paws are on the ground or whether the head is above or below the level of the shoulders may provide reasonably objective criteria. The most suitable ones to use will depend on experience: the observer must be confident that the features he is scoring do, in fact, provide a measure of intensity. This is most clearly so where more and more minor elements get added on to the same basic pattern in a predictable fashion, or where the behavior is goal-directed and follows a fairly set sequence (e.g. sniffing—mounting—intromission—ejaculation in a sexual encounter). Here a score may be assigned to the behavior depending on how fully developed it is and these scores may then be compared between individuals and situations.

Some final difficulties should be mentioned. Watching animals for the length of time required to gather adequate quantitative data can be tedious, especially if they are doing rather little. It is thus tempting to use automatic recording for behaviors, like feeding and locomotion, that are suited to it. Such methods can be very useful, but the investigator must realize that he, or the designer of his apparatus, has decided upon a category of behavior to measure that may be very different from the behavior that he would have used for direct observation. Automatically recorded "feeding," for instance, is frequently a measure of how often an animal puts its head in a hole, a behavior that need not necessarily involve ingestion. Automatic recording thus involves dangers of categorization which may, in some cases, override its usefulness.

"Locomotion" is a difficult category of behavior to split up, whether or not it is automatically recorded, but any simple measure of it must be recognized as heterogeneous, including as it does running, jumping, walking, rearing, and even flying. Furthermore, even if these different aspects can be recorded separately, each of them is likely to be complex both causally and functionally. Regarding exploration of the cage, moving to the food hopper, and running in a running-wheel as the same is unlikely to give a measure that provides any useful insight into the behavior of the animal under study. There are no simple solutions to this problem, but the important point is that models of behavior based on categories that are heterogeneous in form, function, or causation are less likely to be heuristic than those in which the categories are homogeneous. Results using blanket categories such as "locomotion" should therefore be treated with some reserve.

The choice of categories is not the simple task that it might, at first sight, appear to be. Even with extensive groundwork it may not be possible to be absolutely sure about the categories to choose: the ethologist must always cast his mind back and consider how his results would have been affected had he taken different decisions. As Fentress (1973) has succinctly put it, "Categories of behavior must be formed, but the investigator must not believe them!"

3. SAMPLING METHODS

There are several possible ways of sampling behavior, and it is important that the investigator should choose his method carefully, depending on the sort of analysis he wishes to carry out and the recording apparatus available to him. We will consider a selection of the more commonly used methods here. Altmann (1974) provides a fuller treatment, with particular emphasis on methods useful for the field-worker studying social behavior; many of the points raised here rely on her careful appraisal of the problems involved.

Before dealing with the methods themselves, we must consider briefly when samples should be taken. The informal observations that are used to decide upon categories of behavior can also be used to devise a timetable for more formal data collection. By carrying out these informal observations at varied times the observer will maximize his chances of seeing the complete behavioral repertoire of the species he is studying. He may discover, for example, that his animals usually sleep in the afternoon and that observations on their feeding behavior at this time of day are therefore unlikely to yield adequate data. By contrast, when it comes to collecting quantitative data, regularly scheduled observation sessions of standard duration are usually preferable. In some studies this will be impracticable as, for example, when a field-worker wants to record the effect of rain on his animals or their behavior round a waterhole. It may not be possible to collect enough information from observations scheduled at regular times and thus be necessary to record when the particular situation arises regardless of time. The regular scheduling of observations has, however, a number of advantages. One of these is that the decision of when to record is too open to observer bias when records are made irregularly: he is, for instance, liable to overestimate the frequency of "interesting" events because he is more likely to start recording when they are happening. Irregularly scheduled observations are also likely to be more inconsistent than regular ones. The fact that animals usually show daily rhythms in their behavior is one reason why this may be the case: observations massed from different times of day are likely to yield more diverse results than those regularly scheduled at the same time.

We will now consider several possible sampling techniques, the recording methods suited to them, their uses, and their restrictions. The form of the record obtained when using each of the main methods discussed is illustrated in Figure 1.

The "Complete" Record

Description. The observer attempts to record the timing and duration of every occurrence of all the categories of behavior in which he is interested.

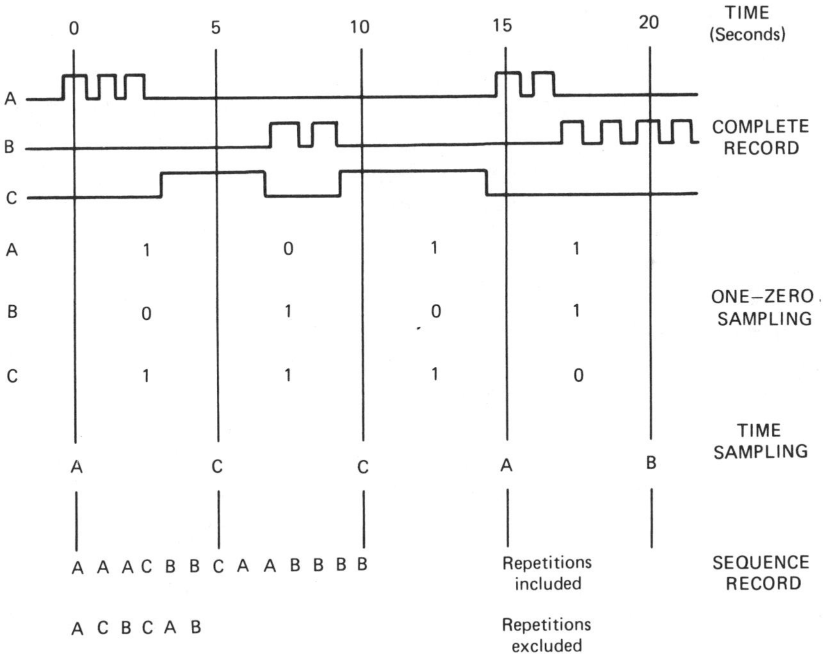

Figure 1. Some ways of recording behavioral data. In this hypothetical example an animal has been watched for 20 seconds and three different behavior patterns (labeled A, B, and C) observed. The complete record at the top would be obtained by using a keyboard and pen recorder to record the timing and duration of each event. The other records show how the same data would have been recorded by alternative methods.

Recording methods. Recording is most accurately done by means of a keyboard linked to a pen-recorder, tape punch, or tape recorder modified in a way compatible with computer analysis (e.g. Dawkins 1971; White 1971). These all provide an accurate time-base not obtainable with the use of check-sheets or dictation onto tape, unless changes in behavior are very infrequent.

Uses. Many sorts of analysis are possible as the information available in the record approaches the amount present in the original behavior. In particular, frequencies and durations of events and of bouts can be assessed and sequence analysis carried out.

Restrictions. The complete record is mainly limited to observations where few categories are involved and the behavior does not take place very rapidly. Watches must also be kept short because of observer fatigue. These restrictions may be overcome by using more than one observer or by watching the session more than once on film or videotape, preferably slowed down. The automatic recording of some categories, where practicable, may help the observer to record more. Despite these possibilities, the method will be impracticable for many purposes, particularly in field-work and where several animals are being observed simultaneously. Where this is so, the investigator must decide which aspects of behavior are most important to his analysis and choose another sampling method suited to their examination.

Sequence Sampling

Description The observer records which acts occur in the correct order without using a time-base. If more than one animal is involved he will also record which animal showed which behavior.

Recording Methods. The methods mentioned in the last section are also suitable here but, as no time-base is involved, the use of check-sheets and dictation into a tape recorder are also possibilities.

Uses. Sequence sampling is purely for sequence analysis, which can be used to assess sequential dependencies both within and between individuals and thus also the associations between acts.

Restrictions. Many types of information are lost in this case, notably any measures of frequency or duration. It is also hard to achieve an accurate record if several animals are involved. Combining this method with the next may help to avoid this difficulty.

Focal Animal Sampling

Description. During observations on the behavior of animals in groups, the observer concentrates for a period of time on one individual, taking either a complete record or a sequence record of his behavior and of his interactions with others.

Recording Methods. As in the previous two sections, the choice depending on whether a complete or sequence record is being made.

Uses. The uses of focal animal sampling vary, depending on the exact recording method used. This technique is particularly useful in studies of social behavior in the wild where the observer cannot watch all animals at once and cannot be sure that the same animals will always be present or remain in view. By choosing the animal to be focused on the observer can make sure that he obtains adequate results on different age and sex classes.

Restrictions. Though this possibility only arises in studies of animals in groups, it is probably the most suitable method for accurate recording of their behavior (see Altmann 1974 for a fuller discussion). It is not, however, ideal for various reasons. If animals are difficult to see, the samples for each may be short and biased towards those which were most obvious. It is also easy to miss events that occur elsewhere in the group and influence the behavior of the focal animal.

"On-the-Dot" Sampling

Description. The observer notes what an animal is doing at fixed time intervals at a particular instant. If a group of animals is involved, he may scan the group at each sampling point recording what each in turn is doing.

Recording Methods. Uşe of a check-sheet or tape recorder will be perfectly adequate. The latter is preferable where several animals are involved, as the observer will not have to look down between individuals. Time-lapse photography provides an alternative that makes longer "watches" possible.

Uses. These depend to some extent on the interval between samples. Some keyboard systems for obtaining complete records do, in fact, involve on-the-dot sampling by a scanner that samples the state of the keyboard at intervals of a fraction of a second (e.g., White 1971). However, most studies using this method sample at intervals of 15 seconds or more. Such samples may be used to assess synchrony between individuals or the percentage of time spent in particular activities. If the latter measure is required, this provides an easy and objective way of obtaining it which may even be preferable to efforts based on the greater information available in the complete record. The latter may involve difficulty in deciding how long the gap between successive events of the same type should be if it is to be recorded as a break in the performance of that behavior.

Restrictions. This is not useful for sequence analysis or for any precise measures of frequency or bout duration. It may also be a rather difficult method to employ in practice. At many sampling points the animal may not, from the strict point of view, be in the act of performing any of the categories being scored, but have just finished one and be about to start another. This method is therefore more suited to recording those behaviors that persist in time (''states'': e.g., sleeping, standing, sitting) than those that are of brief duration and so likely to be missed (''events'': e.g., scratching, coughing, shaking). Furthermore, unless the observer is very careful, or uses time-lapse photography, he will be tempted to score events which occurred just before the sampling point, but not at it, especially if these are uncommon. A difficulty with time-lapse photography is that correct identification of behavior patterns may depend on movement, which, of course, cannot be seen on photographs.

One-Zero Sampling

Description. This is the name given by Altmann (1974) to the collecting of scores that are commonly called Hansen frequencies. The observer simply notes whether or not a particular behavior is present in each of a series of successive time units. Several behaviors may be scored at the same time.

Recording Methods. This method is usually a check-sheet (Hinde (1974) discusses the design of these with this application, amongst others, in mind).

Uses. This method provides a measure of neither duration nor frequencies and Altmann (1974) does not recommend its use under any circumstances. For some forms of analysis it may, however, offer advantages. First, the score that it provides for each behavior pattern (the number of time units in which it occurred) may be a more realistic guide to the level of that behavior than would be a count of the total number of events that occurred or the total time that the behavior occupied (Baerends et al. 1970; Ollason and Slater 1973). The one-zero score for an animal showing a single long bout of a behavior will tend to be less than that for one showing several short bouts even though the time spent in the behavior may not differ between them. Thus, with 5 second sampling units, a 20 second bout will score 4 or 5, depending on how many time units it encroaches upon; twenty 1 second bouts that are spread out in time may achieve a score of up to 20. The one-zero score thus comes closer to indicating how often bouts of that behavior started than does the measure of exact duration. This said, it is of course preferable, where possible, to use a complete record and assess the number of bouts shown in a more objective fashion (see Chapter 4). A second possible use of this method is in looking for associations between behaviors by means of correlations to see whether they tend to occur in the same time unit as each other. Binkley (1973) has shown that clipping data (reducing the frequencies in successive time units to 1 or 0) has little effect on the autocorrelations calculated from them. It seems therefore that one-zero scores provide a source of data suited to correlating within and between behaviors.

Restrictions. Despite the possibilities raised above, the uses of this method are clearly limited. Altmann and Wagner (1970) and Fienberg (1972) have suggested that it can be used to measure rates of performance, but only for the unlikely situation where the behavior concerned follows a Poisson process. Simpson and Simpson (1977) have explored the possibility of using a Markovian model to convert one-zero frequencies to measures of duration. The results were not encouraging probably, again, because bout lengths of behavior do not usually follow a negative exponential distribution as either a Markov or Poisson model would assume. The number of time units in which a behavior occurs depends in a rather complicated way on the temporal pattern of the behavior and on the length of the time unit chosen. If the time unit used is short, the method may provide a reasonable measure of the frequency of rare events, two of which are unlikely to occur in the same unit. The method will always overestimate the duration of states because a behavior that extends into a time unit, no matter how briefly, will be scored as occupying the whole of it. The only case where a reasonable measure of duration may be obtained is where bout lengths are long in relation to the length of the time unit. Generally, however, the relationship of one-zero scores to more precise measures, such as duration and frequency, is not a simple one, and for most purposes their use is best avoided wherever another method of sampling is feasible.

CHAPTER

2

Repertoire Analysis

ROBERT M. FAGEN
University of Illinois at Urbana-Champaign

A behavioral repertoire is a set of mutually exclusive, collectively exhaustive behavioral acts of an animal or species. Like a workperson's toolbox or computer's instruction set, it represents the constraints under which intelligence operates in its interactions with the environment. Were we intelligent plants (for imagination and mathematics may flourish in the same garden, vide Lewis Carroll), animal behavior would no doubt seem a poor substitute for *our* malleable and docile plasticity of form, *our* splendid array of noxious secondary compounds capable of giving even the most casual browser indigestion or a ticket to the next world. And yet, impoverished though they may be, animals behave, survive, and endure.

A behavioral catalogue (S. Altmann 1965; Hutt and Hutt 1970; Kaufman and Rosenblum 1966) is a sample from the repertoire; it is some observer's list of distinguishable behavioral acts. Practical statistical methods are available for use in acquisition and analysis of behavioral catalogues (Fagen and Goldman 1977). In this chapter I give procedures for predicting the number of different kinds of acts that will be contained in a behavioral catalogue as a function of the total number of acts observed, and I present a method for determining whether or not an appropriately defined completeness requirement is satisfied by an existing catalogue relative to a hypothetical repertory. The latter problem appears to have been first posed by S. Altmann (1965). Some interesting theory and observations on repertoire size and its evolution may be found in Moynihan (1970) and in Smith (1969).

The organization of the chapter is as follows. A regression method and an *a priori* rule for predicting the amount of additional sampling needed in order to add a given number of new behavior types to an existing catalogue are outlined and illustrated in Section 1. Several different mathematical definitions of catalogue completeness appear in Section 2. Methods for estimating catalogue completeness using two of these definitions are presented in Sections 3 and 4, and several new examples of the application of these methods to behavioral data given. In Section 5, I discuss these results and additional applications of the methods.

1. LOGARITHMIC REGRESSION

How will the number of types of acts in a sample of behavior depend on the total number of acts observed? By total number of acts, I mean the number of acts counting all repetitions. For example, a catalogue consisting of three instances of "walk," three instances of "sit," and two instances of "run" would contain three types of acts (walk, sit, and run) with a total of 3 + 3 + 2 = 8 acts. Will there be an orderly relation between the number of different types of acts and the total number of acts? Such "type-token" relations are not uncommon in the behavioral and life sciences. For example, the number of different animal species collected by ecologists in a given sample area may be predicted from the size of that area (MacArthur and Wilson 1967) or from the total number of individuals (tokens) in the collection (Preston 1948, 1962). The number of different word types in a literary work may similarly be predicted from the total number of words (tokens) contained in that work (Brainerd 1972, and references cited therein). In both cases the logarithm of the number of types in the collection will be found to depend approximately linearly on the logarithm of the number of tokens. A general description of this linear relation is given by the equation $Y = CX^z$, where Y is the number of types in the collection, X is the number of tokens, and C and z are parameters that depend on the nature of the biological system analyzed.

May (1974) and Webb (1974) show that this regularity of type-token samples is no accident. By analyzing the statistical properties of samples from populations of an infinite number of objects belonging to a finite number of types or categories, they demonstrate that such regularity is a direct consequence of the mode of analysis itself and should be expected whenever the number of different types of objects in a sample is analyzed in relation to the total number of objects in that sample. Accordingly, the logarithm of the number of types of acts in a behavioral catalogue is expected to depend linearly on the logarithm of the total number of acts in that catalogue.

In particular, parameters C and z of this linear relation may be estimated from behavioral data using regression analysis, making it possible to predict the expected number of types of acts in a behavioral catalogue resulting from various amounts of additional sampling. A procedure for carrying out such analyses is outlined below and an illustrative example presented.

An observational study of behavior typically comprises a series of sampling periods, designed according to techniques that serve to minimize sampling bias with respect to occurrence of acts of different types (J. Altmann 1974). Suppose that the ith sampling period, when analyzed, is found to contain X_i total acts. If $\log Y_i$ is plotted against $\log X_i$ for $i = 1, 2, \ldots, n$, where n is the total number of sampling periods, the linear regression of $\log Y_i$ on $\log X_i$ may be calculated as

$$\log Y_i = \log C + z \log X_i$$

An estimate of the number of different types of acts in a catalogue of size X_0 is then $Y = CX_0^z$. Approximate confidence bounds for Y may be derived using standard techniques (see, e.g., Lewis 1960).

As an example, the regression procedure outlined above has been applied (Fagen and Goldman 1977) to a set of observations of maintenance, search, prey capture and handling, and object play in domestic cats, *Felis catus* (Table 1). Regression analysis of these data (Figure 1) yields $Y = 4.01X^{0.29}$. This relation is highly signiifcant ($t = 7.76$ on 33 d.f., $p < 0.001$) with an estimated correlation coefficient of $r = 0.80$. A 95% confidence interval for z, the slope of the regression line, is [0.22, 0.36].

The regression equation predicts that for the 12,211 acts in the data of Table 1, $\hat{Y} = 56$ with 95% confidence bounds for Y of [28, 105]. A behavioral catalogue constructed using the Table 1 data actually contains $Y = 69$ types of acts. This agreement is of course simply a measure of internal consistency, since prediction and observation are based on the same data. For $X = 20{,}000$, $\hat{Y} = 64$ with 95% confidence bounds for Y of [36, 120]. This prediction means that if 20,000 additional acts were sampled, this imaginary new catalogue would contain from 36 to 120 types of acts with 0.95 probability. A 95% upper confidence bound on the number of types of acts present in the new ($X = 20{,}000$) catalogue,

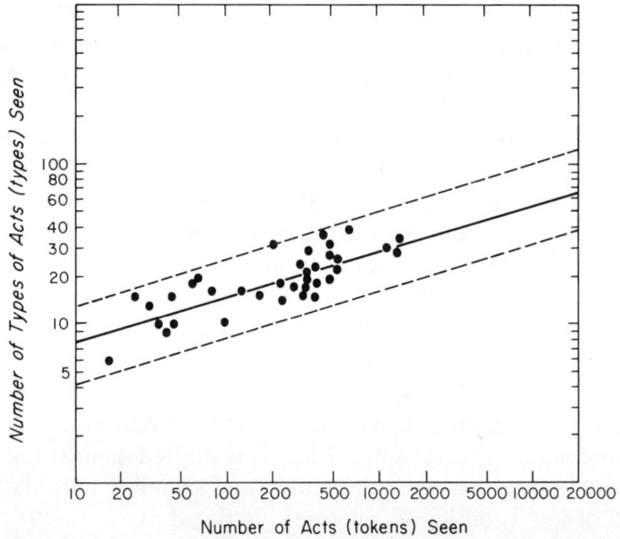

Figure 1. Type-token plot of domestic cat behavior data with fitted regression line $Y = 4.01X^{0.29}$ (solid line) and 95% confidence bounds for regression line (dashed lines).

Table 1. Summary of Behavior Types and Tokens for Domestic Cat Observations

Context of Behavior	Individuals Observed*	Acts	Types of Act
Maintenance and search	D	224	18
Maintenance and search	D	31	13
Maintenance and search	D	80	16
Object play	D	1402	34
Object play	D	426	36
Object play	D	478	32
Object play	D	537	22
Object play	D	359	29
Maintenance and search	A	40	9
Object play	M	60	18
Object play	M	25	15
Maintenance and search	T	336	21
Handling ("play with") prey	T	1154	30
Handling ("play with") prey	T	43	15
Maintenance and search	T	17	6
Handling ("play with") prey	T	1349	28
Maintenance and search	T	331	17
Handling ("play with") prey	T	468	27
Maintenance and search	T	393	18
Maintenance and search	T	237	14
Maintenance and search	T	124	16
Object play	T	64	19
Prey capture	T	35	10
Maintenance and search	T	98	10
Maintenance and search	T	383	23
Prey capture	T	301	24
Handling ("play with") prey	T	645	39
Maintenance and search	T	538	26
Maintenance and search	T	326	15
Maintenance and search	T	168	15
Maintenance and search	T	342	19
Maintenance and search	T	44	10
Maintenance and search	T	488	19
Maintenance and search	T	389	15
Maintenance and search	T	276	17

* Individuals observed: D, male, 4 months old, seen in cage; A, sex unknown, adult, seen in open field; M, female, 5 years old, seen in large room; T, male, 9 months old, seen in large room (object play) and open field (all other observations).

but not present in the old (X = 12,211), is 120 − 69 = 51 types. In other words, by increasing the size of the catalogue from 12,211 to 32,211 we would at most nearly double the number of types contained in the catalogue, but we would expect far fewer new types than this on the average.

A general principle in ecological sampling is that the size of a collection (i.e. the number of individuals contained in the collection) must be increased tenfold in order to double the number of animal species represented (MacArthur and Wilson 1967). This rule corresponds to a slope of $z = 0.301$ in the regression equation $Y = CX^z$. In the above example, the calculated value $z = 0.29$ does not differ significantly from 0.301 ($t = 0.27$ on 33 df, $P > 0.10$). A somewhat different analysis of behavioral type-token distributions (Section 4 of this Chapter) indicates that regression slopes of about 0.3 characterize these distributions in all animal species for which these slopes have been calculated. As mentioned above, May (1974) and Webb (1974) present strong theoretical reasons for the existence of such constancy in nature. I therefore recommend the rule "A tenfold increase in the total number of acts will, on the average, double the number of behavior types in the catalogue" for use as a general *a priori* guideline in behavioral catalogue acquisition, subject to future revision and/or refinement as more data become available. An investigator who questions the applicability of the theory to a particular biological problem would be well advised to carry out an actual regression analysis as outlined above, or to check results using other methods (e.g., Brainerd 1972; Good and Toulmin 1956).

It should be emphasized that the logarithmic regression procedure does not enable an investigator to make statements about actual repertoire size, since the theoretical regression line has no finite asymptote. However, the predictions of the underlying statistical model do seem consistent with several kinds of biological data. For sufficiently large samples one would indeed expect departures from linearity. Such departures in very large samples are predicted by Brainerd's (1972) regression model, which should be used in the event that this problem arises.

Width of the confidence limits on predicted number of types in the above example is understandable in view of the heterogeneity of the data. Analysis of internally homogeneous subsets of these data would be expected to result in more accurate predictions. When, for instance, I reanalyzed the cat data of Table 1, considering play behavior separately from all other behaviors, I obtained $r = 0.86$ for play and $r = 0.86$ for nonplay behavior. (Since different individuals did not contribute equally to the play and nonplay samples, this effect may be attributed either to differences between play and nonplay or to overall differences between individuals, although the former explanation is much more likely.)

As the above mentioned increase in r from 0.80 to 0.86 may possibly have been due to subdivision of the sample into two smaller samples, I subjected the data to a one-way analysis of covariance. Difference between mean number of types in play and nonplay was highly significant ($F = 116.6$, $df = 1/32$, $P < 0.01$). Difference between regression slopes for play and nonplay was not significant ($F = 2.82$, $df = 1/31$, $P > 0.05$).

2. DEFINITIONS OF BEHAVIORAL CATALOGUE COMPLETENESS

Catalogue Fraction

The first and most natural definition of completeness is the fraction of behavior types in the repertoire present in the catalogue. According to this definition, if N = true repertoire size and Y = number of behavior types observed = observed catalogue size, then completeness may be measured by repertory fraction $\phi = Y/N$. (The problem of estimating the unknown value of N is dealt with in Section 4.) If ϕ were greater than some fixed value such as 0.90 or 0.99, then the catalogue would be considered essentially complete. This definition of completeness contains the implicit assumption that all acts in the repertoire are equally important to the observer in evaluating repertoire completeness, so that what matters is the absolute proportion of the repertoire seen.

Sample Coverage

A second definition of completeness weights each type of behavioral act by the probability of occurrence of that act in a hypothetical, very large sample of the animal's behavior. If the N types of acts in the repertory have average probability of occurrence p_i, $i = 1, 2, \ldots, N$, completeness may be defined as follows in a way that places greater emphasis on the more common behaviors. If some p_i is strictly less than 1, there is a chance that the ith behavior will not appear in the sample. This behavior might equally well be represented several times, especially if the sample is large. In general, a sample of an animal's behavior consists of I_i occurrences of the ith behavior type, for $i = 1, 2, \ldots, N$; I, the total number of acts (tokens) in the sample is equal to $\sum_{i=1}^{N} I_i$.

Suppose that behaviors $i_1, i_2, \ldots, i_y$ are present in the sample. If we have not seen all behaviors in the repertoire, can we at least be reasonably sure that the missing behaviors are very rare ones having very low values of p_i? We pose this question mathematically by defining a quantity θ called the sample coverage. θ is defined to be the sum of the true p_i's over all Y observed behavior types

$$\theta = \sum_{i=i_1}^{i_y} p_i$$

The sample coverage is the probability that in a new, independent sample of behavior, a randomly chosen act will belong to a type already represented in the initial sample of I behaviors. Equivalently, $1 - \theta$ is the probability that this next behavioral act will belong to a type not previously seen. In this sense θ measures our ability to predict the composition of the animal's behavior, based on information gained from sampling. If $\theta = 0.99$, for example, then the missing behaviors must have an aggregate probability of only 0.01. In other words, the unseen behaviors are all rare and our coverage is essentially complete. If, on the other hand, θ is 0.1 or 0.2, we have not yet seen most types of acts that the animal is likely to perform.

Note that while the concepts of sample coverage and repertoire fraction are related, they are not identical. For example, if an animal has five common behavior types, each with $p = 0.19$, and five rare categories, each with $p = 0.01$, and our sample contains four different common behavior types and one rare behavior type, we have only seen $5/10 = 50\%$ of the acts in the repertoire, but the coverage of the sample is $4 \times 0.19 + 0.01 = 0.77$, or 77% of the total probability associated with the repertoire.

Perhaps the most fundamental distinction between repertoire fraction and sample coverage as measures of catalogue completeness is that an investigator who uses ϕ emphasizes the significance of rare behavioral acts in ontogeny and evolution, while an investigator who prefers to use θ attaches no special importance to rare events. Chance and Russell (1959), Washburn and Hamburg (1965), and Wilson (1973b) stress the biological significance of rare behavioral acts. However, the rarest behaviors in an animal's repertoire are not important if one is interested only in the gross predictability that high coverage would tend to provide.

Time Coverage

Hutt and Hutt's (1971) discussion of the catalogue acquisition problem implicitly contains a third completeness measure similar to θ. With p_i as before, let t_i = average time taken to perform behavior type i. Then

$$\pi_i = \frac{p_i t_i}{\sum_{i=1}^{N} p_i t_i}$$

would be the fraction of time spent performing behavior type i in a very large sample of behavior. Accordingly, a measure of time coverage would be θ_T, the sum of the π_i over the Y observed behavior types. No statistical estimator of θ_T is now available. It is important to realize that use of θ_T demands a sampling technique in which duration of each act is recorded, in addition to types and frequency of acts.

Information Coverage

A fourth measure of catalogue completeness is the information coverage statistic θ_{cat} defined as follows:

$$\theta_{\text{cat}} = \frac{\min\left(-\sum_{i=1}^{i_y} p_i \log_2 p_i,\ -\sum_{i=1}^{N} p_i \log_2 p_i\right)}{\max\left(-\sum_{i=1}^{i_y} p_i \log_2 p_i,\ -\sum_{i=1}^{N} p_i \log_2 p_i\right)};\quad 0 \le \theta_{\text{cat}} \le 1$$

The p_i are behavioral probabilities of acts in the repertoire, as discussed above. This complicated expression with its minima and maxima is necessary because the value of $-\sum p_i \log_2 p_i$ for a sample may actually exceed that for the population.

Some sampling theory for information indices of the form $-\sum p_i \log p_i$ is derived by Bulmer (1974) and by Engen (1975). I do not discuss this measure further here, primarily because it appears reasonably difficult to estimate.

3. ESTIMATING SAMPLE COVERAGE

Let N_1 be the number of behavior types represented exactly once in the sample. If the individual observations are statistically independent, a distribution-free estimate of the average value of θ is (Good, 1953)

$$\hat{\theta}_g = 1 - \frac{N_1}{I}$$

Although, as indicated, $\hat{\theta}_g$ is not an estimate of θ itself but of the average θ for all samples of I acts from the animal's repertoire (Engen, 1975), $\hat{\theta}_g$

is a useful coverage indicator unless I is very small. For example, S. Altmann (1965), observing free-ranging rhesus monkeys, recorded I = 5507 tokens; N_1 = 32 behavioral types were observed only once. For these data, assuming independence, $\hat{\theta}_g = 1 - (32/5507) = 0.9942$.

4. ESTIMATING REPERTOIRE FRACTION

Repertoire fraction $\phi = Y/N$ will be estimated by $\phi = Y/\hat{N}$, where $\hat{N}$ is an estimate of true repertoire size. $\hat{N}$ is obtained by analyzing the observed *behavioral-abundance distribution* (frequency distribution of the N_i, where N_i = number of behavior types represented exactly i times in the sample). This mode of analysis has been used in ecology (Preston 1948, 1962; Patrick, Hohn, and Wallace 1954), in lexicography (Carroll, Davies, and Richman 1971; Gani 1976), and in ethology (Schleidt 1973). While the behavioral catalogue itself records how many different kinds of acts have been observed, and therefore suggests the question "Why are there so many different kinds of acts?", the tabulation of behavioral abundances records the relative commonness and rarity of the various acts in the repertoire and leads us to ask "Why are some acts common while others are rare?" Ecologists frequently ask questions about the diversity and relative abundance of species (Hutchinson 1953, 1959; MacArthur 1960; May 1974; Preston 1948, 1962; Whittaker 1970, 1972). Schleidt (1973) considers the potential importance of such questions in the study of animal communication.

The form of behavioral-abundance distributions may vary with situation, individual, and species. Three such distributions are shown in Figure 2.

The most direct approach to the problem of estimating N requires a general method for estimating the number of different categories in an infinitely large population with a finite number of types, on the basis of the number of categories represented in a finite random sample. This problem appears to be unsolved at present, and Goldman and I adopted the parametric approach used by him (1973) for the ecological species sampling problem. A behavioral-abundance distribution includes N_1 rare behavior types which appeared only once in the sample, N_2 behavior types which appeared twice, N_3 types which appeared three times, and so on. We observed that the $N - Y$ behavior types present in the repertoire, but missing from the catalog, have abundance zero (by definition, because they did not appear in the sample). Our method fits some particular discrete probability distribution to the behavioral-abundance data and extrapolates the fitted curve back to the origin, obtaining

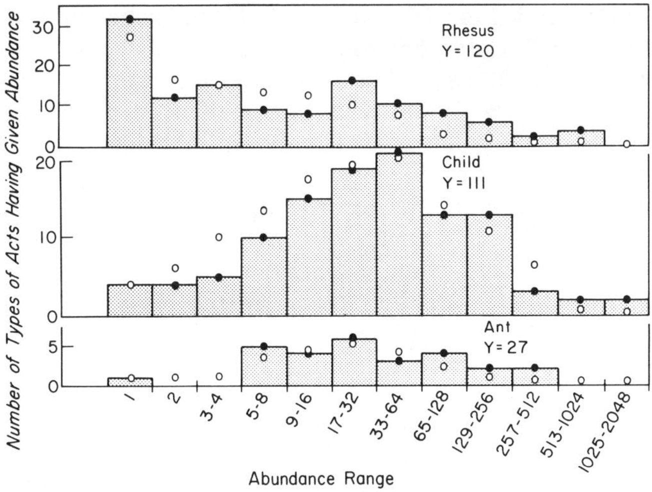

Figure 2. Behavioral-abundance distributions in three species. Filled circles indicate observations; open circles indicate the lognormal Poisson distribution fitted to these data using methods described in the text.

in this way an estimate, $\hat{N}_0$, of $N_0 = N - Y$, the missing zero-abundance class of the distribution. An estimate of repertoire size is then $\hat{N} = Y + \hat{N}_0$. But which probability distribution should be chosen for curve fitting? Many candidates exist (Patil and Joshi 1968). Two different distributions (the lognormal Poisson and the negative binomial) occasionally fit behavioral-abundance data equally well if the catalog is incomplete. However, empirical evidence is accumulating in favor of the lognormal Poisson and against the negative binomial (see below). Perhaps other, unknown distributions now crouching in the mossy undergrowth of probability theory could exhibit even better performance than that of the lognormal Poisson. A way around this problem, as MacArthur (1960) recognized, is to replace the empirical, curve-fitting approach with mathematical models describing the biological processes producing observed differences in abundance. Some theory relevant to this problem will now be discussed.

Theoretically, lognormal distributions of behavioral abundance should occur often in nature, both in relatively constant and in relatively variable environments, for the following reasons (May 1974). In stable environments, lognormal distributions would be expected whenever repertoires contained a large number of behavioral types fulfilling diverse roles. In fluctuating environments, lognormal distributions would occur not only in large but also in small repertoires. Indeed, exceptions to the lognormal "rule" should be found only in animals which both inhabit stable environments and exhibit small repertoires. (This last observation raises a question that ethologists do not yet appear to have considered in detail: What is the effect of different levels of environmental stability and predictability on the evolution of repertoire size and on the number of roles that each behavior is called on to fulfill? A possible analogy with the ecological theory of niche breadth and species packing (May 1973) might be misleading, since natural selection acts on entire repertoires but seldom, if ever, on entire communities.) The practical implication of this theory in the present context is that unless the animal studied lives in a very stable environment and has an exceedingly small repertoire of behaviors as well, use of the lognormal distribution is strongly preferable on theoretical grounds. If, however, the repertoire is in fact very small (20 types of acts or less), a reasonably complete catalogue will easily be obtained, in which case the lognormal and negative binomial estimates of repertoire size will tend not to differ.

As an example, behavioral-abundance data from a number of vertebrate and invertebrate species were analyzed (see Appendix). Good's (1953) estimate of expected sample coverage was computed. The (truncated) lognormal Poisson, (truncated) negative binomial, and (truncated) Poisson distributions were fit to the data using Bulmer's (1973) method (lognormal Poisson), Brass' (in Cohen 1971) and Goldman's (1973) methods (negative binomial), and the method of maximum likelihood (Poisson). Confidence intervals for catalogue size were calculated using Bulmer's (1974) method (lognormal Poisson analysis) and Goldman's (1973) method (negative binomial analysis). Goodness of fit of the estimated distributions to the data was tested using χ^2, chi-square approximation with theoretical cell frequencies 1 or greater (Cohen 1971).

Table 2 summarizes results of catalogue analysis. The truncated Poisson can be rejected as a possible distribution of behavioral abundance in every case. As expected from theory, the truncated negative binomial often fails to provide an adequate fit, while the lognormal Poisson distribution fits existing behavioral-abundance data quite well. N estimates for the relatively complete human and ant catalogues do not depend strongly on the distribution chosen, but these two distributions yield vastly different estimates of N when applied to the relatively incomplete rhesus catalogue. The error of these estimates is very high in the latter case, and further data would obviously be required in order to obtain a reliable repertoire size estimate and to make a final discrimination between the two distributions.

The lognormal Poisson distribution is a discrete version of the

Table 2. Behavioral Catalogue Analyses

Species*	Types	Tokens	Lognormal Poisson	Negative Bionomial	Value of a
Human children *(Homo sapiens)*	111	10789	$\hat{N} = 113 \pm 2.7$	$\hat{N} = 180 \pm 18$	0.308
Rhesus macaque *(Macaca mulatta)*	120	5507	$\hat{N} = 76 \pm 18$	$\hat{N} = 10^4 \pm 10^6$	0.200
Domestic cat *(Felis catus)*	74	3362	$\hat{N} = 102 \pm 28$	Failed to fit, $p < 0.05$	0.215
Timber wolf *(Canis lupus)*	50	7134	$\hat{N} = 50 \pm 0.4$	$\hat{N} = 51 \pm 1$	0.357
Coyote *(C. latrans)*	48	7582	$\hat{N} = 48 \pm 0.04$	$\hat{N} = 48 \pm 0.8$	0.498
Beagle *(C. familiaris)*	47	9408	$\hat{N} = 47 \pm 0.3$	$\hat{N} = 48 \pm 1$	0.345
Wolf-Malemute hybrid *(C. lupus × C. familiaris)*	44	8593	$\hat{N} = 44 \pm 0.1$	Failed to fit, $p < 0.05$	0.437
Ants:					
Cephalotes atratus workers	39	6644	$\hat{N} = 40 + 1$	$\hat{N} = 44 \pm 2.9$	0.269
Leptothorax curvispinosus workers	27	1951	$\hat{N} = 27 \pm 1$	$\hat{N} = 27 \pm 0.1$	0.338
Pheidole dentata minor workers	26	1222	$\hat{N} = 26 \pm 1$	$\hat{N} = 27 \pm 1.5$	0.375
Pheidole dentata major workers	8	204	$\hat{N} = 8 \pm 1$	$\hat{N} = 8 \pm 0.5$	0.372
Zacryptocerus varians minor workers	38	2163	$\hat{N} = 42 \pm 3$	Failed to fit, $p < 0.05$	0.268

* Data from McGrew (in Hutt and Hutt 1970); Altmann (1965); Fagen and Goldman (1977); Bekoff and Fagen (unpublished); Wilson and Fagen (1974); Wilson (unpublished); Corn (unpublished).

lognormal distribution (Aitchison and Brown 1957) and approaches the lognormal very closely for large sample size. The lognormal contains a parameter a, related to the variance of the distribution; estimates of a from observed distributions of ecological abundance tend to cluster around 0.2, as do a estimates from human population size or gross national product distributions (May 1974). The lognormal Poisson distribution "inherits" a from the underlying lognormal (Patil and Joshi, 1968); this parameter may be estimated during the computation of $\hat{N}$. The a values for behavioral-abundance data analyzed to date (Table 2) are all relatively close to 0.2, and few differ significantly from this value. However, the higher a values may represent a biologically interesting deviation from the theory for "rich" versus "impoverished" faunas (MacArthur and Wilson 1967). As May (1974) and Webb (1974) have clearly pointed out, constancy of a in biological data is merely a mathematical property of the lognormal distribution for the kinds of sample sizes and abundance ranges that biologists normally consider. Systematic deviations from the expected constant value, for fixed sample sizes and abundance ranges, may of course contain information of biological significance.

A close relation exists between parameter a of the lognormal distribution and slope z of the type-token regression line (Section 1). MacArthur and Wilson (1967) show mathematically that if the relationship between types and tokens is such that $a \approx 0.2$, then $z \approx 0.3$. My analysis of data from many species indicates that $z \approx 0.3$ is a valid empirical relation for much animal behavior, and one which was expected from theory.

5. DISCUSSION

After a sufficiently long period of sampling, a good estimate of repertoire size ought to be obtained as behaviors continue to accumulate in the sample, thus increasing I and pushing the mode of the observed distribution to the right. (Compare the rhesus distribution, which has its mode at abundance 1 and yields poor estimates of repertoire size, with the human and ant distributions, which exhibit a mode at abundances greater than 1 and yield good estimates of N.) It is interesting to note that the definition of "sufficiently long" appears to depend on the species concerned. A sample of approximately 2000 acts (59 hours' observation time) is adequate for *L. curvispinosus* workers but seems less adequate for rhesus. Indeed, a good estimate of repertoire size in a primate or carnivore might require a sample of size 50,000 acts or more. If we suppose that behavior is recorded once each minute in order to obtain a more nearly random sample, since the assumption of random sampling underlies all of the preceding, 840 hours of sampling would be required to obtain a total of 50,000 acts and an accurate estimate of repertoire size. This figure could be reduced if several independently acting animals could be monitored simultaneously.

These results suggest the not very surprising observation that familiarity with an animal's behavior will tend to require years of experience if the animal is a mammal or bird with a complex repertoire. If, however, the animal's behavior is simple and relatively stereotyped, such familiarity may be gained in a few months.

The method of behavioral catalogue completeness estimation has now been applied to data from a fairly wide variety of species. In all cases, intuition about possible completeness of the working catalogue has been borne out by the analysis. Other applications of the method are now possible. For instance, the behavioral sample sizes required for accurate sequential analyses (Chapter 4), and for information-theoretic analyses (Chapter 3) have been shown to be functions of repertoire size. Methods of completeness analysis therefore make it possible to obtain meaningful estimates of the sample sizes necessary for quantitative study of such important behavioral processes as communication, play, and sequential structuring.

APPENDIX

ESTIMATING REPERTOIRE GIVEN A DISTRIBUTION OF BEHAVIORAL PROBABILITIES

Methods for repertoire size estimation using truncated lognormal, Poisson, and truncated negative binomial distributions are presented; more general methods are given by Harris (1959) and by McNeil (1973). See also Gani (1976) for a preliminary account of Markov-chain techniques for estimating repertoire size.

LOGNORMAL POISSON DISTRIBUTION

The lognormal Poisson fitting procedure of Fagen and Goldman (1977) is statistically correct but inefficient. Bulmer's (1974) method is vastly superior but has the possible disadvantage of requiring use of an automatic computer. A tested FORTRAN IV program for lognormal Poisson catalogue analysis using Bulmer's method has been written by R. Fagen. Bulmer (1974) developed a similar program in the ALGOL language. See also Slocomb, Stauffer and Dickson (1977).

NEGATIVE BINOMIAL DISTRIBUTION

The following procedure is based on methods in Cohen (1971) and in Goldman (1973). Engen (1974) presents an extended negative binomial species-abundance model and a procedure for estimating the parameters of this model; this work is applicable to problems of behavioral abundance and should be considered in addition to the method presented here.

1. Calculate the sample mean $\bar{x}$ and the sample variance s^2.

$$\bar{x} = \frac{\sum_{i=1}^{\infty} iN_i}{Y} \qquad s^2 = \frac{\sum_{i=1}^{\infty} i^2 N_i}{Y} - \bar{x}^2$$

2. Use Sampford's test (Cohen 1971) to determine whether this combination of sample mean and sample variance is consistent with the equations for moment estimation of the parameters of the truncated negative binomial distribution.

By Sampford's test, the conditions for the existence of a solution to the moment-matching equations for the truncated negative binomial are:

$$q_1 < q_2 < q_3$$

where

$$q_1 = 1 - \exp\left[-\left(\bar{x} + \frac{s^2}{\bar{x}} - 1\right)\right]$$

$$q_2 = \frac{\bar{x} + (s^2/\bar{x}) - 1}{\bar{x}}$$

$$q_3 = \log\left(\bar{x} + \frac{s^2}{\bar{x}}\right)$$

If the data pass Sampford's test, estimate the parameters of the truncated negative binomial distribution using the moment-estimation method of Brass (Cohen 1971).

Table 3. Asymptotic Standard Error of Negative Binomial Repertoire Size Estimate for $N = 100$

	$\hat{p}$									
$\hat{r}$	0.001	0.003	0.007	0.01	0.03	0.05	0.15	0.25	0.35	0.45
0.001	17449	22619	22650	32036	48208	61028	121090	193800	294248	443081
0.005	1571	2034	2575	2878	4326	5475	10850	17358	26336	39680
0.009	655	847	1071	1197	1798	2274	4501	7198	10914	16444
0.02	201	260	328	366	548	692	1366	2181	3304	4974
0.04	73.2	94.1	118	132	197	248	487	776	1174	1766
0.15	11.1	14.2	17.8	19.8	29.1	36.4	70.2	111	166	248
0.35	2.78	3.84	5.02	5.66	8.60	10.8	20.8	32.6	48.5	72.0
0.50	1.23	1.88	2.63	3.04	4.92	6.33	12.5	19.6	29.1	43.1
2.5	0	0	0	0	0.014	0.042	0.406	1.12	2.20	3.78

$$\hat{p} = \frac{\bar{x}}{s^2}\left(1 - \frac{N_1}{Y}\right)$$

$$\hat{r} = \frac{\hat{p}\cdot\bar{x} - (N_1/Y)}{1 - \hat{p}}$$

If the data do not pass Sampford's test, maximum likelihood estimation (Goldman 1973), which requires use of a computer, can still be employed.

3. Estimate N_0, the number of behavioral categories not seen, using the equation:

$$\hat{N}_0 = \frac{Y\hat{p}^{\hat{r}}}{1 - \hat{p}^{\hat{r}}}$$

An estimate of repertoire size is $\hat{N} = Y + N_0$, and an estimate of repertoire fraction is $\hat{\phi} = Y/\hat{N}$.

4. The fit of the truncated negative binomial to the data can be measured using the χ^2 test with $k - 3$ degrees of freedom, where the observations are grouped into k cells for the purpose of the test. (See Cohen 1971 for discussion.) Pahl (1969) and Hinz and Gurland (1970) suggest alternative procedures for testing goodness of fit of negative binomial distributions to data. These procedures are recommended for use in place of the standard χ^2 test when expected cell frequencies are small.

5. The asymptotic standard error of $\hat{N}$ is given by

$$\text{SE} = \sigma(\hat{p}, \hat{r})\cdot\frac{\hat{N}}{100}$$

Table 3 gives $\sigma(\hat{p}, \hat{r})$ for ranges of p and r normally encountered by biologists. Confidence intervals for N can now be constructed, since $\hat{N}$ is asymptotically normally distributed. For instance, an (approximate) 95% confidence interval for N would be $\hat{N} \pm 1.96$ SE. The approximation is excellent for $N \geq 50$, but it is not as good for small ($N \approx 10$) repertories; confidence intervals when repertoire size is small may be obtained by simulation.

CHAPTER

3

Information Theory and Communication

GEORGE S. LOSEY, JR.
University of Hawaii

Information theory originated in communication engineering as a method to express information in quantitative terms and describe the behavior of machines (Wiener 1948; Shannon and Weaver 1949). The techniques might have little interest for ethologists except that they described phenomena such as feedback and behavior based only on observed input and output: Informational qualities could be expressed in quantitative terms without knowledge of the mechanics of the machine itself. Ethologists might apply these measures to their four-legged (or otherwise) "black boxes" that have entrancingly complex outputs and usually obscure behavioral mechanics.

Soon after its inception, psychologists weaned information theory by applying it to their living "machines" with hopes of domesticating a powerful analytical tool (Shannon and Weaver 1949; Miller and Fricke 1949; McGill 1954; Garner and McGill 1956), but it did not provide a clear solution to all problems in psychology (Attneave 1959). Ecologists adopted information theory as a tool to describe diversity, but again problems were encountered (Pielou 1966; Lloyd, Zar, and Karr 1968; Fager 1972). Ethologists have more recently laid claim to information theoretical measures for the analysis of behavior. Dingle (1972) has presented a review and critique of past studies.

Past applications of information theory in ethology can be divided into three major areas: indicators of relative diversity, measures of structure such as Markovian relationships, and measurement of communication between individuals as the most popular application. Indicators of diversity are most similar to ecological applications of information theory. Studies include comparison of the diversity of action patterns employed by species (Dawkins and Dawkins 1973; Mock 1976; Kerr 1976), the diversity of species that an animal interacts with in a symbiotic relationship (Losey 1972), and diversity in the utilization of space (Lim in preparation) and resources (Gray 1976). Chatfield and Lemon (1970) applied techniques described in Attneave (1959) for investigating Markovian dependency in the sequential structure of bird song. Many authors (e.g. Hazlett and Bossert 1965) have presented measures of communication between individuals based upon dyadic sequences of behavior (signal-response pairs). The most interesting results have come from studies that dealt with correlations between the amount of communication and independent variables such as dominance (Dingle 1969), sex (Baylis 1975), the effects of initiating a sequence (Steinberg and Conant 1974), and the effects of the length of the sequence and position within the sequence (Hazlett and Estabrook 1974a, b; Rubenstein and Hazlett 1974; Rand and Rand 1976). While dyadic sequences have been the most popular application, higher order sequences have also been investigated by Altmann (1965). It is frequently profitable to compare the degree of "internal structure" in the sequence of actions employed by an individual with the structure that results from the need to respond to changing signals from another individual (e.g. Dingle 1969, 1972). Baylis (1975) presented a novel form of the arrangement of dyadic contingency tables that favor such comparison. Moehring (1972) investigated a communication complex among several species in a symbiotic relationship.

While information theory has not been altered in its basic structure through these various applications, both the demands and popular fads of the various fields of study have imposed somewhat different forms on the basic methods. Psychologists concentrated on multivariate applications. But their interest in the use of information theoretical measures as analytical tools has declined as other more robust multivariate techniques have become widely available (see Chapters 4 to 8). Ecologists concentrated on problems that resulted from sampling techniques (Pielou 1966; Lloyd et al. 1968) and the nonlinear relationship between information measures of diversity and subjective impressions of what community diversity really is (Fager 1972). Ethologists have used techniques to help elucidate the sources of pattern in observations of communication (Hazlett and Bossert 1965; Steinberg and Conant 1974; Baylis 1975) but have paid little attention to problems of statistical inference and sampling bias (Losey 1975; Fagen ms.).

In this chapter, I attempt to bring together some of these different methods and explore their possible applications to ethology. I also introduce a few structural modifications of my own in an attempt to adapt more fully information theory as a behaviorist's tool. I follow the symbols and terminology in Attneave (1959) whenever possible, and strongly suggest reading of Attneave for further understanding of the techniques.

1. ASSUMPTIONS

The most basic and troublesome assumption that concerns us stems from the fact that information theory was devised to describe the behavior of communicative devices such as codes, languages, and machines. The theory assumes that there are no unknown inputs to this machine that alter its basic operation during our observations. For example, if the definitions of code symbols continue to change, information theory will not give an accurate description of the coding process: The machine does not exhibit stationarity. Similarly, we might observe a species in which a dominant individual always "howls" and a subordinant always "whimpers" regardless of the howls and whimpers of the other. If we have an observation record that includes a reversal of dominance due to unobserved causes, and we record only howls and whimpers, information measures would be very high: Howling and whimpering would be indicated as communicative behaviors. While such extreme cases would probably be obvious, Hazlett and Estabrook (1974b) have pointed out that lumping observations with different response motivation can lead to inflated estimates of the degree of communication. I suggest some methods for at least reducing some of the influences of the lack of stationarity.

Fortunately, information theory does not require scaling or ordering of measurements. Counts of behavioral action patterns are natural units for this technique. However, the distribution of these counts can become troublesome as illustrated in the discussion of sampling bias below.

2. METHOD

Information Content and Transmission

I use the term "event" as the unit of measurement to indicate that some condition has been observed such as an action pattern occurred, a contextual variable such as food was present, or a switch was actuated. The basic measure for information theory is the uncertainty or diversity of these events in a set of observations. Uncertainty is expressed as information content in units of bits per act. (In this chapter, all logarithms are to the base two and information is measured in "bits.") If there are many events that are all common, uncertainty, diversity, and information content of the set are high. If only a few events are common and the remainder are rare, information content is low. This information content is represented by H. Symbols, terminology, and formulas for calculating H, and other information measures are presented in Table 1.

The first step in the use of information measures is to define the dimensions of analysis. While most applications in ethology have employed only two dimensions, three or more are possible and can be extremely valuable. Dimensions might be chosen to be "signals" (any action pattern), "responses" (any action pattern following a signal), and so forth. These definitions set the conditions that must be observed in order for each cell of the matrix to receive a count. The number of each dyad (two act sequence), triad (three act sequence), and so on, that was observed is entered into its respective cell. The total count in each cell is the n in Table 1. Most values are calculated directly from these counts. The formulas are straightforward and any of the H or T values is simple to calculate with a pocket calculator. One important convention is that the *log of zero equals zero*! Programmers must remember to insert this convention. Many calculators and computers do not have base-two logarithms available in their languages, but base-two should be used so that all workers have the same units. The confidence limit calculations below demand base-two logarithms. Conversion is made by multiplying base-ten logarithms by 3.3219 and natural logarithms by 1.4427.

Table 1 includes all of the information measures that appear to be of general applicability to ethology, but others, and suggestions for expansion to the fourth dimension, can be found in Attneave (1959). Transmission efficiency or *TE* (*D* in Attneave) may have many possible forms depending on choice of T and H. I feel that it is best merely to state which T and H estimates were used in calculation of *TE* instead of creating a complex series of subscripts to cover all situations.

Miller's Chi-square

Miller and Madow (1954) reported on the relationship between the χ^2 distribution and information measures. They suggested that this relationship could be used to test the following null hypotheses:

$$\text{To test for } T = 0, \qquad \chi^2 \approx 1.3863N\hat{T}$$

$$\text{To test for } H = H(\max), \qquad \chi^2 \approx 1.3863N[\hat{H} - H(\max)]$$

The appropriate degrees of freedom are given in Table 1. Significant χ^2 values can be found in standard χ^2 tables and would indicate that the population sampled had a T greater than zero or H less than H(max), respectively. As with a standard χ^2 test, there should be few or no zero or low frequency cells for this test to be effective. But, unlike the standard χ^2, this test becomes *more conservative* as this guideline is violated. This test can be very useful, but my research has revealed several guidelines that must be followed (see below).

Bias Correction

All $\hat{H}$ and $\hat{T}$ values are biased as a function of sample size. Since rare events may not be sampled, $\hat{H}$ is generally an underestimate of H. Since bias increases with degrees of freedom at a given sample size, $\hat{H}(x, y)$ is generally more severely biased than $\hat{H}(x)$ or $\hat{H}(y)$. This results in a bias

Table 1. Information Measures and Terminology

Terms and Formulas	Degrees of Freedom	Descriptions
i, j, k		Position in x, y, and z dimensions of an array
P_i		Probability that event i in the x dimension will occur
P_{ij} or P_{ijk}		Probability that dyad ij or triad ijk will occur
n_i, n_{ij}, or n_{ijk}		Frequency that event i, dyad ij, or triad ijk was observed to occur
$N = \sum^{i} \sum^{j} \sum^{k} n_{ijk}$		Sample size
Nx, Ny, Nz		Number of rows, columns, and layers (i, j, and k dimensions) in the matrix
$\hat{H}(x) = \log N - \frac{1}{N} \sum^{i} n_i \log n_i$	$Nx - 1$	Estimate of information present in the x dimension (similar formulas are used for the y and z dimensions)
$\hat{H}(x, y) = \log N - \frac{1}{N} \sum^{i} \sum^{j} n_{ij} \log n_{ij}$	$NxNy - 1$	Estimate of information present in the ij dyads (similar formulas are used for other x, y, and z pairs)
$\hat{H}(x, y, z) = \log N - \frac{1}{N} \sum^{i} \sum^{j} \sum^{k} n_{ijk} \log n_{ijk}$	$NxNyNz - 1$	Estimate of information present in the ijk triads

$\hat{H}_{xy}(z) = \hat{H}(x, y, z) - \hat{H}(x, y)$	$NxNyNz - NxNy$	Estimate of information present in the z dimension that is unique to that dimension (not shared with x or y)
$H(\max) = \log Nm$		Maximum information possible in the m dimension (m can be x, y, xy, etc.)
$\hat{T}(x; y) = \hat{H}(x) + \hat{H}(y) - \hat{H}(x, y)$	$(Nx - 1)(Ny - 1)$	Estimate of the information shared by the x and y dimensions
$\hat{T}(x; y; z) = \hat{H}(x) + \hat{H}(y) + \hat{H}(z) - \hat{H}(x, y, z)$	$(Nx - 1)(Ny - 1)(Nz - 1)$	Estimate of the information shared by the x, y, and z dimensions
$\hat{T}(x, y; z) = \hat{H}(x, y) + \hat{H}(z) - \hat{H}(x, y, z)$	$(NxNy - 1)(Nz - 1)$	Estimate of the information in z that is shared with the x and y dimensions (all combinations possible)
$\hat{T}_x(y; z) = \hat{H}(x, y) + \hat{H}(x, z) - \hat{H}(x) - \hat{H}(x, y, z)$	$Nx(NyNz - Ny - Nz + 1)$	Estimate of the information shared by y and z with the x dimension held constant or partialled out (all combinations possible)
T or H		True value of estimates, $\hat{T}$ and $\hat{H}$
$TE = T \div H$		Transmission efficiency for any values of T or H
$R(x) = H(x) \div H(\max \text{ for } x)$		Relative entropy of x dimension

of $\hat{T}$ toward overestimation of T. Miller and Madow (1954) suggested that the same χ^2 relationship could produce bias corrected measures, H' and T', as

$$H' = \hat{H} + (\text{degrees of freedom} \div 1.3863N)$$

$$T' = \hat{T} - (\text{degrees of freedom} \div 1.3863N).$$

Restriction as to the number of zero or low frequency cells also applies to the bias correction. If the guideline is violated the relationship will be an overcorrection. These problems are discussed below.

Confidence Limits

David and others (1956) reported that an arcsine transformation of information measures approximated a normal distribution. This enables the calculation of statistical confidence limits for H' and T'. While their methods appear to be well suited for ethologists until a more satisfactory solution is found, they should not be uncritically applied. The user should consult Fagen (ms.) and the results of my research (below).

The procedure for calculating "David method" confidence limits is presented as in David and others (1956) with my correction of what appear to be editorial errors and instruction for expansion to three-dimensional measures. All logarithms must be base-two and all trigonometric functions must be in radians. Programmers should note that these calculations frequently demand exponentiation of zero values, which is precluded by many computer languages. Preventative steps should be taken. In addition, since many computers lack an arcsine function, a handy relationship is

$$\arcsin x = \arctan \frac{x}{\sqrt{1 - x^2}}$$

The first step in the calculation of all confidence limits is to convert the counts for each cell to estimates of the probability of each event (p) by dividing each count by the sample size (N). Then choose a "Student's t" value (t) from an appropriate table. This choice must be guided by the discussion presented below. Degrees of freedom are as in Table 1.

To calculate upper ($\bar{H}$) and lower ($\underline{H}$) confidence limits for H', compute

$$H(p) = \sum^{i} p_i \log p_i$$

$$SH(p) = \sum^{i} p_i (\log p_i)^2$$

$$A = \left[\arcsin \frac{H(p)}{SH(p)}\right] + \frac{t}{N}$$

$$B = \left[\arcsin \frac{H(p)}{SH(p)}\right] - \frac{t}{N}$$

$$D = \frac{[\text{degrees of freedom}/2] \log e}{N}$$

$$\bar{F} = D + [SH(p) \sin A]$$

$$\underline{F} = D + [SH(p) \sin B]$$

Then for the upper confidence bound

1. If A is less than $\pi/2$, $\bar{H}$ is the smaller of $\bar{F}$ or $H(\max)$.
2. If A is greater than or equal to $\pi/2$, $\bar{H} = H(\max)$

The lower confidence bound is computed as follows:

1. If B is greater than zero, $\underline{H} = \underline{F}$.
2. If B is less than or equal to zero, $\underline{H} = 0$.

To calculate confidence bounds for T' ($\bar{T}$ and $\underline{T}$) (see Table 2), compute

$$T(p) = -\sum^{i}\sum^{j} p_{ij}(\log p_i + \log p_j - \log p_{ij})$$

$$ST(p) = \sum^{i}\sum^{j} p_{ij}(\log p_i + \log p_j - \log p_{ij})^2$$

$$A = \left[\arcsin \frac{T(p) - (D/N)}{ST(p)}\right] + \frac{t}{N}$$

$$B = \left[\arcsin \frac{T(p) - (D/N)}{ST(p)}\right] - \frac{t}{N}$$

$$D = \left[\frac{\text{degrees of freedom}}{2}\right] \log e$$

Then compute the upper confidence bound as follows:

1. If A is less than $\pi/2$ compute

$$G = ST(p) \sin A$$

and $\bar{T}$ is the smallest of G, $H'(x)$ and $H'(y)$.
2. If A is greater than or equal to $\pi/2$, $\bar{T}$ is the smaller of $H'(x)$ and $H'(y)$.

The lower confidence bound is computed as follows:

1. If B is greater than zero

$$\underline{T} = ST(p) \sin B$$

2. If B is less than or equal to zero

$$\underline{T} = 0$$

Formulas for use with any of the H or T configurations in Table 1 can be easily derived. The formulas from Table 1 are first changed into p notation in order to use probability estimates. For example, $T_x(y;\, z)$ changes to

$$T_x(y;\, z) = -\sum^{i}\sum^{j}\sum^{k} p_{ijk}(\log p_{ij} + p_{ik} - p_i - p_{ijk})$$

This is the $T(p)$ function for this information measure. The $ST(p)$ function is obtained by squaring the bracketed portion of the expression. The remainder of the calculations are the same as in the two dimensional case except that the appropriate degrees of freedom must be used for each case.

Partialized "Chi-square"

Many authors have followed the example of Hazlett and Bossert (1965) in examination of the sources of shared information in a two dimensional matrix for signals and responses. A standard χ^2 calculation is computed for the matrix. Then, to locate the major sources of departure from randomness, the contribution that each row makes to the total χ^2 is examined to indicate whether the row event directs, inhibits, or has no apparent effect on the respective column events. I do not feel that this method is particularly illuminating for ethologists (see below) and suggest a more logical alternative. The null hypothesis in question is that the distribution of dyads in a single row is not different from random, given that the row event has occurred. A standard χ^2 calculation is performed but the expected values are calculated such that each row of the matrix contributes equally to the expected distribution. To calculate expected values $e_{j|i}$, compute as follows:

$$p_{j|i} = n_{ij} \div \sum^{j} n_{ij}$$

$$D_j = \sum^{i} p_{j|i}$$

$$E_j = D_j \div \sum^{j} D_j$$

$$e_{j|i} = E_j \sum^{j} n_{ij}$$

The relative magnitude of the standard χ^2 values for each cell and row can be examined to indicate sources of information transmission. A χ^2 table of statistical probability might be used to guide judgement as to the importance of differences in magnitude, but this is only a guide. Probability levels of getting this result from a random distribution must not be inferred (see below).

Other methods exist for locating sources of nonrandomness in contingency tables (Goodman 1968; Chapter 6 of this volume). These techniques are valuable in describing sequences of an individual's behavior (Slater and Ollason 1972; Slater 1973) and for statistical evaluation of randomness. But their application to evaluating sources of communication is questionable (see below).

Partialized Information Measures

Steinberg and Conant (1974) applied a technique suggested by Blachman (1968) for exploring sources of information transmission that is based on the information measures themselves. Probability estimates are used in this calculation as above.

$$J_i \doteq \sum^{j} P_{i|j} \log (P_{j|i} \div P_j)$$

$$I_i = p_i J_i$$

The relative magnitude of the *I* values are then used to indicate the degree to which each row (*i*) is an effective signal.

3. EXAMPLES

Communication and Dimensions of Analysis

Ethologists should be most attracted to information theory as a measure of shared information. It can reveal pattern in sequences of behavior, communication, and effects of contextual variables. Examination of communication is a basic example. In the simplest approach, you define a matrix in the dimensions of: (1) "Signal" dimension is any act of animal A. (2) "Response" dimension is any act of animal B that follows a given signal by animal A. A matrix such as the hypothetical case in Table 2 results.

In this case, H'(signal) and H'(response) both indicate the presence of much information. You conclude that this set of observations could reveal communication. The T'(signal; response) of 0.07 bits per act indicates that this amount of information is shared by signals and responses in the observations. Miller's χ^2 indicates that this set of observations would be very unlikely if there was actually no shared information. The *TE* estimates indicate that T' was much lower than it

Table 2. A Two Dimensional Matrix of Signal (*S*) Response (*R*) Dyads. In this hypothetical communication system, I have pretended that act A is an agonistic threat, act B is a comfort movement, act C is a submissive behavior. Information measures are given below the matrix for discussion. "χ^2" is Miller's χ^2.

		Responses			
		A	B	C	Σ
Signals	A	38	14	38	90
	B	40	28	22	90
	C	65	15	10	90
	Σ	143	57	70	

Information Value	Estimate	χ^2	Degrees of Freedom
$H'(S)$	1.58	0	2
$H'(R)$	1.47	44	2
$H'(S, R)$	2.98	79	8
$T'(S; R)$	0.07	30	4
$TE[T'(S; R) \div H'(S)]$	0.04		
$TE[T'(S; R) \div H'(R)]$	0.05		

David method 95% confidence limits for $T' = 0.06$ to 0.1

could have been; little of the total information was shared; communication was inefficient.

This T' value is a *minimum* estimate of communication. One reason that it is not a true measure of communication is that many important variables have been ignored. The context of the signal, preceding signals, and the motivational state of the responding animal are all important variables. Some authors have examined additional matrices from the same set of observations to address this problem (e.g. Hazlett and Bossert 1965; Dingle 1969, 1972). You can examine two act sequences of an individual's behavior to indicate the importance of "internal constraint" over behavior. Define a "preponse" dimension as any act that precedes a response and generate a matrix of preponse-response dyads. The matrix in Table 3 is from the same set of observations as Table 2. T'(preponse; response) indicates that the last act performed holds more information about the response than does the signal.

The difficulty is that effects of signals are included as "noise" in the latter analysis; and internal constraint masks communication in the first analysis. Both analyses must be combined to partial out the effects of one variable while considering the effects of the other. The three dimensional matrix that results contains preponse-signal-response triads for the same set of observations (Table 4). The response now gains 0.33

Table 3. A Two Dimensional Matrix of Preponse (*P*) Response (*R*) Dyads. These data are from the same hypothetical data set as in Table 2.

		Responses			
		A	B	C	Σ
Preponses	A	75	8	7	90
	B	38	34	18	90
	C	30	15	45	90
	Σ	143	57	70	

Information Value	Estimate	χ^2	Degrees of Freedom
$H'(P)$	1.58	0	2
$H'(R)$	as in Table 2		
$H'(P, R)$	2.87	120	8
$T'(P; R)$	0.18	71	4
$TE[T'(P; R) \div H'(P)]$	0.11		
$TE[T'(P; R) \div H'(R)]$	0.12		

Table 4. A Three Dimensional Matrix of Preponse (P) Stimulus (S) Response (R) Triads. These data are from the same hypothetical data set as Tables 2 and 3.

	Preponse A			Preponse B			Preponse C		
S\R	A	B	C	A	B	C	A	B	C
A	20	5	5	13	4	13	5	5	20
B	25	3	2	5	20	5	10	5	15
C	30	0	0	20	10	0	15	5	10

Information Value	Estimate	χ^2	Degrees of Freedom
As in Tables 2 and 3 with additional values:			
$H'(S, P)$	3.17	0	8
$H'(P, S, R)$	4.32	183	26
$T'(P; S; R)$	0.34	135	8
$T'(P, S; R)$	0.33	138	16
$TE\ [T'(P, S; R) \div H'(R)]$	0.22		
$TE\ [T'(P, S; R) \div H'(S)]$	0.21		
$TE\ [T'(P, S; R) \div H'(S, P)]$	0.10		
$T'_P\ (S; R)$	0.15	67	12
$TE\ [T'_P(S; R) \div H'(S)]$	0.09		
$T'_S\ (P; R)$	0.26	109	12
$TE\ [T'_S(P; R) \div H'(P)]$	0.16		

bits of information per act from preponse and signal combined. T'_{preponse} (signal; response) gives a more accurate estimate of communication by partialling out the effects of preponse. T'_{signal}(preponse; response) estimates preponse-response dependency with the signal held constant. The various *TE* values give added insight. T'(preponse, signal; response) $\div$ H'(response) indicates that 22 percent of the information in the responses is shared with preponses and stimuli combined; much of the response variability has been located. T'_{preponse}(signal; response) $\div$ H'(stimulus) indicates that 9 percent of the information present in the stimuli was communicated to the other animal when its preponses were accounted for; communication was twice as efficient as that indicated by the two dimensional analysis.

I obviously "rigged" these data to demonstrate the value of combined analysis, but the power of multiple dimensions of analysis are well known (see Chapters 5 to 9). Information theory must be considered with other correlation testing techniques such as analysis of variance or multiple correlation statistics. Garner and McGill (1956) and Attneave (1959) discuss some of the similarities such as

$$T(x; y) = \log \frac{1}{\sqrt{1 - r^2}} - 0.25$$

Correlation statistics such as r can be transformed directly into T estimates. (The reverse transformation can not be made.) T is a correlation statistic that uses counts of unordered, unranked events.

The use of counts of events as a measure makes this a valuable tool for ethologists. Many of our variables cannot be measured or even ranked. These events can be literally any class that can be defined. One dimension of analysis could be presence or absence of a particular feature of the coloration of a signal sender. T'_{signal}(coloration; response) tests whether the presence or absence of this color affects communication. T'_{signal}(treatment; response) evaluates the effects of some treatment such as "winning a fight" on subsequent communication. The possible applications are endless, but there are practical limitations on the number of dimensions. Each dimension increases the degrees of freedom, and the relationship between degrees of freedom and sample size is critical (see below). But, so long as the dimensions include only a few events, a fourth or even fifth dimension might be valuable. Contextual behaviors such as piloerection, and contextual measures such as distance between opponents, are logical choices for added dimensions.

A few words of caution are needed at this point:

1. A significant $T'(x, y; z)$ does not demand that z share information with both x and y. It might be correlated with only one. You can explore this by examining both $T'_x(y; z)$ and $T'_y(x; z)$.

2. Interaction between dimensions is as troublesome in information theory as in analysis of variance. $T'(x, y; z)$ includes information that z shares with unique combinations (interaction) of x and y. Attneave (1959) discusses this interaction that, in some cases, can lead to negative estimates of shared information. For our purposes it is sufficient to realize that estimates of T may not be additive; $T'_x(y; z)$ plus $T'_y(x; z)$ may not equal $T'(x, y; z)$.

3. In the discussion of communication, a direction of information flow is implied. But the definition of the dimensions of analysis determines the direction. Information measures do not have a directional component. $T(x; y)$ equals $T(y; x)$.

Communication and Stationarity

So far we have discussed one set of observations on one pair of individuals. David method confidence limits enable comparisons between individuals or observations. But usually the investigator wants to make some inference about a population of individuals. To make such an inference you must combine observations that include different sessions with different test subjects and different opponents. This raises the problem of stationarity: Was communication to different opponents similar enough to justify lumping of observations with different opponents? Was communication by different test individuals similar enough to justify lumping them?

I am using information theory to guide this type of session lumping. I am interested in differences in the communication of species A with different species. Each test individual is matched against a variety of opponents of each species. For any one species, with any one test individual, T'_{signal}(opponent; response) indicates whether communication was similar with different opponents; if so the data from different opponents are lumped. If not, partialized χ^2 is used to indicate the source of the difference, and similar opponents are lumped in subgroups. The lumped data for several test individuals are then joined to form a matrix and T'_{signal}(test individual; response) indicates whether they can be lumped. This is repeated for each species until one or more matrices of species-signal-response triads is obtained.

Such step-wise lumping guards against gross violation of stationarity due to differences between individuals, but other differences may remain. Communication shows clear changes as dominance is established (e.g., Dingle 1969): The early stages of an encounter session may differ from later stages. On a finer time scale, Hazlett and Estabrook (1974a, b) found changes over the time course of a single interaction. By applying concepts of character analysis, they calculated information transmission for the first, second, . . . *N*th response in an interaction. Transmission estimates changed as a function of position in the sequence. Calculation of T'_{signal}(position; response) would be a valuable companion to such character analysis and indicate the relative importance of order effects. T'_{position}(signal; response) would give a corrected estimate of communication.

Hazlett and Estabrook (1974a, b) also applied character analysis to a variety of contextual or static variables such as size, sex, and color phase. They calculated information measures for each condition of these characters. The reasons for these methods are the same as those for multidimensional analysis. Researchers could profit by using character analysis and multidimensional analysis of the same data.

Rand and Rand (1976) adopted a different approach to the problem of position within a sequence. They examined the frequency distribution of action patterns that were the first, second, . . . *N*th to occur during an encounter. Multiple χ^2 analyses of these distributions indicated changes in behavior due to position in the sequence. They found three distinct groups, intruder's initial acts, intruder's later acts, and all resident's acts. They then examined information transmission between signal-response and preponse-response for all possible combinations of these three groups. Their interpretation of the results profited by this division. While their approach did not examine stationarity of communication as a basis for grouping, it did provide an objective basis for subdividing heterogeneous observations.

Definition of Dimensions

Definition of the dimensions of analysis is the most critical step in applying information theory. A "simple" choice such as signal-response can be defined in many ways. Nelson (1964) indicated the importance of specifying some minimum time interval between signals and responses, or preponses. After the interval is chosen, is it realistic to use only the first response, or should all responses during this interval be used? What is recorded if the animal does nothing after a signal or changes its behavior with no detectable signal?

Choice of definition must be guided by the biological questions that are at stake. I recommend the application of more than one definition to the same set of observations. After choice of a set of definitions, they must be carefully examined during interpretation of information measures. For example, setting a minimum time interval between signal and response may result in some signals not having a response if "do nothing" is not a possible response. Thus, H'(signal) will not be a true representation of signal diversity, and *TE* demands different interpretation. If a single response such as "arm up" is maintained during a set of signals, do you record a single "signal-arm up" dyad, add on a series of "signal-do-nothing" dyad, or record a series of "signal-arm up" dyads. I prefer the final alternative, but any choice will effect all of the information measures. If you include all responses that follow within a minimum time interval, sample size and information values are both difficult to interpret.

I particularly advise against use of a "no change" category for signals or responses. This appears to solve problems of maintaining the same signal or response while the opponent changes behavior, but it only creates confusion. "No change" could mean both "continue aggression" and "continue submission" in the same sample. It is best to tabulate this behavior as if it were a unique event when the opponent changes behavior. But remember that this choice will alter the apparent diversity of that dimension.

General Application

Information theory appears to be an analytical tool of broad generality. It reduces correlation to terms of information transmission. We can test for the presence of transmission and compare values from different analyses. Unfortunately, ethologists, like psychologists, will find that information theory is not an answer to all problems. In many cases, interpretation and statistical inference are crude. The very generality of information theory can be its greatest drawback: It is tempting to compare T' values from very different kinds of observations when such comparisons have little biological validity. Different biological phenomena may have similar effects on information measures: The T' values can be the same while their biological causes are different. Second, Miller's χ^2 and David's confidence limits are only approximations. Statisticians tell us that the approximations are fairly good, but I find it difficult to interpret this judgement. What is our real probability of error? Third, Fager (1972) reported that, as he altered a hypothetical community of animals from maximum to minimum diversity, H(community) at first decreased slowly and then very quickly: H was not a perfect measure of what he considered to be biological diversity. Is the same true of communication? If we show a difference of 0.2 bits per act, does this have significant biological impact?

I address these problems in the remainder of this chapter. I urge the reader not to apply these techniques before evaluating my suggestions as to interpretation and statistical error rates.

4. INTERPRETATION AND LIMITATIONS

I could not gain any insight as to the statistical power or reliability of information measures by study of their derivation or the theoretical basis for Miller's χ^2 and David's confidence limits. I resorted to calculating information measures for a variety of simple matrices to observe the behavior of information measures. I also observed the relationship between statistical error suggested by the above techniques and actual error rates resulting from random sampling of these matrices. "Monte Carlo" studies of this sort provide some general rules of thumb and can be applied to specific data when more detailed rules are desired.

My goals were: (1) To determine the sample size required for acceptable alpha and beta error rates. Type I or alpha error is the probability of rejecting a null hypothesis when it is true. Type II or beta error is the probability of accepting a null hypothesis when it is false. High alpha error rate is nonconservative while a high beta error rate suggests that the technique lacks discriminatory power. (2) To explore the accuracy of Miller's bias correction. (3) To estimate the effects of low $R(x)$ that occurs when some events are far more common than others.

Monte Carlo Methods

Two dimensional matrices of transitional probabilities were chosen to represent different levels of communication. These were used by a computer program that performed the following functions:

1. Calculate true population information measures $H(x)$, $H(y)$, $T(x; y)$, and so on from transitional probabilities p_{ij} where

$$\sum^{i} \sum^{j} p_{ij} = 1.0$$

2. Take a random sample from this population of probabilities by:

   a. Transforming the p_{ij} array to cumulative p_{ij} such that the last p_{ij} = 1.0.
   b. Choose N random numbers from 0 to 1.0.
   c. Record each n_{ij} as the number of random numbers that were less than or equal to the corresponding p_{ij} and greater than the preceding p_{ij} in the cumulative series.

3. Calculate information estimates [$T'(x; y)$ etc.] for the random sample of size N.

4. Repeat steps 2 and 3 100 times, with a different random number series each time, resulting in 100 estimates of information measures from the same population. (The limitation of the number of repetitions to 100 appeared adequate. In all 14 cases where a given matrix was again sampled 100 times with a different series of random numbers, the results were nearly identical.)

5. Repeat steps 2, 3, and 4 for different sample sizes.
6. Repeat steps 2 through 5 for various types of matrices.

By analogy, the x dimension represents signals from animal A and the y dimension responses by animal B. The computer technique is analogous to making 100 different sets of observations of interaction between A and B with N signal-response pairs in each set. The matrix used as input to the program represents the relationship between the responses of B and signals from A. In this study, the signals from A did not depend on preceding events.

Matrices were constructed as if each signal was a different "message" and was usually followed by a unique response. For a 10 by 10 matrix, there were 10 different messages. Sampling started with nearly perfect communication; the unique response for each signal was 900 times more common than any other response to that signal. This was accomplished by having relative transitional probabilities of 0.9 in all cells on the major diagonal and 0.001 in all other cells. Subsequent matrices were constructed by altering the diagonal and off-diagonal cells in steps to represent less efficient transmission of the same number of messages. Six different types of matrices were used (Table 5) representing from near perfect information transmission to zero transmission.

In most behavioral studies some action patterns are far more common than others. The effect of such imbalance was investigated by "weighting" some rows or signals more heavily than others. For instance, in a 10×10 matrix, the transitional probabilities in one row can be made two orders of magnitude greater than all others; 10 percent of the cells are made 100 times greater. Row one would have 90.0 in the major diagonal and 0.1 in the off-diagonal while all other rows would have 0.9 in the major diagonal and 0.001 in the off-diagonal. All such relative probabilities must be so altered to "actual probabilities" P that their sum equals one before calculating the true population information measures.

Monte Carlo Results

I concentrated my efforts on results that were within a meaningful range in order to conserve computer time. In many cases the results (Tables 6, 7; Figures 1–8) are indicated as approximate ranges or beyond the scale of the figure. Estimates that were outside of the bounds of reasonable difference were only approximated.

Miller's Chi-square. Miller's χ^2 was dangerously nonconservative when applied as suggested by Attneave (1959). More than 10 percent of the

Table 5. Types of Transitional Arrays Tested in the Monte Carlo Study and Total Number of Samples Made from each Array. For a 10 × 10 matrix, there are 10 cells on the major diagonal and 90 cells off the major diagonal. A 4 × 4 matrix of type II is presented for explanatory purposes both as relative and absolute probabilities. See text.

Type	Relative Transitional Probabilities: On Major Diagonal	Off Major Diagonal	Number of Times Sampled
I	0.9	0.001	18,200
II	0.9	0.01	32,675
III	0.8	0.13	8,675
IV	0.7	0.25	4,175
V	0.6	0.38	1,775
Even	0.5	0.5	1,032

Type II Relative Matrix				Type II Absolute Matrix			
0.9	0.01	0.01	0.01	0.24	0.003	0.003	0.003
0.01	0.9	0.01	0.01	0.003	0.24	0.003	0.003
0.01	0.01	0.9	0.01	0.003	0.003	0.24	0.003
0.01	0.01	0.01	0.9	0.003	0.003	0.003	0.24

Table 6. The Percentage of 100 Iterative Estimates, $T'(x; y)$, That Were Significantly Greater than Zero When the True Population $T(x; y)$ = 0.0. P = probability level of the test. t = Student's t value.

		Percent of $T' > 0$		
	N	By Miller's χ^2		By David Method
Number of Acts	Degrees of Freedom	$P = 0.05$	$P = 0.01$	$t = 1.96$
4	9	16	5	2
8	1.3	12	2	5
8	3.8	11	2	1
10	1.2	32	7	11
10	2.5	31	6	6
10	4.9	23	8	0

Table 7. Estimates and Population Values of Information Measures for Selected Matrices. All estimates are medians of 100 values. Signal weighting is the percentage of signals that were either 10 or 100 times more common than the other signals. Blanks in a column indicate that the value has not changed from that given above. $TE(m)$ is $T(x; y) \div H(m)$.

Matrix Type	Signal Weighting	N	$T(x; y)$	$T'(x; y)$	$TE(x)$	$TE'(x)$	$TE(y)$
I	None	400	3.20	2.10	0.96	0.94	0.96
		800		3.16		0.95	
	30% are 10×	400	2.42	2.29	0.95	0.91	0.96
		800		2.35		0.92	
		1600		2.40		0.94	
	10% are 100×	200	0.62	0.30	0.84	0.47	0.91
		400		0.47		0.65	
		800		0.57		0.76	
		1600		0.58		0.80	
		3200		0.60		0.86	
	30% are 100×	400	1.70	1.56	0.93	0.87	0.96
		800		1.64		0.91	
		1600		1.68		0.91	
		3200		1.69		0.94	
II	None	100	2.59	2.26	0.78	0.69	0.78
		200		2.51		0.76	
		400		2.59		0.78	
	10% are 10×	100	1.91	1.50	0.72	0.55	0.76
		200		1.78		0.63	
		400		1.86		0.73	
		800		1.90		0.72	
	30% are 10×	100	1.94	1.55	0.73	0.65	0.77
		200		1.79		0.67	
		400		1.87		0.73	
		800		1.90		0.72	
	10% are 100×	100	0.44	0.0	0.37	0.0	0.65
		200		0.20		0.23	
		400		0.32		0.30	
		800		0.39		0.32	
		1600		0.42		0.37	
	30% are 100×	400	1.41	1.31	0.66	0.63	0.79
		800		1.37		0.62	
		1600		1.39		0.65	

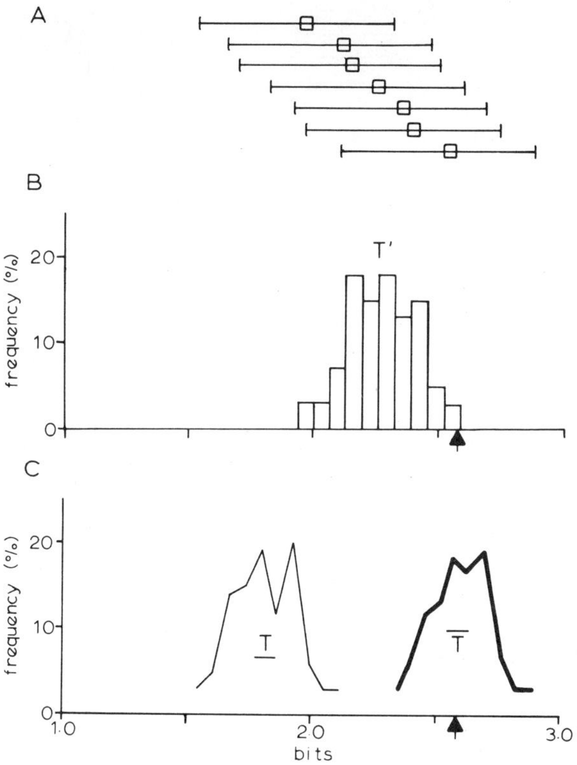

Figure 1. Information estimates for a Type II, 10 × 10 matrix with sample size of 100 for each of the 100 estimates: *A*. *T′ (x; y)* and 95% David confidence limits in rank order for the 3rd, 12th, 25th, 50th, 75th, 87th, and 98th ranks out of 100 estimates. *B*. Frequency distribution for 100 estimates, *T′(x; y)*, of the same transitional array. The true *T(x; y)* for the array is indicated by the arrow. *C*. Frequency distributions for the limits of the 95% confidence bounds for the *T′(x; y)* above.

$\hat{T}(x; y)$ drawn from even matrices [$T(x; y) = 0$] were indicated as significantly greater than zero in each case (Table 6). The alpha error rate was clearly greater than the 5 percent suggested by the χ^2 tables. But, when 1 percent probability was used as the significance level from the χ^2 table, the alpha error rate of Miller's χ^2 was only 2 to 8 percent.

The beta error rate was indicated by observing the number of $T(x; y)$ indicated as significantly greater than zero when $T(x; y)$ was nonzero. If only 50 of 100 samples were significant, beta error would be about 50 percent. Fifteen matrices were examined with $T(x; y)$ ranging from 0.01

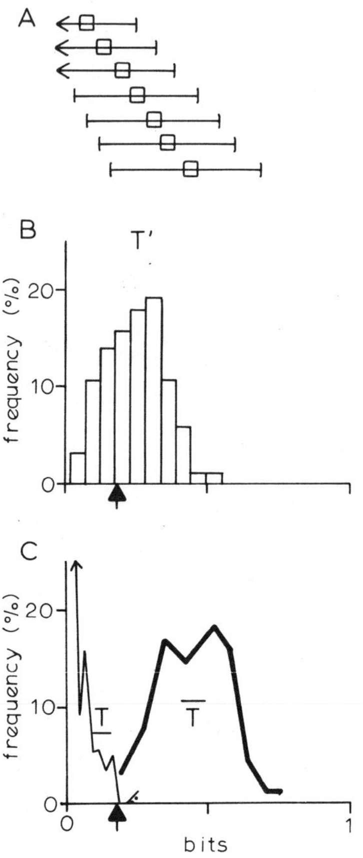

Figure 2. Information estimates for a Type IV, 10 × 10 matrix with sample size of 100 for each of the 100 estimates. See legend of Figure 1.

to 3.20 bits per act and 9 to 18 degrees of freedom. When feasible, N was increased until 95 percent of the $\hat{T}(x;\ y)$ were significantly different from zero.

A simple answer was elusive, since beta error was a function of T, N, and degrees of freedom. To provide a conservative guide, I considered only the worst cases, those with a signal weighting of 10 and 100. Parametric regression of the results of these cases suggested that, to have a beta error rate less than 5 percent

$$\frac{N}{\text{degrees of freedom}} \geq 4 - 4.7\ ln\ T(x;\ y) \tag{1}$$

But, if $T(x; y)$ is greater than 1.8, a sample size equal to degrees of freedom should suffice. The sample size required for a low beta error rate becomes astronomical at $T(x; y)$ less than about 0.1 bits per act. But this expression is only a crude guide for choice of sample size and, in some cases, may be overly pessimistic.

Bias Correction. $\hat{T}(x; y)$ was frequently a gross overestimation of $T(x; y)$. At low values of $T(x; y)$, $\hat{T}$ minus T was up to 98 percent of $\hat{T}$. At near maximum T, bias was slight. Conversely, T' was a good estimate of T except at near maximum values of T (Figures 1 to 6, Table 7). Signal

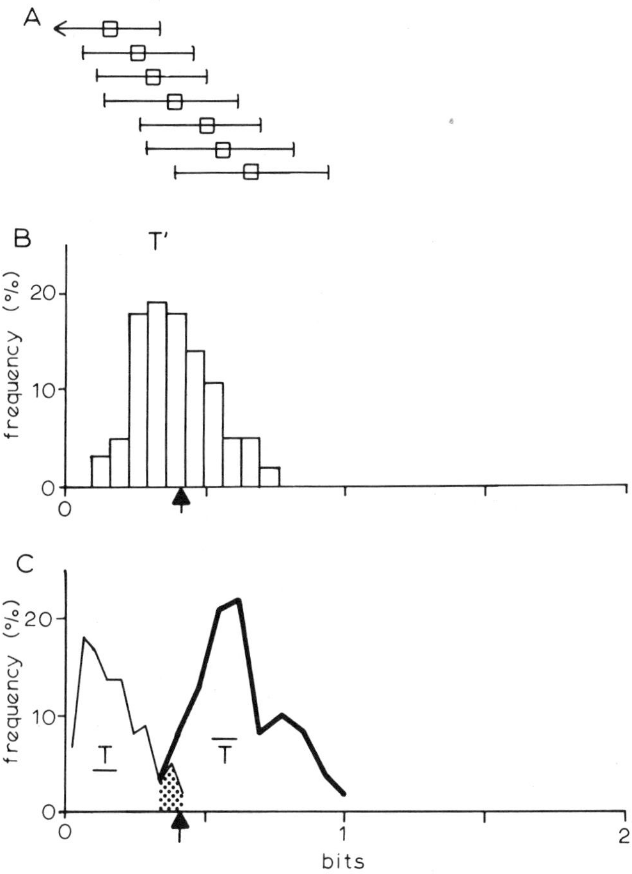

Figure 3. Information estimates for a 4 × 4 matrix similar to Type III with sample size of 80 for each of the 100 estimates. See legend of Figure 1.

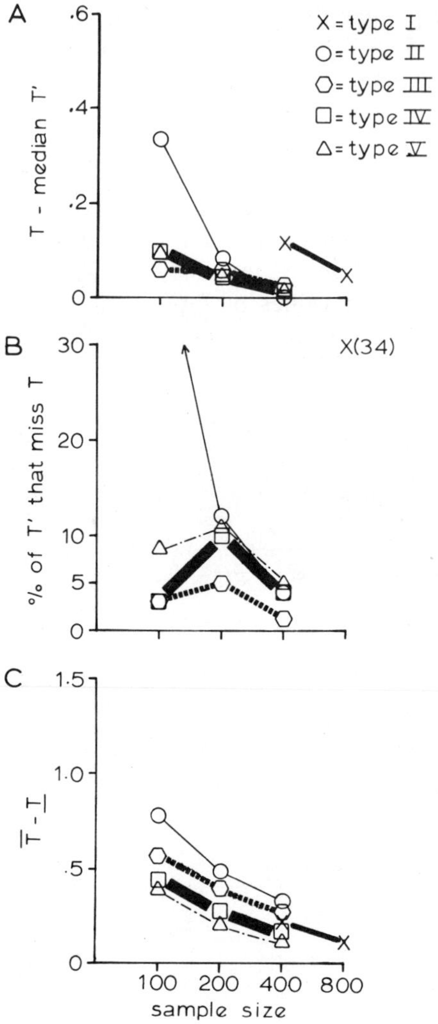

Figure 4. Evaluation of information measures for 10 × 10 matrices where all signals had equal weighting: *A*. Bias indicated as the true population *T(x; y)* minus the median of 100 estimates, *T'(x; y)*; *B*. The percentage of the 100 estimates with either lower or upper confidence limits that did not include the population *T(x; y)*; *C*. The average difference between upper and lower confidence bounds.

weighting also had a strong effect on bias (Figures 4*A*, 5*A*, 6*A*, Table 7): Strong signal weighting increased bias.

I found no consistent relationship between the degree of bias and indicators such as the number of zero cells in the matrix. In order to gain some numerical guide for bias, the results from all Type II, 10 × 10 matrices were examined by parametric regression. The results suggest that

$$T' \text{ bias} = 0.32 - 0.14\, lnS \qquad (2)$$

where T' bias = $|T(x;\, y) - \text{median } T'(x;\, y)|$ and $S = N \div$ degrees of

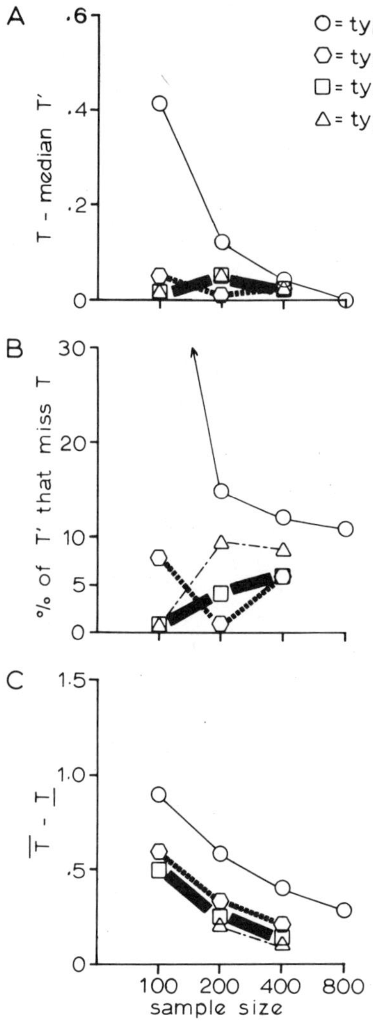

Figure 5. Evaluation of information measures for 10 × 10 matrices where 10% of the signals were weighted to be 10 times more common. See legend of Figure 4.

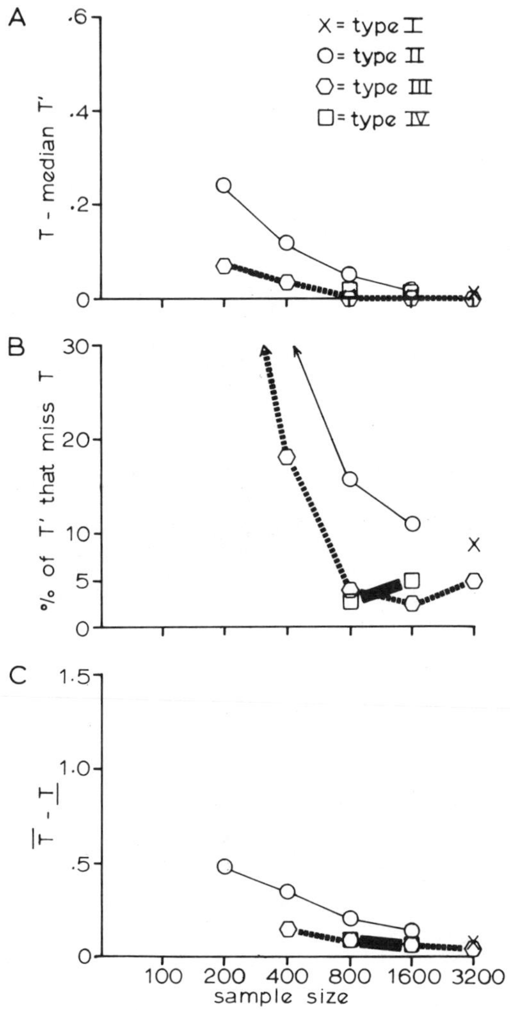

Figure 6. Evaluation of information measures for 10 × 10 matrices where 10% of the signals were weighted to be 100 times more common. See legend of Figure 4.

freedom is a fairly good guideline for bias if S is less than 12 (r for regression = 0.8, N = 11). The results from 4 × 4 matrices fit this relationship fairly well.

Confidence Limits. The David method was conservative as a test of the null hypothesis that T equals zero. Its 95 percent alpha error rate was approximately the same as the 99 percent Miller's χ^2 (Table 6). The beta error rate appeared to be slightly higher than Miller's χ^2 for every matrix tested.

The limits of the David method confidence bounds were asymmetrical around T' (Figures 1*A*, 2*A*, 3*A*). This asymmetry was similar to the skewed distribution of T' that was particularly marked for values of T less than about 0.5 bit (Figures 1*B*, 2*B*, 3*B*), and suggests that the arcsin transformation might be accurate.

The David method was much less conservative as an indication of T and for comparing samples of different sizes. The alpha error rate was different for tests of different null hypotheses. The simplest null hypothesis is that $T(x; y)$ indicated by one sample is the same as the $T(x; y)$

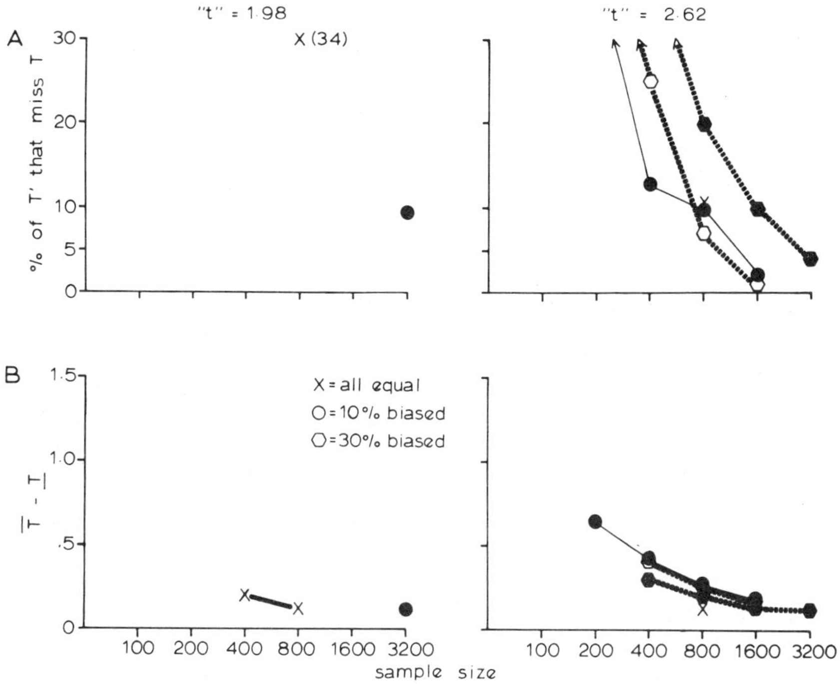

Figure 7. Evaluation of information measures for Type I, 10 × 10 matrices where all signals had equal weighting or were weighted to have 10% or 30% of the signals to be 100 times more common. Open and closed symbols were weighted by a factor of 10 and 100 respectively. *A*. The percentage of 100 estimates, $T'(x; y)$, that were indicated as significantly different from the population $T(x; y)$ by the David method; *B*. The average difference between upper and lower confidence bounds. Results are given where t values of both 1.98 (95%) and 2.62 (99%) were used.

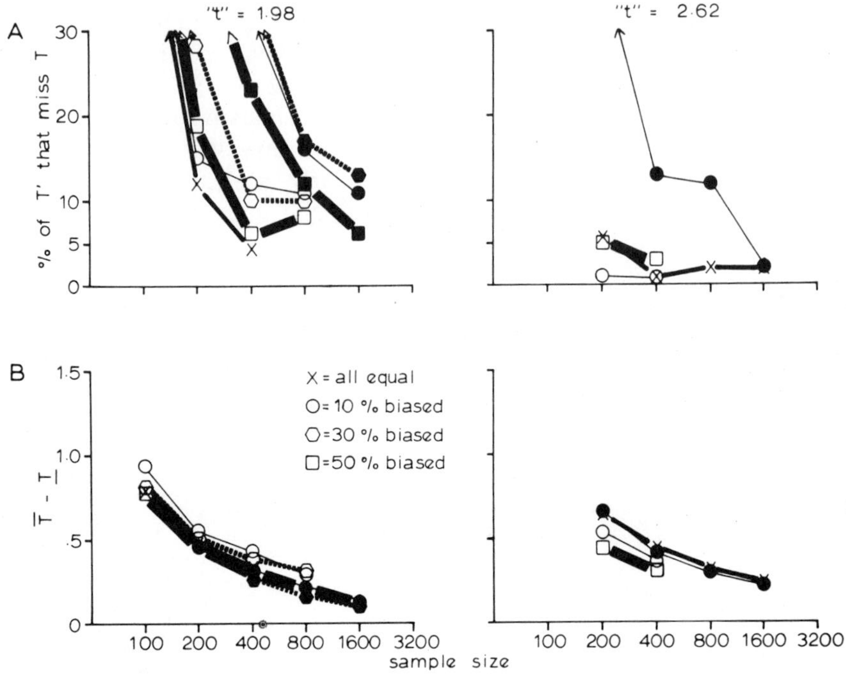

Figure 8. Evaluation of information measures for Type II, 10 × 10 matrices. See legend of Figure 4.

estimated by another with equal sample size. An indication of beta error rate for this hypothesis is gained by examining the distributions of the upper and lower bounds of estimates drawn from the same matrix. Confidence limits are calculated for 100 estimates of T from the same distribution. The frequencies of occurrence of the upper limit and lower limit of the confidence intervals are plotted separately (Figure 1*C*, 2*C*, 3*C*). Overlap of these distributions indicates that the upper limits for some estimates were less than the lower limits for some others: The estimates were indicated as statistically different.

The stippled areas in Figures 2*C* and 3*C* indicate the percentage of extreme bounds that overlap. In 107 trials of different matrices with 100 T' in each trial, I found only nine trials that indicated an alpha error rate greater than 5 percent (more than 5 of the 100 T' indicated as different from each other). There was little chance of obtaining statistically different estimates of T when samples of the same size were drawn from the same matrix.

This result is encouraging but does not reveal the alpha error rate when samples are of different sizes. To evaluate this source of error, I took samples of N equals 100, 400, 800, and 1600 from various 10 × 10 matrices with 100 repetitions of each sample size. I then examined the number of extreme values for T' from one sample size that were significantly different from the extreme values of T' at a different sample size.

The David method had a high alpha error rate when comparing T' values of different sample size (Table 8). For example, a Type II matrix with 10 percent of the signals weighted by 100 was sampled at all sizes. All of the T' where N equals 400 and 800 were significantly different from all of the N equals 100 estimates when a Student's t value of 1.98 was used. When comparing samples where N equals 800 with estimates

Table 8. The Percentage of $T'(x; y)$ of Greater Sample Size That Were Significantly Different (David Method) from Estimates of Sample Size 100 and 400 in 10 × 10 Tables. Dash indicates no samples made. Percentages are approximate ranges due to the technique used.

Matrix Type	Signal Weighting	$T(x; y)$	t	Percent That Differ from N = 100 Sample		Percent That Differ from N = 400 Sample	
				N = 400	N = 800	N = 800	N = 1600
I	None	3.20	1.98	—	—	3–12	—
	None	3.20	2.35	—	—	3	—
	None	3.20	2.62	—	—	3	—
	30% are 10×	2.42	2.62	—	—	3–12	13–25
	10% are 100×	0.62	2.62	—	—	13–25	26–50
	30% are 100×	1.7	2.62	—	—	3	51–75
II	None	2.59	1.98	50–75	—	—	—
	None	2.59	2.62	—	—	3	3
	10% are 10×	1.91	1.98	50–75	100	3–12	—
	10% are 10×	1.91	2.62	—	—	3	—
	30% are 10×	1.95	1.98	88–97	100	3–12	—
	50% are 10×	2.15	1.98	75–88	100	3–12	—
	50% are 10×	2.15	2.62	—	100	3–12	—
	10% are 100×	0.44	1.98	100	100	13–25	50–75
	10% are 100×	0.44	2.62	—	—	3	3–12
	30% are 100×	1.41	1.98	—	—	13–25	25–50
	50% are 100×	1.92	1.98	—	—	3–12	13–25
III	None	0.45	1.98	3	—	—	—
	10% are 10×	0.33	1.98	13–25	—	—	—
	10% are 100×	0.07	1.98	3–12	—	—	—

where N equals 400, error rates of 13 to 25 percent were indicated with a Student's t of 1.98. Increasing the Student's t to 2.62 decreased the error rate to about 3 percent. Samples where N equals 1600 could not be reliably compared with N equals 400 samples at either choice of Student's t.

This result is discouraging but still does not evaluate the alpha error rate of T' as an indicator of T. I examined the error rate as the number of T' that were significantly different from the T of the sampling matrix (T'that miss T, Figures 4*B*, 5*B*, 6*B*, 7*A*, 8*A*).

The short upper bound of the confidence limits was frequently less than T indicating a high alpha error rate. The worst cases were again at high ranges of information transmission and signal weighting. Without signal weighting, a sample size of about four times the degrees of freedom produced a satisfactory alpha error rate of about 5 percent (Figure 4*B*). But for the worst cases, alpha error rate approached a minimum of about 10 percent (Figures 5*B*, 6*B*); additional increases in sample size were profitless. Increasing Student's t to 2.62 usually resulted in satisfactory alpha error rates, but sample sizes of 10 to 40 times the degrees of freedom were required (Figures 7*A*, 8*A*).

The beta error rate is more encouraging. The mean range of the confidence limits ($\bar{T}-\underline{T}$) indicates the power of the technique to discriminate between values of T. These values are relatively consistent for different types of matrices (Figures 4*C*, 5*C*, 6*C*, 7*B*, 8*B*). Parametric regression of the data in Figure 8*B* for a Student's t of 1.98 suggests that

$$\bar{T}-\underline{T} = 1.8-0.23 \ln N \qquad (3)$$

(r for the regression = 0.9, N = 19). This relationship was calculated for the worst cases so it should be a conservative guide toward the prediction of confidence intervals. (The calculation of $\bar{T}-\underline{T}$ excludes all confidence bounds that are truncated by $\underline{T}$ equals zero. Thus this measure only applies to confidence intervals that do not include zero.) Note that increasing Student's t to 2.62 does not seriously alter the confidence intervals (Figures 7*B*, 8*B*).

Conclusions on Measures of Statistical Probability

My most important conclusion from the Monte Carlo study is that if you want to be certain of alpha and beta error rates, perform a similar Monte Carlo study of your matrix. Neither Miller's χ^2 nor David's confidence limits displayed what I would call reasonable stability or statistical power on all types of matrices.

The Monte Carlo approach is not as formidable as it sounds. My study was expensive and time-consuming because I was searching for rules that might have general application. Study of a specific case can be relatively cheap and easy to perform. To check the alpha error rate for Miller's χ^2, sample repeatedly from an even matrix of the desired sample size and dimensions. The estimate of alpha error rate is the percentage of T' significantly greater than zero. The beta error rate can be estimated by sampling from a matrix of transitional probabilities that represent what you expect from your animals, and finding the proportion of the samples that are significantly different from zero. Similar runs could be made for testing the effectiveness of the confidence limits. Monte Carlo trials are particularly important for multidimensional applications that I have not examined.

For many applications, the 99 percent Miller's χ^2 is effective. It should be particularly valuable for multidimensional analysis of the relative importance of context, signals, and preponses on communication. Chatfield and Lemon (1970) discussed the similarity between χ^2 and measures of information. They concluded that Miller's χ^2 should not be used unless all of the expected values are greater than one and about 80 percent are greater than five, and providing the degrees of freedom are not too large. In studying the sequence structure of bird song, they concluded that the only advantage gained over a standard χ^2 was the opportunity to judge significance from a graphical representation. But I found that, at least up to the degrees of freedom encountered herein, Miller's χ^2 could be a valuable test at sample sizes that violated their suggested guidelines.

The David method for calculating confidence limits should probably be limited to comparing estimates with similar sample sizes. You should be suspicious of "significant" differences between T' values that differ by only a few tenths of a bit per act. Considerable caution should be used when comparing T' values estimated by samples of different size or when comparing T' with a hypothesized value of T.

Sampling bias was the most difficult problem. Uncorrected estimates of T overestimate information transmission. Fortunately, both bias correction and confidence limit estimation were fairly well behaved at low to moderate values of T such as an ethologist frequently encounters. Corrected estimates of T underestimate information transmission but approach the true value as sample size increases. Fagen (ms.) performed a Monte Carlo study on matrices selected from the behavioral literature. He concluded that bias correction and the David method were adequate for description of these matrices.

Most studies in ethology have used $\hat{T}(x; y)$ and are thus very likely to be overestimates of actual information transmission. This is particularly true where sample sizes were only a few times the degrees of freedom or where some signals were far more common than others.

Failure to correct measures may be particularly troublesome in applications such as character analysis (Hazlett and Estabrook 1974a, b; Rubenstein and Hazlett 1974). Hazlett and Estabrook (1974b) examined the effects of using an entire sample as opposed to subsamples of lesser size from the same data. They found higher estimates of information transmission from the lower sample sizes. They drew attention to the fact that overestimates of transmission can result from spurious patterns that appear when qualitatively different sequences of behavior are lumped together. While this is certainly important, their estimates could have been improved by correcting for sample size. This may be particularly important when one of the characters is position in a behavioral sequence. Unless all sequences are of the same length, fewer dyads will be sampled later in the sequence resulting in decreased sample size and increased bias. They discuss the importance of separating long from short sequences but their reasons are based upon biological as opposed to statistical problems.

Biological Interpretations

Degree of Communication. It is critical to examine what is meant by information sharing. A nonzero T'(signal; response) indicates that, given the diversity of signals and responses in the observations, the subjects were able to demonstrate transfer of so many bits of information per act. This may be representative of the individual's capability to communicate if: your definition of a signal-response dyad was adequate; the sample size was large enough for statistical purposes; you did not ignore relevant signals, responses, or contextual variables; you did not fail to distinguish between signals of different meaning; and the diversity of signals was representative of the individual. This is a large set of assumptions.

Comparisons are relatively safe if we stay within the same or very similar species that are observed in the same context and with the same definition of dimensions. But I doubt the biological validity of comparisons between dissimilar species and dissimilar methodologies. Even if we ignore the statistical difficulties, the biological problems are immense. Failure to account for any of the above problems can lead to large differences in T values. Biological inference must be carefully qualified as to the conditions under which T was estimated.

The values in Table 7 were selected from the results of the Monte Carlo study to demonstrate the variability of the measures even when all of these assumptions are satisfied. $T(x; y)$ is reduced if:

1. Less information is offered by the signal sender [low $H(x)$];
2. If some responses become uncommon in the responder [low $H(y)$]; and/or
3. If the "importance" of the signals, or the response constraint is reduced [higher $H(x, y)$].

It is difficult to draw conclusions about phenomena such as changes in communication as dominance is established if only $T(x; y)$ is known. Do the signals now mean more or less to the subordinate animal? Has the dominant animal shifted its pattern of display?

Problems of interpretation are evident in the literature. In discussing changes in $T(x; y)$, Dingle (1969) concluded that displays changed in how meaningful they were. This wording is correct if "meaningful" is interpreted in the context of the sequence. That is, a signal sequence can become less meaningful merely by having much repetition of a few signals resulting in a low H(signal). Brown (1975) may have made a common error in interpretation of Dingle's results or at least offered the chance of misinterpretation to his readers. He states (p. 280) that as $T(x; y)$ decreases, the effect of a "given act" is reduced. This implies that the signal is given a less precise interpretation by the receiver. Examination of Table 7 clearly indicates that this is not necessarily true. Much confusion could be avoided if authors would provide more precise wording of their interpretation of such measures.

Transmission efficiency (TE) is commonly used to correct information measures for the relative frequency of occurrence of the action patterns. The usual definition of $TE = T(x; y) \div H(x)$ is not sufficient to document changes in the strength of the effect of a given signal (Table 7). However, $T(x; y) \div H(y)$ shows little change with the relative frequency of the signals when the signals show no change in the strength of their effects on responses.

I suggest that any interpretation of information measures should include consideration of a variety of T', TE', and R' values. Rand and Rand (1976) carefully considered H and T values to interpret their observations. As a result, while bias correction should have been applied, their discussion was clear and not misleading.

After the sources of communication have been examined, one final and critical judgment remains: How much biological importance should be inferred when two measures differ by x number of bits per act. My Type I and Type II matrices were very similar but both TE measures indicate a difference of about 0.2 bit (Table 8). In both cases, 10 messages were communicated with a high degree of accuracy: The differences between the matrices appear to have little biological reality.

T may not be an ideal measure of what we consider to be biological communication, constraint of another individual's behavior. If a perfect communication matrix is made less perfect, *T* at first drops rapidly and then decreases more slowly. Fager (1972) found a similar difficulty in using *H* to indicate diversity of a community. Users of information theory should keep this in mind while evaluating the biological significance of information measures.

Sources of Communication. You can gain insight into the sources of correlation in a contingency table by examining partialized χ^2 values, *I* values, or by using more powerful iterative techniques (Goodman 1968; Slater 1973; Chapter 6 of this volume). But the value of the technique depends on the biological questions that are posed. Slater and Ollason (1972) and Slater (1973) discuss problems associated with testing such tables and generating expected values. But their methods are tailored to reveal pattern in a sequence of behavior by an individual (see Chapter 3), not to reveal the significance of particular signals in a communication system. Their methods, and those used by other ethologists (e.g., Hazlett and Bossert 1965) all test the null hypothesis that the observed values are unlikely given the observed distribution of signals and responses: Expected values depend both on the distribution of signals and responses.

If we want to gain insight as to the meaning of a signal, expected values should not depend on the frequency of other signals. Expected values should reflect the response tendencies of the animal in that context regardless of the signal given. This goal is usually impossible so we are forced to use our observation data itself to generate expected values. The method I suggest provides a somewhat better approximation to our desired distribution by correcting for the relative frequency of signals before calculating column sums. By this method, sampling abnormalities have less effect on our conclusions.

Final Remarks

I judge information theory to be a useful, but not omnipotent, tool for behaviorists. It should be employed as one of several analytical and descriptive tools. At times, it may be the only applicable technique due to its generality. At other times it should accompany other, more robust and reliable techniques.

CHAPTER
4

Temporal Patterns of Behaviors: Durations, Intervals, Latencies, and Sequences

ROBERT M. FAGEN and DONALD Y. YOUNG
University of Illinois at Urbana

Temporal patterns of behavioral acts shed light on the motivation and adaptation of animals, and perhaps even on the evolution and function of behavior itself. Elements of this concern with temporal patterns are of course present in daily life. One may wait, perhaps for some time, for a friend or contributor to write (latency of response). Likewise, to cite a relevant but unrelated literary parallel, Jane Austen's Emma was intrigued to learn of the progression of Mr. Elton's courtship, from the initial stage of accidental rencontres and evenings with others at Mrs. Brown's, to later interesting blushes and palpitations (sequence of responses). Like Emma and other arch Austen maidens, ethological analysis of temporal patterns of behavior reflects its historical context. The ethological study of behavior sequences began with Tinbergen's classic analysis of stickleback courtship, and sequential analysis of courtship in a variety of species has continued. But concern has also broadened, and it would be difficult to find a major area of behavior in which at least one form of temporal pattern analysis has not been performed. The goal of this chapter is to introduce a variety of techniques for analyzing temporal patterns of behavior, stressing biological informativeness of such techniques, whether these be simple graphical methods or complex algorithms requiring a medium- to large-sized digital computer. In discussing temporal patterns of behavior, we examine duration and interval measures as well as the more traditional forms of sequence analysis. We adopt this approach and terminology because, as Nelson (1964) has pointed out, "descriptions of behavior should include the temporal as well as the sequential relations between the behavioral events being described." Use of correct sampling techniques is a crucial first step in sequential analysis; since appropriate techniques have been presented in depth by J. Altmann (1974), there is no need to repeat that material here.

The organization of our chapter in support of the above goal is as follows. Section 1 treats temporal patterns of single types of behavioral acts, the log survivor function approach, and some general models that aid in interpreting empirical log survivor functions. In Section 2 we outline the ethologist's standard Markov chain technique for analyzing behavioral sequences in discrete time. We then present a number of currently accepted methods for analyzing continuous-time sequences (i.e. sequences in which the durations of or intervals between acts, as well as the identities of the acts themselves, are measured). In Section 3 we discuss relationships between temporal pattern analyses and certain quantitative methods presented in a more general context elsewhere in this volume. Section 4 offers some ideas on possible futures for ethological temporal pattern analysis.

1. TEMPORAL PATTERNS OF SINGLE TYPES OF BEHAVIORAL ACTS

Most problems in animal behavior may be approached from at least five distinct subdisciplinary points of view. Morphology and physiology help answer "how" an organism responds to stimuli. "Where" an animal exists is basically an ecological question, while "why" is often an evolutionary (and occasionally a teleological) question. Ethology, however, generally involves the questions of "what" and "when." We address the latter question in this section by describing concepts and methods useful in examining the time dimension of behavioral acts. Our approach deals with onset and cessation of acts, as well as with sequential order of several successive acts. Wolf vocalizations, whale respiratory "blows," fiddler crab courtship, starfish feeding—all these behavioral phenomena lend themselves to temporal and/or sequential analyses.

Durations, Intervals, Latencies

Before discussing time analyses of behavioral data it would be appropriate to define the terms duration, interval, and latency. The *duration* of a particular act is the amount of time elapsed between the onset of the act and its end. The length of time that a Japanese macaque spends uninterruptedly rinsing the sand off a single sweet potato, for example, may be thought of as wash duration. A single wolf howl is said to have a duration of 0.5 to 11 seconds (Theberge and Falls 1967). Not all acts, however, have measurable duration; those acts with duration of zero length are considered to be point events. It can be argued that techniques are available to measure the time elapsed while a chicken lowers its head, strikes a food item, and raises its head, as well as the time it takes an odontocete spiracle to open and close. However, the duration of such events may be sufficiently small relative to other acts to permit the investigator to treat them as point events.

The time between the end of one act and the beginning of another (whether or not of the same type) defines the *interval* between two acts. Depending upon the question being asked, the second act may be of a different type than the first, or simply another "token" of the same type. To investigate interval length between successive fin displays (tokens) of a single male *Betta splendens*, one would start the chronometer at the end of one fin display and record the amount of time elapsed until the commencement of the next fin display, regardless of what acts had transpired in the interim. Since animals rarely do "nothing" during the interval between any two acts, this interval can be equivalent to the duration of interim behaviors. For instance, the interval between two successive respirations in whales is equivalent to the duration of the dive separating the two surfacings, since most whales submerge between "blows."

The quantity of time that elapses between the presentation of a stimulus and the initiation of a response is known as the *latency* or latent period. It often represents the time during which the stimulus information is being neurophysiologically processed. As an example, Archer (1973a) examined the effects of testosterone upon young male chicks' latencies to resume movement after presentation of a loud auditory stimulus.

The concepts of durations, intervals, and latencies are schematically summarized in the time lines of Figure 1 where the horizontal bars represent the occurrence of act A (open) and act B (solid) and where

D_A = duration of act A
D_B = duration of act B
I_{AA} = interval between two act As
I_{AB} = interval between act A and act B
I_{BA} = interval between act B and act A
L_A = latency of act A if stimulus were applied at $t = 0$
L_B = latency of act B if stimulus were applied at $t = 0$

Log Survivor Functions

Under many circumstances, neither the duration of individual acts nor minor variations in the form and intensity of these acts are of interest,

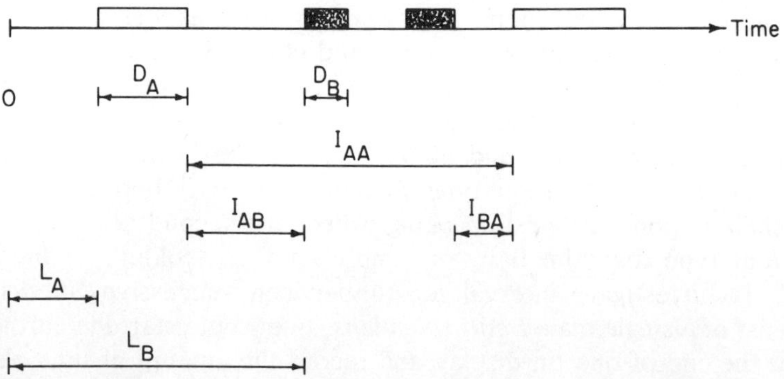

Figure 1. Behavioral time lines, illustrating durations, intervals, and latencies. See text for notation.

and the occurrences may be viewed as identical point events (see above). One may then examine the temporal pattern established by these points along a one dimensional time line. Often the points are arranged in such a pattern that several tokens of a single type occur in rapid temporal succession, with relatively long periods of time demarcating these "bursts" of activity. The ethologist may wish to organize behaviors having this pattern into groups or bouts, with long periods of time (gaps) separating the bouts. For many purposes it is important to define an objective criterion for determining whether a particular interval is an intrabout space or an interbout gap. In other words, what determines whether a single point event is still part of a preceding bout or simply the beginning of the next bout? Ideally, the criterion chosen reflects the organization or structure of the behaviors themselves (Slater 1974). Criteria have been based on arbitrary decisions (Baerends 1970; Magnen and Tallon 1966; Thomas and Mayer 1968; Rowell 1961), on frequency histograms of intervals (Ewing 1969; Isaac and Marler 1963; Lemon and Chatfield 1971; Levitsky 1970), and on log survivor functions (Delius 1969; Machlis 1977; Nelson 1964; Ollason and Slater 1973; Slater 1974a, 1974b; Slater and Ollason 1972; Wiepkema 1968).

A log survivor function is displayed by plotting the intervals between behavioral events semilogarithmically. On the graph of a log survivor function, the abscissa corresponds to time t and the ordinate displays the log of the number of intervals whose lengths are greater than time t. Construction of a log survivor function typically begins with a frequency histogram of the data to be plotted. Figure 2 represents a frequency histogram of interblow (respiration) interval lengths of a particular blue whale (*Balaenoptera musculus*). Figure 3 is the log survivor plot of the same data. Both X-axes have identical scales. Whereas the Y-axis of the frequency histogram represents the number of intervals of X length, that of the log survivor plot denotes, on a log scale, the number of intervals greater than X length. For instance, the point on the log survivor function designated by the coordinates (1, 28) corresponds to the total number of intervals greater than one minute, as well as the shaded region in the frequency histogram of Figure 2. A log survivor plot uses a tabulation, for each x, of the number of intervals in a frequency histogram to the right of successive values of X.

A frequency histogram of intervals between randomly occurring point events (arrivals in a time-homogeneous Poisson process) will be described by a negative exponential distribution (Cox and Lewis 1966; Duncan et al. 1970). This distribution describes many sets of ethological time pattern observations and has been used in an interesting estimation procedure for characteristics of behavior data (Chow 1975). Its intriguing

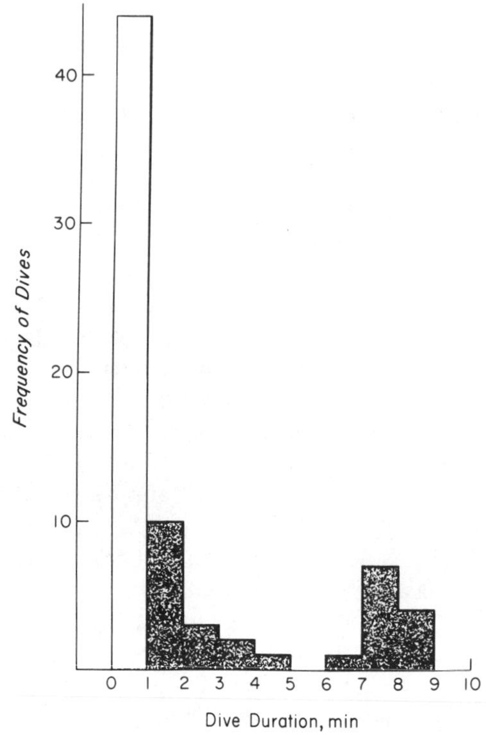

Figure 2. Frequency histogram of interrespiration interval lengths in a blue whale.

theoretical implications are discussed below in connection with evolutionarily stable strategies. A log survivor plot of intervals between these events may be fitted by a straight line whose slope is proportional to the probability of an event occurring at any given time after the last event. This technique is widely used in mortality, reliability, and risk analyses, since it is considerably easier to compare such a plot with an expected straight line or lines than it is to compare a corresponding frequency histogram with a negative exponential curve. Moreover, since the slope of the log survivor function is proportional to the probability of an event occurring at any given time after the last event, a change in probability will be signaled by a break in the curve, as illustrated in Figure 3. Such a break may be used as the bout criterion to separate intrabout intervals from interbout gaps.

Slater (1974a) describes a minor problem with such an analysis of intervals. Because observation periods are often relatively short, long behavioral intervals are infrequently observed or recorded because these intervals tend not to be complete within a single observation period; they either begin or end outside the limits of an observation session. As a result, the observed number of complete long intervals is generally less than would be expected. To lessen the impact of such an effect, Slater (1974a) constructed a data point by adding the time elapsed between the onset of the observation session and the first event to the time between the last event and the end of the session.

Machlis (1977) developed an objective computer procedure for determining the value of the bout criterion. She then applied this procedure to experimental data on interpeck intervals in chicks. Bouts of pecks were identified. In turn, these bouts were themselves found to cluster into superbouts. These clusters of bouts seemed to indicate periods of ongoing attention to the stimulus. Some experiments, however, were difficult to interpret quantitatively in a bout-superbout framework.

Objective characterization of a behavior bout may sometimes be quite difficult. While a stable and objective bout criterion can be defined, this criterion by itself fails to provide a reliable means of classifying intervals

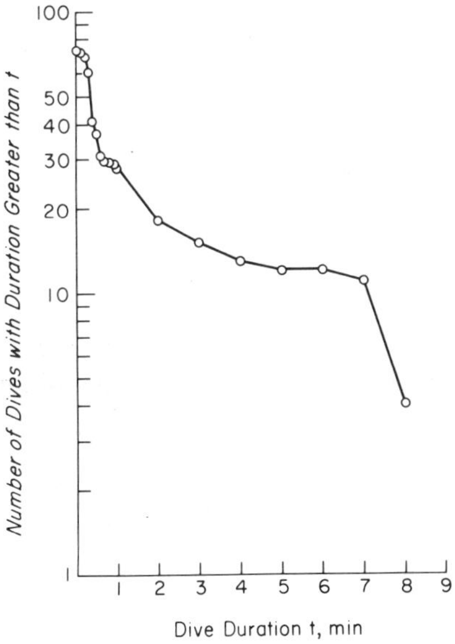

Figure 3. Log survivor plot of data of Figure 2.

into short "within-bout" and long "between-bout" categories. The reason for this failure is that these intervals are produced by (at least) two distinct behavioral processes. While the mean interval lengths for the two processes are distinct, the two probability distributions overlap. The motivational process generating between-bout intervals will sometimes by chance produce a short interval, which would then be misclassified as within-bout by existing statistical procedures subsequently applied to the data. As a result, the true mean within-bout interval length would be overestimated, the true mean between-bout interval length would be underestimated, and a substantial proportion of between-bout intervals would go undetected.

A possible solution to the problems identified above is as follows. Use a maximum-likelihood statistical estimation procedure (cf. Bard 1974; Hogg and Craig 1965; Weiss and Wolfowitz 1974) to ascertain the rate parameters and mixing ratios of a mixed exponential distribution. If behavior is clustered into bouts, but superbouts do not exist, estimate p, a_1, and a_2, the parameters of the mixed distribution

$$f_t(t_0) = pa_1 \exp(-a_1 t_0) + (1 - p)a_2 \exp(-a_2 t_0), \quad t_0 \geq 0$$

The bout criterion, T, is the value of t_0 for which the $(p, 1 - p)$-weighted log survivor functions of the component distributions intersect: T is the solution to the equation

$$\ln p - a_1 T = \ln(1 - p) - a_2 T$$

implying

$$T = \{\ln[p/(1 - p)]\}/(a_1 - a_2).$$

The probability density function of interval lengths in the corresponding superbout model is

$$f_t(t_0) = p_1 a_1 \exp(-a_1 t_0) + p_2 a_2 \exp(-a_2 t_0) + (1 - p_1 - p_2)a_3 \exp(-a_3 t_0), \quad t_0 \geq 0$$

Once the behavioral intervals are partitioned into bouts and gaps, separate, but similar, analyses of both may proceed. If the bouts are statistically independent of one another the probability of beginning a new bout (or equivalently, ending a gap) is constant with the passage of time since the previous bout (Slater 1974b). In addition, the log survivor function of the gap lengths should form a straight line. Failure to approximate a straight line may be due to a poorly chosen bout criterion and/or a tendency for the gap lengths to cluster around a single value. This clustering would be the result of recurrence of the behavior after a particular interval length. To test the observed distributions of gap lengths against a straight line or random model, one may apply a χ^2 test for goodness of fit, outlined in the sequel.

The bout criterion, the total number of gaps, and their mean length allow one to calculate the expected distribution of gaps under the assumption that the gap length is a random variable having the negative exponential distribution, $f(t) = ae^{-at}$, where a is equal to the reciprocal of the mean gap length and t is the gap length. By integration, it is possible then to calculate the probability that a gap length t will be within specified limits (t_1, t_2). This probability is

$$P(t_1 \leq t \leq t_2) = \exp(-at_1) - \exp(-at_2)$$

The time between the bout criterion and infinity is then partitioned into four (Slater 1974b) to eight (Slater 1974a) blocks, within each of which an equal number of gap lengths should fall. The observed number of gap lengths in each of these time periods is then compared with the corresponding expected number using the χ^2 test for goodness of fit. Defining the temporal limits of the blocks involves setting $P(t_1 \leq t \leq t_2)$ equal to 0.25 (for four blocks), specifying t_1, and solving for t_2 as follows:

$$\begin{aligned}
P(t_1 \leq t \leq t_2) &= \exp(-at_1) - \exp(-at_2) \\
0.25 &= \exp(-at_1) - \exp(-at_2) \\
e^{-at_2} &= \exp(-at_1) - 0.25 \\
\ln e^{-at_2} &= \ln(\exp(-at_1) - 0.25) \\
t_2 &= \frac{\ln[\exp(-at_1) - 0.25]}{-a}
\end{aligned}$$

For the first block, t_1 is equal to the bout criterion; for each successive block t_1 is equal to the preceding block's t_2. For example, using the data upon which Figures 2 and 3 are based, we find that

bout criterion = 0.7 minute
total number of gaps = 29
mean gap length = 4.506 minute

It follows that $a = 0.222 (=1/4.506)$ and that the temporal limits of the theoretical blocks are

0.7 — 2.26 minutes
2.26 — 4.65 minutes
4.65 — 10.11 minutes
10.11 — ∞ minutes

Within each block we expect 7.25(=29/4) gaps, in contrast to the observed numbers of gaps of 11, 6, 12, and 0, respectively. A χ^2 test for goodness of fit shows that the observed data differs significantly from the theoretical distribution ($P \leqslant 0.01$, 2 d.f. = number of blocks − 2). Hence, for this whale, the gap length appears to be nonrandom and clustered; the intervals between bouts tend to have typical lengths.

Alternatively one may calculate a jackknife estimate of the exponential parameter *a* and its 95 percent confidence bounds, and then plot three negative exponential curves using:

1. Jackknife estimate of *a*
2. Upper 95 percent confidence limit of the jackknife estimate of *a*
3. Lower 95 percent confidence limit of the jackknife estimate of *a*

The observed log survivor function may then be evaluated in light of these three curves. (It is beyond the scope of this chapter to present a detailed discussion of the jackknife technique. This method is discussed by Miller (1974) and by Mosteller and Tukey (1968).)

Once a value for *a* is obtained (either the jackknife estimate, the upper limit, or the lower limit) the straight line describing the corresponding negative exponential distribution may be plotted by setting *a* equal to the slope of the line and the *Y* intercept as the total number of gaps. Once the theoretical lines are plotted, the shape of the actual log survivor function can be visually assessed.

If events in a time series are independent of one another, there will be no significant correlations among the times of their occurrences. Autocorrelation coefficients may be used to determine the existence of such correlations. An autocorrelation coefficient for lag 1 is the correlation coefficient between the length of each interval and that of the immediate subsequent interval. Similarly, an autocorrelation coefficient for lag 2 is the coefficient between each interval length and the second succeeding interval length, and so forth. Delius (1969) used Spearman's Rho (r_s) as the autocorrelation coefficient in his examination of skylark behavior. In their study of merganser displays, van der Kloot and Morse (1975) first calculate the autocovariances of interval lengths (C_i) for a series of different lags, j:

$$C_i(j) = \frac{1}{n} \sum_{i=1}^{n-j} (X_i - \bar{X})(X_{i+j} - \bar{X})$$

where $j = 0, 1, \ldots n - 1$

$$\bar{X} = 1/n \sum_{i=1}^{n} X_i$$

n = total number of intervals

The autocorrelation coefficients $r_i(j)$ are then calculated by

$$r_i(j) = \frac{C_i(j)}{C_i(0)}$$

At an α level of 0.05, the critical value for the autocorrelation coefficients is given by

$$r_{i0.05} = \frac{t_{0.05}}{(t_{0.05})^2 + n - 2}$$

where $t_{0.05}$ is Student's t at the 0.05 level (d.f. $= n - 1$) and n is again the total number of intervals.

Van der Kloot and Morse (1975) point out a possible problem when using autocorrelation coefficients, for lags other than 1, to assess the independence of events in a time series. If a number of $r_i(j)$ for several lags (j) are found to be significant, caution must be applied during interpretation, since the $r_i(j)$ for different lags (j) are not independent of one another. Their approach to this problem is to construct a spectrum, or periodogram, $In(\omega_p)$, using the autocovariances as coefficients to a Fourier series and transforming the variances from a time scale to a frequency scale:

$$In(\omega_p) = \frac{1}{2\pi} C_i(0) + 2 \sum_{j=1}^{l} C_i(j)\cdot\cos(j\omega_p)$$

where $\omega_p = \pi P/l$
$P = 1, 2, 3, \ldots l$
$l = (n - 1)/2$
n = total number of intervals

If the events in a given time series occur at random, then the values of $In(\omega_p)$ are exponentially distributed with a mean of var$(x)/2$. By multiplying each spectral estimate by 2/var(x), the mean becomes $1/\pi$.

Figure 4 is a periodogram of intervals between consecutive salute curtsies in red-breasted mergansers as presented by van der Kloot and Morse (1975). A distinct peak along the ω_p axis at 0.375 indicates an excess of variance for intervals spaced approximately 2.67(=1/0.375)

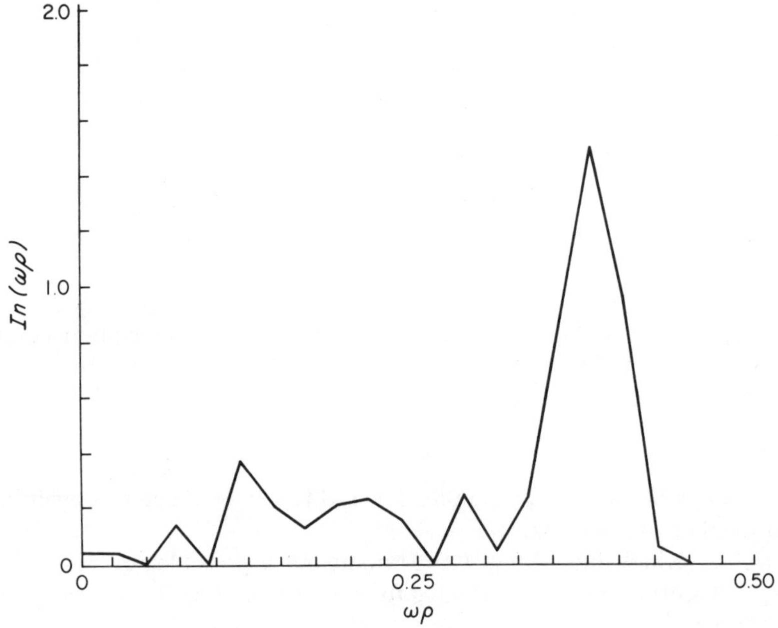

Figure 4. Periodogram of intervals between consecutive salute curtsies in red-breasted mergansers.

lags apart. In other words, if the periodogram can be shown to deviate significantly from an exponential distribution, significant interactions are occurring between intervals that are separated by about 2.67 lags. A periodogram can be compared with an exponential distribution by calculating a series of U_i's where

$$U_i = \sum_{i=1}^{l} \frac{In(\omega_p)}{l/\pi}$$

If one were to plot the U_i's for an exponential distribution as a function of i/n, a straight line from the origin to (1, 1) would be obtained. The Kolmogorov-Smirnov test could then be used to evaluate the goodness of fit of the calculated U_i's to the predicted straight line (van der Kloot and Morse 1975). In similar fashion, it is possible to analyze bout lengths and the occurrences of events within bouts using the methods described above for gap lengths.

Given a log survivor function, can we extract from it any biologically meaningful information other than that of possible existence of bouts and gaps? The answer to this question is an optimistic, if cautious, "yes." Durations of behavioral acts, of intervals between acts, and of latencies between stimulus and response can indicate much about processes controlling behavior. Duration data have stimulated quantitative analyses at least since Peirce (1873) measured and characterized response latencies of human observers, but only recently have duration, latency, or interval data served to test explicit models of behavioral control. Behavior durations seem to reflect strength of reward, a primary determinant of feeding behavior (Sibly and McFarland 1975); analysis of intervals between behavioral acts of a given type reveals varying degrees of control by internal and external factors (Slater 1974a, 1974b; Halliday 1975). In the following subsection we present an explicit mathematical model for duration, interval, and latency data. The advantage of this approach over earlier, empirical approaches is that the model raises the possibility of measuring and comparing relative strengths of different behavioral control processes from observational data alone.

Temporal Patterns of Behavior

Intervals between behavior bouts may exhibit very different types of probability distributions, each type corresponding to a particular kind of behavioral control (Spurway and Haldane 1953; Metz 1974; Slater 1974a). In some cases, the probability of a new bout starting remains constant between bouts, and the intervals are purely random; their lengths have a negative exponential distribution. In other cases, bouts of behavior are overdispersed in time. They are separated by intervals having a characteristic length. Overdispersion would be expected "if causal factors for a particular behavior rise during nonperformance so that its probability of being elicited by the appropriate stimulus becomes greater" (Slater 1974a).

The same sort of reasoning may be applied to behavioral durations. Some types of acts tend to terminate independently of the time at which performance begins. This could be the case, for instance, if occurrence of the event that caused the act to terminate were equally likely at any time during the act's performance (i.e. the act is "interrupted"). Durations of acts of this type would have a negative exponential probability distribution. But other types of acts may tend to terminate with increasing probability as performance time increases (i.e. the act is "finished"). Durations of such acts would not be negative exponentially distributed. Instead, their probability distribution will be positively skewed, will exhibit a nonzero modal value, and will be characterized by an increasing hazard rate (known in the actuarial sciences as "force of mortality'' and in reliability theory as ''wearout'' if it is increasing) (Barlow, Marshall, and Proschan 1963). Thus the probability that the act will terminate in a small interval after t seconds of performance have elapsed will increase as t increases.

Previous studies of temporal patterns of behavior have related (deterministic) causes to (stochastic) outcomes, using probabilistic methods of analysis. We now extend this approach to the ultimate causation of behavior and to mechanisms through which ultimate causation proximately determines behavioral outcomes. Our model is neither the first nor the only approach to this problem. For some interesting alternatives, see Metz (1974) or Schleidt (1964a, 1964b, 1965, 1974).

Whatever the proximate (physiological, motivational) causes of an act of behavior, natural selection determines that its performance will begin when the benefit resulting from the act is seen to outweigh its cost, and likewise performance will end when cost is seen to exceed benefit. The mathematical forms of these benefit and cost functions, and the properties of the behavioral mechanism by which information on costs and benefits is translated into response tendencies, completely determine the temporal properties of a given behavioral act.

Thus, to specify the probability distribution of behavioral durations, intervals, or latencies, we must specify three functions: benefit, cost, and response. If

$$J(t) = B(t) - C(t)$$

is the difference between benefit and cost at time t, and

$$R[J(t)]dt$$

is the probability that the animal will first exhibit the response between time t and $t + dt$ if the difference between benefit and cost at t is $J(t)$, then if $P(t)$ is defined as the probability that the response will occur at some time greater than or equal to t, it follows that

$$P(t + dt) = P(t)\{1 - R[J(t)]dt\}$$

This difference equation is equivalent to the differential equation

$$\frac{dP}{dt} = -R[J(t)]P$$

whose solution is

$$P(t) = P(t_0)\exp - \int_{t_0}^{t} R[J(t)]dt$$

The probability density function of t

$$f(t) = \frac{d}{dt}[1 - P(t)]$$

may then be derived

$$f(t) = R[J(t)]P(t_0)\exp - \int_{t_0}^{t} R(J(t))dt \quad \text{for } t_0 \leq t < \infty$$

$$= 0 \text{ elsewhere.}$$

In the simplest case, benefit and cost would bear a constant relation to one another over time, and $B(t) - C(t)$ is constant. The benefit of beginning performance might increase with time linearly or in some other way. For instance, if each successive minute of nonperformance depleted the animal's reserves of some substance by some fixed amount, performance would be more and more beneficial as the time elapsed since the last performance increased. For physiological processes such as heartbeat or respiration, such benefit ought to increase very rapidly, perhaps exponentially, with time. Biological intuition will suggest theoretical forms for benefit and cost functions that are reasonable *a priori*, and these hypotheses may then be tested against data by fitting the theoretical probability distribution to observations on the animal or system of interest.

Omniscience regarding benefits and costs is not to be expected in animals. Perceived benefits and costs may differ from time benefits and costs for a number of reasons: the animal may lack necessary information or experience, (stochastic) thresholds may exist at the cellular level, a positive J may not be detected if it is sufficiently small, or deception of some sort may occur. Signal detection models are used in psychology (Swets 1973) to analyze phenomena of this type. These models specify a criterion value for response (in the context of the current discussion, this criterion is simply J) and probability distributions for signal and noise. The simplest response model assumes that judgmental errors regarding J are normally distributed about 0 with variance σ^2, so that the tendency to respond correctly increases in a sigmoid manner as J increases:

$$R[J(t)] = \int_0^{\infty} \frac{1}{\sigma\sqrt{2\pi}} \exp\{-[x - J(t)]^2/2\sigma^2\}dx$$

As an example, consider two simultaneous control processes, J_1 = constant and J_2 = linear. Suppose that J is small compared to $B - C$, so that benefits and costs are perceived correctly in the majority of cases.

If the first process J_1 (interruption) is relatively strong compared to the second process J_2 (finishing) then the animal's behavior will appear highly distractible and would be interpreted as an instance of external control (interruption), while if the second process is dominant then the animal would appear relatively persistent when performing the act, and internal control (finishing) would be suggested. Most behavioral acts would be expected to reflect both external and internal control in proportions that depend proximately on the situation and on the individual and ultimately on evolutionary processes (i.e. on the outcome of selection for the most advantageous balance between internal and external control).

The interrupter-finisher model is formulated as follows. Define, again, $P(t)$ = probability that the duration of a given act is greater than or equal to t units of time, given that it began at $t = 0$. If the probability that an act terminates due to interruption in time interval dt is a dt for all $t \geq 0$, and if the probability that the act is finished in time interval dt is $bt\, dt$ for all $t \geq 0$, and if interruption and finishing are statistically independent events, then

$$P(t + dt) = P(t)[1 - (a + bt)dt]$$

This equation is equivalent to the differential equation

$$\frac{dp}{dt} = -(a + bt)P(t) \tag{1}$$

whose solution, with initial condition $P(0) = 1$ because the act is defined to begin at $t = 0$, is

$$P(t) = \exp[-(at + 0.5bt^2)] \tag{2}$$

and the probability density function for t, $f(t) = -\dfrac{dP}{dt}$, is

$$f(t) = (a + bt)\exp[-(at + 0.5bt^2)] \tag{3}$$

If $b = 0$ and $a \neq 0$ (pure interrupter), $f(t)$ is a negative exponential density with mean $1/a$; if $a = 0$ and $b \neq 0$ (pure finisher), $f(t)$ is the so-called Rayleigh density (Drake 1967, p. 276). Parameter a measures the animal's distractibility when performing the act in question, parameter b measures persistence of the animal when performing this act, and the dimensionless ratio $ab^{-1/2}$ measures the relative strengths of distractibility and persistence for this particular animal and act.

By changing "duration" to "interval" or "latency" in the above derivation, and substituting "begins" for "terminates," a model for act latencies or for intervals between acts may be derived.

Important assumptions in the above derivation are that causes of interruption and finishing are independent of one another (independence), that parameters *a* and *b* do not change over time (stationarity), that *a* and *b* depend only on the particular act—and possibly particular individual—studied (homogeneity), and that response errors do not occur.

The first assumption (independence) might fail if the animal's tendency to respond to an interrupting stimulus depended strongly on time already spent in performance of the act in question, but in this case the tendency to respond would itself follow a "finisher" process, and the model, once this second process had been taken into account, would still offer a valid description of the behavior. The latter two assumptions are more onerous. Response tendencies may change complexly over time. Distractibility during the fiftieth consecutive repetition of some act may greatly exceed distractibility early in the same sequence. Furthermore, the probability distribution of time intervals between consecutive acts of different types may sometimes depend on the identity of the following act (Halliday 1975). Where the model fails to describe data adequately, the next step would be to identify potential sources of nonstationarity and/or heterogeneity, and to conduct separate analyses as required.

As argued above, simple graphical analysis of interval distributions by use of log survivor functions often indicates an additional type of nonuniformity in the temporal organization of behavior. Behavioral acts frequently occur in bouts, during which several acts of the same type are performed in rapid succession; such bouts are separated by long intervals of time during which no acts of this type occur. A sharp change in the slope of the log survivor function at a particular point on the time axis indicates the temporal boundary separating within-bout intervals and between-bout intervals (Slater 1974a).

Factors determining length of within-bout intervals may be entirely distinct from factors determining length of between-bout intervals. For instance, gaps occurring during a bout of prey-catching behavior are likely to depend mainly on the handling and swallowing time of each prey item, while gaps between bouts may be determined by other behavioral and ecological factors (e.g., drinking, resting, territorial defense, distance between patches of habitat in which prey may be found).

Clearly, the model only applies to gaps of a given type and is not intended or expected to describe entire interval distributions. For this reason, data on behavioral latencies, intervals, and durations (since durations may also exhibit this sort of nonuniformity) should first be plotted using the log survivor function method in order to separate the raw data into elementary components, each of which may be analyzed separately using the model.

We give three examples of the use of the interrupter-finisher model. The first example involves intraspecific, observational data on zebra finches and can be viewed as a novel approach to the ethology of individual differences and/or of developmental processes. The second example involves intraspecific, experimental data on mice and illustrates how a change in behavior resulting from drug administration (or from some other experimental manipulation) can be analyzed and interpreted in terms of basic control processes in the context of this model. The third example is an analysis of field data on whale behavior. In all three examples we assume that behavior wholly under internal control could be described by linear J_2 (rather than quadratic or exponential J_2, for example) and that both interruption and finishing could terminate the acts or gaps analyzed. As explained above, data were first separated into elementary components by visually inspecting log survivor function plots. Fit of the pure interrupter model (negative exponential probability distribution) was assessed using chi-square. Data sets not adequately described by a pure interrupter model (if the χ^2 test showed $P < .05$, the fit was deemed inadequate) were analyzed with the full interrupter-finisher model. In both cases, a computer program using the method of maximum likelihood served to generate the fitted curves and parameter estimates. Results of the three analyses appear in Table I.

The first analysis dealt with data of Slater (1974b) on gap lengths between feeding bouts in nine zebra finches, *Taeniopygia guttata*. Six birds were found to be interrupters, one (Bird 27) exhibited a mixture of interrupter and finisher components, and two (Birds 30, 31) were entirely finishers. These marked individual differences in feeding patterns raise a number of interesting theoretical questions. Might these differences reflect the shifting locus of control felt by some authorities to characterize behavioral development? Do they represent alternate optimal feeding strategies? Were the interrupters exhibiting feeding strategies consistent with Maynard Smith's (1974a) argument on negative exponentiality of certain persistence time distributions under evolutionarily stable strategies of behavior?

The second analysis was suggested by Wiepkema's (1968) finding in CBA mice that one injection of goldthioglucose (GTG) changes the log survivor function of intergnawing intervals from linear to quadratic. As expected, the interrupter model adequately fit data on a control mouse but failed to fit data on a GTG-treated mouse. Feeding behavior of the GTG mouse exhibited both interrupter and finisher components, and this interpretation was confirmed in successful model fitting.

Table 1. Interrupter-Finisher Models: Three Analyses

	No. data points	Interrupter			Interrupter-Finisher			
		Parameter *a*	χ^2(d.f.)	*P*	Parameter *a*	Parameter *b*	χ^2(d.f.)	*P*
1. Zebra Finches (time unit: sec)								
Bird 6	65	0.0012	9.1(9)	N.S.				
9	120	0.0021	14.5(16)	N.S.				
10	59	0.0011	15.3(8)	N.S.				
13	183	0.0037	8.9(14)	N.S.				
14	78	0.0016	10.4(10)	N.S.				
22	46	0.00079	10.5(6)	N.S.				
27	41	0.00068	27.5(5)	<0.005	0.00018	5×10^{-7}	5.6(4)	N.S.
30	81	0.0013	31.9(11)	<0.005	0.0	2.3×10^{-6}	7.2(11)	N.S.
31	56	0.0009	34.8(8)	<0.005	0.0	1.2×10^{-6}	5.3(6)	N.S.
2. Mice (time unit: min)								
Control	67	0.025	11.2(8)	N.S.				
GTG-treated	50	0.020	13.7(5)	<0.025	0.0014	0.00047	2.9(5)	N.S.
3. Whales (time unit: min)								
Minke	55	0.580	0.66(2)	N.S.				

The interrupter-finisher approach is as useful in studying whales as it is in studying mice. We analyzed D.Y. Young's data on long dive (interblow) durations in a baleen whale, the Minke whale (*Balaenoptera acutorostrata*), observed in the Gulf of St. Lawrence near Les Escoumins, Quebec, Canada. (Long dives could reliably be separated from short dives using the log survivor function method.) Log survivor function analysis indicated that Minke whales exhibited pure interrupter behavior, and subsequent model fitting confirmed this hypothesis. Data on durations of blue whale (*B. musculus*) long dives clearly indicate a finisher pattern (Figure 3). The interrupter model failed to fit these data. Given only 22 intervals, we were unable to test the fit of the interrupter-finisher model. This latter model contains one extra parameter, and d.f. (= number of categories − number of parameters − 1) would have been 0 for the fit to the blue whale data.

In the same environment, Minke and blue whales exhibit qualitatively different temporal patterns of long dives. The ecological significance of this difference is not known. There is need for additional formal modeling and theoretical interpretation of the ecology of temporal patterns of behavior, including feeding behavior.

Log Survivor Functions and Evolutionarily Stable Strategies: Ethology Meets the Red Queen

Analysis of behavioral durations and of intervals between behavioral acts, as described in the above subsection, indicates the surprising fact that these behavioral times are often described by simple or compound negative exponential probability distributions. This fact is surprising because, as we showed above, one interpretation of negative exponentiality is that the time that a given act stops is independent of the time that it starts (durations purely random), or that behavioral acts are equally likely to be initiated at any time following the end of the previous act (intervals purely random). Ubiquity of negative exponential distributions means that animals seldom if ever finish what they are doing but rather are always interrupted by something else. Could animal behavior be little more than a stream of random interruptions? An analogous paradox is known in paleontology, where taxa seem to become extinct (i.e., disappear permanently from the fossil record) at a random time independent of their time of origin (Law of Constant Extinction, Van Valen 1976).

Several explanatory hypotheses have been suggested (Maynard Smith 1976), and these discussions bear on the ethological problem as well. The first hypothesis, that of constant extinction probability (interruption), termed by Van Valen the "Field-of-Bullets Hypothesis," corresponds to the view that animal behavior is a stream of random interruptions. Since we have already described above any number of reasonable patterns of behavioral causation that do not result in negative exponential probability distributions, a second hypothesis, the so-called Hypothesis of Triviality (Van Valen 1976) cannot apply to behavioral data.

Another hypothesis is the "Red Queen's Hypothesis" (Van Valen 1974). Freely interpreted and rephrased in behavioral terms, this hypothesis states that at any time the sum of absolute fitness values for performing all behavioral acts is constant: that the effect of behavior as a whole on fitness is constant over the period of time in which time measurements are taken. Is this statement consistent with negative exponential probability distributions?

In fact, the ethologically crucial assumption of the Red Queen's Hypothesis involves a zero-sum game of a particular type already modeled by Maynard Smith (1974a). Using this game-theory model (see Chapter 11), Maynard Smith showed that negative exponential distributions of behavioral durations should be expected in certain forms of animal conflict behavior where time, and not the possibility of injury, is the primary component of fitness cost. (See also Chapter 10 and Krebs 1976). Maynard Smith's theory of evolutionarily stable strategies in games of endurance explains negative exponentiality directly in the social context for which it was originally formulated, and indirectly in a wider variety of contexts more or less remote from social behavior. The indirect explanation follows directly from a simple reformulation of Maynard Smith's theory in which the two participants are not individual animals trying to maximize fitness but rather are different behavioral acts competing for access to effectors (i.e., for access to the "final common path" of McFarland and Sibly 1975). The paradox that the net value of the game to the organism, whatever the outcome, is zero can be explained if the paradox that the expected value of the game to each player is zero can be explained. Maynard Smith (1974a) offers an explanation of the latter and therefore of the former paradox.

Quantitative ethology, through use of the log survivor function technique, has rapidly progressed from an initial descriptive phase to discovery and possible resolution of a major empirical paradox in animal behavior. Our ubiquitous negative exponential probability distributions, like negative exponential extinction curves in paleontology, cannot be explained without recourse to moderately deep theory.

We suggest that experimental manipulation of behavioral durations and intervals be used to test the assumptions of the Red Queen's

Hypothesis as they apply to ethology. Some progress has been made along analogous lines by McFarland and others (Sibly and McFarland 1975), but this work assumes a deterministic model and sheds little direct light on the current problem. We would expect that some entirely novel theoretical explanations of negative exponentiality might come out of these experiments, and that at least some of these explanations would be sufficiently general to aid in resolving the above mentioned current controversy in paleontology as well.

2. BEHAVIORAL SEQUENCES

Before discussing methods for analyzing behavioral sequences, we would be well advised to recall just what a sequence is. By definition, a sequence is an (ordered) succession of named items. The alphabet is a sequence, as are the integers. In ethology, we often study sequences of behavioral acts. We may describe and interpret behavioral sequences of a single individual (intraindividual) or of two individuals (dyadic). We may simply list the acts in succession ("orients; approaches; sniffs; manipulates; pushes away; runs after; mouths; head-shakes; flings away") or we may also tabulate the starting times of each act along with the name of the act itself. In the former case, discrete-time analysis could be employed, and in the latter case continuous-time analysis could be used to extract time information from the sequence.

In ethology, for various reasons, the term "sequential analysis" has become synonymous with a restricted number of statistical techniques traditionally used by ethologists for analyzing sequences of behavior. This limitation of usage is arbitrary, for indeed the existing variety of quantitative approaches to behavioral sequences is virtually unlimited. The first-order Markov model to be discussed below is only one such technique; another commonly used technique, information theory, is a stepchild of first-order analysis rather than an alternative to it, for it shares all of the assumptions and limitations of the Markov model, as well as additional pitfalls less evident to casual users. Information theory and its drawbacks as a method for studying animal behavior are discussed in Chapter 3.

Sequences in Discrete Time

Thus far we have discussed occurrence through time of a single type of act; the point events considered in the previous section were assumed to be qualitatively identical. Of course, no living animal possesses a behavioral repertoire of only one type of act, and it is of interest to examine sequential relationships among different types of acts by analyzing the temporal patterning of all acts displayed by an individual. Under the assumption that behaviors that occur closely together in time may have the same underlying causal factors, sequential analysis of behavior can potentially yield information on causation of behavior (Slater 1973). Moreover, sequential analysis may indicate possible communication between two individuals. That is, to the degree that actions of one individual depend on the immediately preceding behavior of the other, the first animal's behavioral probabilities have altered in response to behaviors performed by the other, a necessary condition for the occurrence of communication (Wilson 1975).

The simplest relationship between two types of acts would consist of one type always being preceded by the other: "push" always precedes "shove." But these deterministic chains of behavior are rare in nature, for it much more often happens that "push" will precede "shove" with a definite probability, but not with certainty. It must be noted that even in a deterministic sequence there need not be symmetry or one-to-one correspondence between acts that follow and those that precede. In other words, while "push" may always precede "shove," "push" might not always be followed by "shove"—"push" could also be followed by some other act. Conversely, while one act may be deterministically followed by the second, the second may only be probabilistically preceded by the first.

Behavioral sequences in discrete time may be thought of as a series of discrete trials or observations, each of which has R possible mutually exclusive outcomes or states. S. Altmann (1965), Nelson (1964), and later many other ethologists modeled such sequences of behavior by using a family of discrete-time random processes known as Markov chains. Generally speaking, Markov chains exhibit temporal dependence in that the identity or probability of future events may depend on past events (Drake 1967). A first-order Markov chain model assumes that the probability of a given act depends only on the identity of the act immediately preceding that act. All other past history of the process is immaterial if behavior is really described by a first-order Markov chain model. On the other hand, this model will inadequately describe the sequence if the probability of the act depends jointly on the identities of the two acts immediately preceding it, but not on the third or on any more remotely preceding act. In such cases, a second-order Markov chain model (implying dependence on exactly two previous acts) would be appropriate. But if, on the other hand, the behavioral process were purely random, that is if acts occurred randomly and independently in time with no dependence at all on the past, then the sequence could be considered to be a zero-order Markov chain, the so-called Bernoulli process (Drake 1967).

The first step in sequential analysis using Markov models is to determine the order of the appropriate model for the data. The (conditional) transition probability p_{ij}* is the conditional probability that given the animal is currently $(n-1)$ performing act A_i, it will perform act A_j during the very next trial or observation n as follows:

$$p_{ij} = P[A_j(n) \mid A_i(n-1)] \quad \text{for } 1 \leq i \leq R,\ 1 \leq j \leq R$$

where R is the repertoire size. The transition probabilities are assumed to be independent of the trial (n) or position in the sequence and constant for all trials (stationary). Since the acts in the repertoire are mutually exclusive and collectively exhaustive by definition, and because probabilities are always positive and sum to one over their sample space (the repertoire in this case)

$$\sum_j p_{ij} = 1 \quad \text{for } i = 1, 2, 3 \ldots R$$

The χ^2 goodness of fit test is used to compare observed transitions with expected ones; hence, transition frequencies are examined in order to test particular hypotheses about transition probabilities. Assuming that any behavior may in principle follow any other one, transition frequencies may be displayed in the form of a contingency table. In this mode of tabulation, the observed frequency with which the jth type of act immediately followed the ith type of act appears in cell (i, j), the table entry in the ith row and j^{th} column. Therefore, the ith row lists the frequencies with which each of the different types of acts in the repertoire immediately *followed* the ith type, and the jth column lists the frequencies with which each of the different types of acts in the repertoire immediately *preceded* the jth type.

We give two examples using real data. The first presents a very small sample of behavior, solely in order to illustrate appropriate tabulation procedures. The second involves a large data set whose size is sufficient to support statistical analysis.

A young male domestic cat, age six months, plays with a ping-pong ball. He looks at it, runs, walks, paws it, turns, walks, creeps toward it, looks at it, creeps, looks at it, creeps, ducks down, sways from side to side, springs, dribbles the ball, paws it, turns, walks, turns, looks at it, creeps, paws, noses, paws, rears, turns. . . . A catalog of all types of

* This conditional probability p_{ij} is less ambiguously written as $p_{j\mid i}$. However, to remain consistent with current usage and Chapter 6 of this volume we are using the p_{ij} notation.

acts included in this short sequence is:

1. Creep
2. Dribble
3. Duck down
4. Look at
5. Nose
6. Paw
7. Rear
8. Run
9. Spring
10. Sway
11. Turn
12. Walk

(These behavioral acts should be familiar to all, but for more specific details consult Leyhausen's (1973) authoritative study of field behavior, or take time to play with a kitten or cat. Of course, both exercises are highly recommended.)

Table 2 is a first-order transition table containing the above data. It was constructed as follows. The first transition in the sequence given above is Look at→Run, or 4→8 using the numbering system specified by the catalog. Therefore, the value in the cell occupying the fourth row and eighth column of the contingency table is incremented by 1. The next transition in the sequence is Run→Walk, or 8→12. Cell (8, 12) is incremented by 1. Note that "Run" was counted twice, first as a following act and then as a preceding act. This procedure involves some

Table 2. First-Order Contingency Table for Cat Play Behavior Sequence*

		Following acts												Row sums
		1	2	3	4	5	6	7	8	9	10	11	12	
Preceding acts	1	0	0	1	2	0	1	0	0	0	0	0	0	4
	2	0	0	0	0	0	1	0	0	0	0	0	0	1
	3	0	0	0	0	0	0	0	0	0	1	0	0	1
	4	3	0	0	0	0	0	0	1	0	0	0	0	4
	5	0	0	0	0	0	1	0	0	0	0	0	0	1
	6	0	0	0	0	1	0	1	0	0	0	2	0	4
	7	0	0	0	0	0	0	0	0	0	0	1	0	1
	8	0	0	0	0	0	0	0	0	0	0	0	1	1
	9	0	1	0	0	0	0	0	0	0	0	0	0	1
	10	0	0	0	0	0	0	0	0	1	0	0	0	1
	11	0	0	0	1	0	0	0	0	0	0	0	2	3
	12	1	0	0	0	0	1	0	0	0	0	1	0	3
Column sums		4	1	1	3	1	4	1	1	1	1	4	3	
	Grand sum (total number of transitions) = 25													

* See text for actual sequence.

statistical complications discussed below, and the actual methods used to analyze the resulting table deal correctly with these subtleties by using the theory of random processes. It happens that the method for calculating the expected frequencies in a test of sequential dependence in behavior is exactly that used for testing independence between rows and columns in a contingency table (row sum × column sum/grand sum). Correspondingly, the calculated χ^2 value is given by

$$\chi^2 \text{ calc} = \sum_{i=1}^{R} \sum_{j=1}^{R} (x_{ij} - m_{ij})^2 / m_{ij}$$

where x_{ij} = observed value in cell (i, j), m_{ij} = expected value in (i, j) = (row sum i × column sum j/grand sum).

The above sample is not sufficiently large for statistical analysis. Assuming a total repertoire of 12 acts, a total of $10R^2 = 1440$ acts would have been required (see below). The second example (Table 3), taken from Baylis (1975), includes a somewhat larger data set (total 2939 acts, repertoire 20 acts; with $10R^2 = 4000$, $2939/4000 = 7.3R^2$ acts are probably just adequate, since we know $10R^2$ is entirely adequate while $5R^2$ appears to be borderline). It was extremely difficult to locate real data that met all of our requirements for this example. We wished to illustrate a catalog of realistic size, a data set sufficiently large to allow statistical analysis, and a conceptually correct formulation of an interindividual behavioral sequence. This table exhibits an unusual feature, namely the use of separate categories for male and female behavior. This conceptualization of interindividual sequences, while not standard, deserves to be better known and more widely used. Baylis (1975) convincingly argues that only in this way can ethologists correctly analyze the four equally important classes of behavioral transitions present in courtship sequences (male act→male act, male act→female act, female act→male act, female act→female act). Moreover, Baylis cites relevant theoretical arguments demonstrating that information-theoretic analysis of interindividual behavior is incorrect unless the contingency table is set up in this way.

The χ^2 technique appears to be quite straightforward. As long as the expected values are large, that is, none is less than one and not more than 20 percent of the expected values are less than 5 (Cochran 1954, Siegel 1956), the χ^2 approximation should be valid. If certain cells contain expected values of insufficient magnitude, then the size of the transition matrix may be reduced by pooling low-frequency acts with related acts. Some authorities suggest use of the original degrees of freedom when this procedure is adopted.

The second assumption that the investigator should consider is the

Table 3. First-Order Contingency Table for Cichlid Courtship Sequence

OBSERVATION PERIOD 1		CICHLASOMA ZALIOSUM — FOLLOWING ACT																			
		MALE										FEMALE									
PRECEDING ACT		A	B	C	D	E	F	G	H	I	J	AA	BB	CC	DD	EE	FF	GG	HH	II	JJ
MALE	A. BITE	7 2.5	1 3.9	0 16.1	0 .3	0 8.5	0 9.3	0 0	0 .8	0 .8	5 6.4	0 .4	2 10.8	2 7	1 1.8	0 .5	7 2.4	0 0	57 3.9	0 1.2	4 9.5
	B. FRONTAL	7 3.9	2 6	16 24.9	0 .5	5 13.1	2 14.3	0 0	0 1.2	1 1.2	4 10	1 .6	11 16.7	56 10.8	2 2.7	1 .8	3 3.8	0 0	14 6	1 1.9	7 14.6
	C. LATERAL	1 16.1	9 24.9	39 103.3	3 1.9	126 54.4	61 59.4	0 0	11 5.1	0 4.9	38 41.2	3 2.4	72 69.4	89 44.6	4 11.2	0 3.2	4 15.6	0 0	6 24.9	3 7.9	82 60.6
	D. DIG	0 .3	0 .5	0 1.9	3 0	1 1	2 1.1	0 0	0 .1	0 .1	0 .7	1 0	1 1.3	0 .8	0 .2	0 .1	0 .3	0 0	0 .5	0 .1	2 1.1
	E. TAILBEAT	0 8.5	10 13.1	29 54.4	0 1	114 28.6	21 31.3	0 0	0 2.7	1 2.6	17 21.7	2 1.3	43 36.5	17 23.5	6 5.9	2 1.7	0 8.2	0 0	2 13.1	2 4.1	24 31.9
	F. QUIVER	0 9.2	0 14.3	14 59.2	1 1.1	7 31.2	152 34.1	0 0	2 2.9	10 2.8	19 23.7	0 1.4	20 39.8	9 25.6	16 6.5	0 1.8	0 8.9	0 0	0 14.3	10 4.5	56 34.7
	G. SKIM	0 0	0 0	0 0	0 0	0 0	0 0	0 0	0 0	0 0	0 0	0 0	0 0	0 0	0 0	0 0	0 0	0 0	0 0	0 0	0 0
	H. YIELD/AVOID	0 .8	1 1.2	6 5.1	0 .1	2 2.7	2 2.9	0 0	0 .2	0 .2	1 2	0 .1	6 3.4	3 2.2	0 .6	0 .2	0 .8	0 0	0 1.2	0 .4	6 3
	I. NIP OFF	0 .8	0 1.2	0 4.9	0 .1	0 2.6	5 2.8	0 0	0 .2	9 .2	2 1.9	0 .1	0 3.3	1 2.1	3 .5	0 .2	0 .7	0 0	0 1.2	1 .4	5 2.9
	J. APPROACH	19 6.5	45 10	28 41.6	0 .8	1 21.9	4 23.9	0 0	0 2	0 2	7 16.6	0 1	19 27.9	23 18	6 4.5	0 1.3	12 6.3	0 0	39 10	1 3.2	18 24.4
FEMALE	AA. BITE	1 .4	0 .6	1 2.4	0 0	2 1.3	0 1.4	0 0	2 .1	0 .1	0 1	4 .1	2 1.6	1 1.1	0 .3	0 .1	0 .4	0 0	0 .6	0 .2	0 1.4
	BB. FRONTAL	0 10.8	43 16.7	258 69.4	2 1.3	10 36.5	16 39.9	0 0	11 3.4	1 3.3	2 27.7	2 1.6	10 46.6	4 30	0 7.6	0 2.1	0 10.4	0 0	4 16.7	1 5.3	6 40.7
	CC. LATERAL	3 6.9	8 10.7	98 44.4	0 .8	11 23.4	14 25.6	0 0	0 2.2	1 2.1	17 17.7	0 1	21 29.8	8 19.2	6 4.8	4 1.4	2 6.7	0 0	9 10.7	1 3.4	34 26
	DD. DIG	1 1.8	3 2.7	8 11.2	0 .2	1 5.9	7 6.5	0 0	0 .6	0 .5	13 4.5	0 .3	8 7.6	0 4.9	13 1.2	0 .3	0 1.7	0 0	0 2.7	3 .9	3 6.6
	EE. TAILBEAT	0 .5	0 .8	3 3.2	0 .1	1 1.7	0 1.8	0 0	0 .2	0 .2	4 1.3	0 .1	1 2.1	1 1.4	0 .3	6 .1	0 .5	0 0	0 .8	0 .2	1 1.9
	FF. QUIVER	8 2.4	0 3.8	3 15.6	0 .3	0 8.2	2 9	0 0	0 .8	1 .7	15 6.2	0 .4	1 10.4	1 6.7	1 1.7	0 .5	33 2.3	0 0	1 3.8	1 1.2	16 9.1
	GG. SKIM	0 0	0 0	0 0	0 0	0 0	0 0	0 0	0 0	0 0	0 0	0 0	0 0	0 0	0 0	0 0	0 0	0 0	0 0	0 0	0 0
	HH. YIELD/AVOID	39 3.9	1 6	4 24.7	0 .4	2 13	1 14.2	0 0	0 1.2	1 1.2	27 9.9	0 .6	1 16.6	4 10.7	0 2.7	3 .8	12 3.7	0 0	0 6	0 1.9	37 14.5
	II. NIP OFF	0 1.2	0 1.9	3 7.9	0 .1	3 4.1	3 4.5	0 0	0 .4	1 .4	7 3.1	0 .2	0 5.3	2 3.4	2 .9	0 .2	3 1.2	0 0	0 1.9	18 .6	0 4.6
	JJ. APPROACH	0 9.5	10 14.7	41 60.7	1 1.1	4 32	25 34.9	0 0	1 3	0 2.9	42 24.3	0 1.4	152 40.8	17 26.2	0 6.6	1 1.9	7 9.2	0 0	1 14.7	0 4.6	22 35.6
	TOTAL:	86	133	551	10	290	317	0	27	26	220	13	370	238	60	17	83	0	133	42	323
	FREQUENCY:	.029	.045	.187	.003	.099	.108	0	.009	.009	.075	.004	.126	.081	.020	.006	.028	0	.045	.014	.110
	N = 2939																				

supposition that any act can follow any other act. In some cases, it may be impossible for certain transitions to occur. In harbor seals, for example, "food gathering" cannot occur immediately following "hauling-out." "Entering water" must intervene between the two actions. Slater and Ollason (1972) discuss a parallel case, in which feeding could neither immediately precede nor be followed immediately by drinking. In this study, food and water were spatially separated and locomotion had to take place between feeding and drinking.

If any act can follow any other act, and if acts occur in repetitious bouts, then transitions between identical acts are more likely than transitions between different acts. The descending diagonal ("hop-to-hop," "skip-to-skip," etc.) will dominate the analysis and hide more interesting associations. In such cases, it may be more fruitful to eliminate homogeneous transitions and enter zeros in those cells (Slater and Ollason 1972). The calculation for the expected values in these incomplete contingency tables must be modified so that the row and column sums for the resultant expected values still equal those of the observed values. Chapter 6 of this volume gives methods for calculating the expected values for such incomplete tables.

Finally, the animal being observed has been assumed to be in a steady state: it is assumed that the sequence is stationary and the probabilistic structure of the events remains constant through time (Slater 1973). If the observation sessions are short, such an assumption is reasonable, but at the same time it becomes more difficult to obtain adequate sample size during short observation sessions. As the length of the observation sessions increases, the risk of nonstationarity generally increases as well. Lemon and Chatfield (1971) actually tested for stationarity before applying Markov chain analysis. They did so by testing the hypothesis that there were no significant differences in the probabilities of different song types between the first and second halves of the observation session (Slater 1973).

Stationarity may also be evaluated using three-way contingency tables where the three variables are preceding act, following act, and half of observation session. If the third variable were associated with either one (or both) of the other two variables, then the data are not stationary. Appropriate methods for analyzing multidimensional contingency tables are given by Colgan and Smith (Chapter 6 this volume) and Fienberg (1970, 1972b).

Multidimensional contingency tables are also used to analyze higher order Markov chains where the variables are acts during the series of trials or observations (following act, first preceding act, second preceding act, etc.). The χ^2 approximation becomes progressively less accurate as the order increases. If the possible number of different types of acts is three or less, Chatfield and Lemon (1970) feel that the χ^2 goodness of fit test is valid until third-order dependency.

Significant Transitions

If Markov chain analysis of behavioral sequence data in discrete time indicates dependence of current acts on past acts, it is natural to pursue further this analysis of temporal dependence in order to identify those specific acts or pairs of acts that account for such temporal dependence. A method for breaking down a transition table in this way exists and has been tested and found to be statistically valid (Fagen ms.). The procedure is as follows: Let x_{ij} = observed frequency in cell (i, j) of the transition table, m_{ij} = expected frequency calculated as indicated in the previous section, and $Y = (x_{ij} - m_{ij})/(m_{ij})^{1/2}$. If $|Y| > \sqrt{(\chi^2_{0.05}, \text{d.f.})/R^2}$ (Bishop *et al.* 1975), then the transition (act $i \rightarrow$act j) is occurring at a frequency that differs significantly ($P < 0.05$) from chance expectation. If $Y > 0$ and the transition is found to be significant, act i may be said to direct act j, while if $Y < 0$ and the transition is found to be significant, act i may be said to inhibit act j. All individual cells may be analyzed using this method. (There is no need to restrict m_{ij}, although for $m_{ij} \ll 1$ moderate caution is advised.) Sample size (total acts in the table) should be at least $10R^2$, where R = repertoire size. The method is expected to detect only one-third to two-thirds of those statistically significant behavioral transitions actually present in the animal's ethogram, and this relative insensitivity means that significant transitions between rare acts, and those significant transitions deviating only slightly from chance, will frequently go unnoticed. Therefore, this method does not completely specify the true transition structure of behavior, but rather identifies a few particularly significant behavioral transitions. Further study of these selected transitions would be expected to be especially productive, since suggestive evidence of motivation or communication was furnished by the significant transitions analysis.

Sample Size Considerations

Recent simulation studies by R. M. Fagen determined the minimum sample size required for use of two standard techniques of discrete-time sequential analysis: (1) significance assessment of single cells in a transition table and (2) information theory. A first-order Markov model was assumed. Results of these studies agreed: defining R = repertoire size, a sample of $2R^2$ behavioral acts (individual acts or tokens including repetitions) is too small, while $10R^2$ acts is sufficiently large. Intermediate sample sizes were not investigated systematically, but $5R^2$ is probably a borderline value, judging from a few simulation runs made at this sample size. Presumably, $10R^{n+1}$ acts would be needed for n^{th}-order analysis. These results suggest that in a typical first-order sequential analysis of behavior using a repertoire of 10 types of acts, total sample size should be at least $5 \times 10 \times 10 = 500$ acts, and more likely $10 \times 10 \times 10 = 1000$ acts. These recommendations agree with current ethological practice. But it is nonetheless reassuring to have such numerical guidelines at hand, especially in the initial, planning phases of a study of behavior. One note of caution is necessary: when acts occur in sets or clusters, resulting in holes in the transition table, these guidelines tend to break down.

In continuous-time transition analysis (see below), where we record the duration of each act as well as its identity, necessary sample sizes are not known. Hazlett and Bossert (1965) found it almost impossible to gather enough data to conduct a meaningful transition analysis in continuous time, but their sample was adequate for discrete-time analysis. If 30 or more observations are required in order to define the shape of a duration distribution, as we have found to be the case in our own analyses, and if a first-order transition analysis is to be carried out in continuous time, so that a separate duration distribution must be calculated for each of the R^2 possible transitions in the animal's repertoire, then at least $30R^2$ acts must be observed. Moreover, since all transitions will not occur with equal frequency, the sample size actually required in order to obtain a reliable quantitative description of the continuous-time transition structure may greatly exceed $30R^2$. As stated above, exact recommendations are not possible at this time, but all indications are that continuous-time transition analyses might require prohibitively large samples unless the repertoire of the animal were exceedingly small. To obtain samples whose size is adequate for other types of continuous-time analysis may, of course, be feasible, but this question must be addressed separately in the context of each method.

Sequences in Continuous Time

When time information, and not merely order information, is recorded for a behavioral sequence, additional possibilities for analysis may be cited. These possibilities may be roughly divided into two broad classes of techniques: "naive" techniques, or those requiring relatively few assumptions about behavioral mechanisms; and "structured" techniques that rest on an explicit, falsifiable formalism or model of behavior. Naive techniques are best used for description and can seldom be used in hypothesis testing, while structured techniques may be used both for description and for analysis of continuous-time behavioral sequences. However, naive techniques are valuable. They provide simplified descriptors of a very complex set of data, and if properly applied and interpreted can be of valuable service to ethological intuition.

Naive Techniques: Time Series Analysis

Time series analysis (Anderson 1971; Brillinger 1975; Box and Jenkins 1970; Jenkins and Watts 1975) is a method of describing temporal regularities in a sequence of behavior (or in any kind of sequence at all). In essence, this method offers a mathematical and graphical picture of the likelihood that an animal will be performing a specified act in its repertoire at some time in the future, given that the animal is performing a particular, specified act at the current time. In considering a single behavioral act now and in the future, we can derive various quantitative measures of temporal dependence. It should come as no surprise that the type of measure used will depend on precisely how we define (1) time, (2) persistence, and (3) a single behavioral act. The important point is that all such measures capture the very same intuition: if we know that an animal is presently behaving in a particular way, how well can we specify the animal's future behavior? Of course, we may consider various combinations of different types of acts now and in the future using a precisely analogous formalism.

Given such a quantitative measure, in whatever terms the constraints of a particular study demand, what might we learn from it that we would not know otherwise? Slatkin's (1975) field study of feeding behavior in two species of primates presents one answer to this question. Slatkin computed autocorrelation functions for feeding in yellow baboons (*Papio cynocephalus*) and in geladas (*Theropithecus gelada*). He found that these functions reflected important ecological differences in the animal's habitats, including differences in environmental characteristics pertinent to predictions generated by ecological and social theory (resource path size and interpatch distance; presence or absence of conspecifics). Other recent field ethological applications of methods of continuous-time sequential analysis include an innovative study of shorebird behavior by Baker (1973) and a thorough analysis of red-breasted merganser displays by van der Kloot and Morse (1975). These methods are equally applicable in the laboratory.

Heiligenberg (1973, 1974a, 1974b) analyzed the behavior of a variety of animals by using continuous-time analysis methods. Dawkins and

Dawkins (1973, 1974) analyzed pecking sequences in the chick in continuous time and obtained some novel and intriguing results on the structure of the sequence of decisions by the animal that results in the observed sequence of behavior.

Naive Techniques: The Periodogram and Spectral Analysis

When behavior occurs cyclically in time, due either to endogenous or to exogenous factors, or when it is expected to do so, methods of naive analysis that focus on this potential cyclicity may be particularly useful. Spectral analysis techniques provide such methods. The end product of spectral analysis is a plot of the observed relative abundance of all possible temporal cyclicities in the behavior. This plot is an empirical estimate of the unknown power spectrum, which would display the true relative abundances but can only be estimated approximately from finite samples of behavior. From the estimated power spectrum we may read directly the various frequencies at which the behavior recurs. A plot showing a peak at about 0.042 (= 1/24) hr^{-1} would indicate a circadian rhythm. Lower or higher frequencies might also be present. The empirical power spectrum does not reveal the biological reasons for the existence of such peaks, but only indicates that periodicities of behavior do in fact exist.

Endogenous rhythms in behavior are of major biological interest in a number of contexts. The self-selection technique, a recent experimental method for the study of endogenous rhythmicity, is particularly compatible with the spectral analysis approach. For instance, Colgan (1975) applied spectral analysis to self-selection data and was able to demonstrate existence of diurnal rhythmic components in pumpkinseed sunfish (*Lepomis gibbosus*).

Structured Techniques: Random Processes in Continuous Time

The first quantitative statistical analyses of behavioral sequences in ethology involved continuous-time data. Spurway and Haldane (1953) analyzed temporal patterns of breathing in a newt. Cane (1959, 1961) reexamined an intriguing empirical relationship between mean duration and proportion of time spent performing a given behavioral act and derived an interesting random-process model of behavioral mechanisms in order to explain this relationship. Cane's explanation rests on certain explicit assumptions about behavior, discussed below. Her model fits not only the data that originally suggested it, but other data as well (see

Nelson 1964). We wish to discuss Cane's results and their implications for ethology.

If we view a particular act (such as feeding) in the stream of ongoing behavior, we can quantify time characteristics of its occurrence in two distinct ways: (1) its mean duration; (2) the proportion of total time spent performing this act. Plots of measure (1) as a function of measure (2) often exhibit a surprisingly regular, simple straight-line relationship.

Why is act duration linearly related to relative behavioral probability? Indeed, why should there be any relationship at all? Cane showed that a number of simple stochastic models of behavior, including a continuous-time (Markov) analog of the ethologist's standard discrete-time Markov chain model, failed to explain the observed relationship. She then proposed a new model, her so-called semi-Markov model, which, under certain interesting assumptions, did adequately explain the observed regularity.

For concreteness, the following example, in the context of feeding behavior, will be used to illustrate the concepts underlying semi-Markov models. (Of course, these models are general, and should not be viewed as applying only to a single behavioral context.) (See S. Altmann (1974) for the original presentation of this example which we paraphrase here.) An animal is eating discrete food items. Each item takes a particular amount of time to find, handle, and eat, but this time varies from item to item, so that total time spent on each item is a random variable. (The reader may demonstrate this as an exercise; the only supplies required for this exercise are a stopwatch, a normal appetite, and small chunks of palatable food.) After finishing each item, the animal must decide whether to eat an additional item or to do something other than eat. (In its ecological foraging theory context, this is the well-known giving-up time problem: Krebs 1974; Krebs, Ryan and Charnov 1974.)

This choice (eat more or do something else) is itself a random variable (described by probabilities whose values are presumably set to some degree of precision by natural selection). Therefore, the time spent uninterruptedly eating before the animal stops eating and does something else is the sum of a random number (this number is one more than the total number of times the animal decided to go on eating) of random variables. Each of these random variables represents the time allocated to a particular food item during this uninterrupted sequence of feeding behavior.

If the conditional decision probabilities $\pi_{j|i}$ (where $\pi_{j|i}$ = probability that, given the animal is performing act i, the next act performed will be j) remain constant for the duration of the observation period, and if the probability distribution for the total duration of a given act (e.g., time spent on a single food item) depends at most on the identities of that act and of the act immediately following in the sequence (thus permitting the animal limited "look-ahead" capabilities), then a semi-Markov model of the behavioral sequence is appropriate. If the animal does not possess the ability to look one act ahead in time, the semi-Markov model will still hold, but it is worth noting that the semi-Markov approach, unlike most other sequence analysis techniques, allows for this fairly complex and realistic kind of behavioral capability. Cane (1959) showed that a semi-Markov model with these assumptions did in fact predict the observed linear relation between mean duration and percent time for a given act. The slope and intercept of this relation may be calculated from the parameters of the semi-Markov model (and vice versa).

3. NEW METHODS OF TEMPORAL PATTERN ANALYSIS

In observing and analyzing animal behavior, one frequently notes complex temporal associations between behavioral acts. These complex contiguities may stretch the simple formalism of Markov chain theory to the breaking point. To cite a hypothetical example, one monkey may react differently to another monkey's display depending on presence or absence of a third monkey. There appears to be a need for naive (in the sense of Section 2) methods of detecting significant clusters of behavioral events, clusters related not merely by immediate temporal precedence (as in the case of first-order Markov theory) but possibly in other ways that we would like to specify (social context, for example). Such a method is cluster analysis (Chapter 5). This approach invokes relatively few assumptions about the structure of behavior and therefore is probably more appropriate for analysis of behavioral sequences than the Markov-chain methods that have come to dominate the field.

A second, potentially useful naive approach to the general temporal structure of behavior sequences is character analysis (discussed in connection with information theory in Chapter 3), a method of analysis that relates sequential structure directly to environmental and contextual variables. In the traditional first-order Markov approach, on the other hand, environmental and contextual information can only enter the analysis in a *post hoc* manner through comparison of computed transition probabilities under different conditions.

A third interesting method is multidimensional scaling (Chapter 7). Kruskal (1964) states that "the problem of multidimensional scaling, broadly stated, is to find *n* points whose interpoint distances match in some sense the experimental dissimilarities of *n* objects." The meaning of this problem in the context of ethology is as follows. Suppose we wished to ascertain which displays of a particular species resembled each other most closely and which were most dissimilar; or suppose it were of interest to determine whether the behavior of offspring of dominant individuals changed to a lesser extent in development than that of offspring of subordinate individuals. The advantage of the (nonmetric) multidimensional scaling technique is that it does not use quantitative (numerical) data, but rather a list of the similarities or dissimilarities of the points, objects, behaviors, or individuals to be ranked, resulting from pairwise comparisons. For instance, a complete social preference network for a given individual could be constructed merely from pairwise assessments of proximity to or time spent with. These networks could then be compared for all individuals. Or data on food preference in a pairwise choice situation could be used to create a general picture of a food item space (i.e. which items an animal might respond to as being equivalent, or nearly so). Unlike cluster analysis, which requires numerical data in order to assess similarity, nonmetric multidimensional scaling only requires preference data and in doing so alleviates the difficult problem of quantifying preference measures. The binary choice paradigm allows the animal to construct its own scale and frees the ethologist from the burden of imposing arbitrary quantitative measures. Accordingly, Golani (1973) used nonmetric techniques to analyze sequences of social behavior in jackals (*Canis aureus*).

We conclude this section by briefly citing three additional analysis techniques. These methods can be applied to sequences whose complexity is such as to defy the simple quantitative approaches presented here. They should be particularly useful in the study of vertebrate behavior. Nelson (1973) and Soucek and Vencl (1975) developed two novel quantitative approaches to the study of bird song. Golani (1976) analyzed interactive movement patterns using frame-by-frame film analysis, a novel notational system (the Eshkol-Wachmann movement notation), and some interesting theoretical and quantitative concepts on the structure of complex behavior patterns.

4. POSSIBLE FUTURES

As the foregoing exposition no doubt indicates, we are frankly skeptical about the value of many quantitative methods currently in use in ethology, and especially of the more complex formalisms. Quantification of perfectly adequate verbal descriptions of behavior is not what we mean by "quantitative ethology." Is any insight really gained by reducing a data set of 10,000 behavioral acts of 10 types to a first-order transition matrix of 100 numbers of unknown accuracy? Grave reservations have recently been expressed about certain aspects of the current trend towards quantification in ethology. By contrast, we strongly support the modeling approach, and we feel that judicious quantification is an essential aspect of the ethological approach to behavior. For example, to carry the above example one step further, suppose that only 10 of the observed 100 transitions had been found to be statistically significant. Analysis of an initial collection of 10,000 items of data would then have resulted in identification of 10 particularly salient behavioral phenomena that could be described, modeled, and analyzed in depth. Here quantitative methods might indeed make a valuable contribution to ethological insight.

If quantitative analysis of temporal patterns of behavior is to serve ethology, it will do so on the basis of intuitive transparency, accessibility, and effectiveness in theoretically motivated empirical investigations. Use of naive approaches and graphical methods (log survivor functions, autocorrelation, cluster and character analysis) and use of straightforward hypothesis testing techniques are indicated. The future may well belong to those who can use simple methods effectively to test a theoretical hypothesis, or who are prepared to construct original models of behavior should no simple technique be available.

CHAPTER

5

Hierarchical Cluster Analysis

V. J. De GHETT
State University of New York at Potsdam

The purpose of this chapter is to intoduce hierarchical cluster analysis to the ethologist. Rather than making an exhaustive survey of hierarchical cluster analysis, this chapter attempts to present the topic in a form that will be most useful for the practicing ethologist. Many ethologists may actually be familiar with the general topic of cluster analysis because of its use by taxomomists in some forms of numerical taxonomy. From this point of view several find textbooks dealing with cluster analysis may already be familiar to ethologists (Cole 1969; Jardine and Sibson 1971; Sneath and Sokal 1973; Sokal and Sneath 1963). For those who would like to explore the topic in more detail and would prefer to stay within the biological context the approach through numerical taxonomy may represent the most familiar path. To this list of books one could also add Bailey (1967, Chapter 7), and Sokal (1966, 1974). However, the taxonomist has some unique problems to deal with that the ethologist does not face (see Hull 1970). The less familiar but more direct approach may be more pragmatic. In recent years several fine books dealing with cluster analysis have appeared (Anderberg 1973; Binjnen 1974; Duran and Odell 1974; Everitt 1974; Tryon and Bailey 1970). Each has its own particular approach and of these Anderberg (1973) may be the most useful.

Cluster analysis is a general term for a large variety of analytic techniques that seek to find: "natural groupings," "natural partitions," "category structures," et cetera that exist in quantitative data. Cluster analysis is not always multivariate. However, the utility of cluster analysis is probably best seen in multivariate applications. In such a multivariate situation, cluster analysis can be a tool for discovering relationships that exist in the data (Anderberg 1973; Andrews 1972; Edwards and Cavalli-Sforza 1965; Fleiss and Zubin 1969; Johnson 1967; Meisel 1972; Zahn 1971). One usually starts with a data set consisting of N entities and one ends with a patterned set of interrelationships among and between the N entities. The original data set could be considered to be lacking any "apparent" structure. The result of cluster analysis will show the "inherent" structure of the data. The resulting structure is derived completely from the relationships that exist in the data set and without any reference to an externally imposed classification (Overall and Klett 1972). In cluster analysis one does not usually specify the desired number of clusters that should emerge.

The specific type of cluster analysis dealt with in this chapter is hierarchical cluster analysis. Sneath and Sokal (1973) and Williams (1971) present a similar kind of organization of various clustering techniques. According to Williams' (1971) system, I deal with exclusive, intrinsic, hierarchical, agglomerative cluster analysis. According to Sneath and Sokal (1973), I deal with sequential, agglomerative, hierarchical, nonoverlapping cluster analysis.

In Williams' (1971) system the cluster analysis is exclusive because an entity can be included in only one cluster (one category or one class). It is intrinsic because there is no externally imposed classification of the data. It is hierarchical because the clustering provides a series of ranks (Sneath and Sokal 1973) that have an optimum route between the total set containing the *N* entities and each of the *N* entities (Williams 1971).

The hierarchical representation is a branched hierarchy because there exists a series of ranks or levels that at one extreme contains all of the *N* entities as a single cluster and at the other extreme contains each of the *N* entities as a separate cluster (*N* clusters for *N* entities) (Hartigan 1967; Johnson 1967; Meredith, Frederiksen, and McLaughlin 1974). The *N* entities are nested when represented in a hierarchical system (Anderberg 1973).

The cluster analysis is agglomerative in Williams' (1971) system because it achieves the hierarchical structure by progressively fusing the individual entities into a complete population. Agglomerative techniques start with *N* entities (*N* sets) and terminate in a single set consisting of all of the *N* entities (Sneath and Sokal 1973). The term sequential is used by Sneath and Sokal (1973) to describe the cluster analysis that is discussed in this chapter. A sequential method employs a recursive definition to proceed from one step to the next during clustering. The term nonoverlapping used by Sneath and Sokal (1973) is similar to the term exclusive used by Williams (1971). From this point on, I use the phrase "hierarchical cluster analysis" to refer to the specific kind of cluster strategy described above.

Hierarchical strategies are not new to ethology. They are part of our heritage. The concept of the hierarchical organization of behavior (eg., Baerends 1941; Hinde 1953; Kortlandt 1955; Räber 1948; Tinbergen 1942, 1950, 1951; Weiss 1941) has not been without its problems. The problems appear to be what is hierarchically organized and not with hierarchical organization itself. Dawkins (1976) has an excellent chapter extolling the virtues and the necessities of employing hierarchical concepts in ethology. His case is well made and I will not dwell on the utility of such concepts.. However, it must be pointed out that hierarchical cluster analysis gives *a* hierarchical organization to behaviors and not necessarily *the* hierarchical organization of behavior. In Tinbergen's (1951, page 104) illustration of the hierarchical organization of the reproductive behavior of the three-spined stickleback (see Figure 1) the

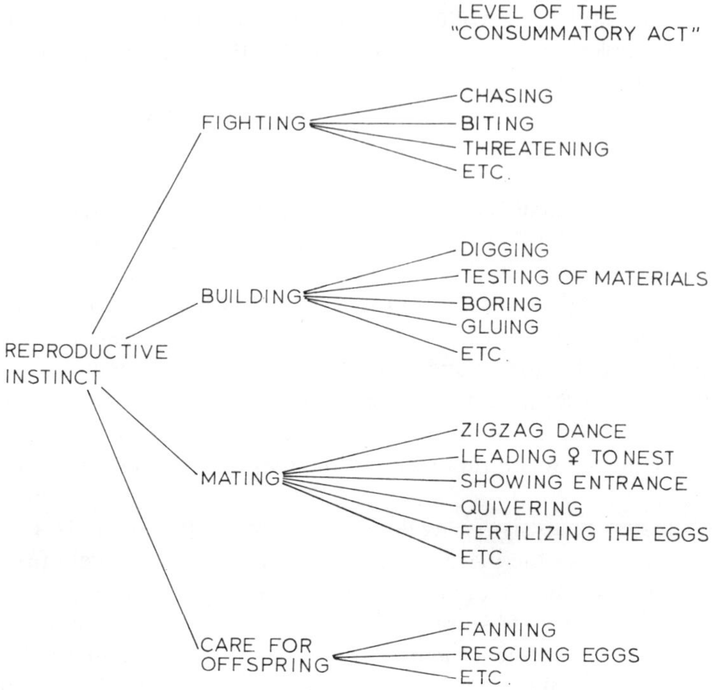

Figure 1. An example of the hierarchical organization of behavior as illustrated by Tinbergen's organization of the reproductive behavior of the male three-spined stickleback. (Redrawn with permission from N. Tinbergen 1951. The study of instinct. Oxford: Clarendon Press.)

clusters are at the level of the consummatory act. Clustering is achieved by organizing behaviors under more molar labels (fighting, building, etc.) which are in turn organized under a yet larger label, reproductive instinct. The nodal points at one level of clustering are labeled (fighting, building, etc.) and imply causation.

Hierarchical cluster analysis can suggest causal mechanisms but does not provide any quantitative mechanism for labeling nodal points. Typically, the variables in the data set are at about the same level on a molar to molecular scale. Therefore, variables that appear at one level of clustering cannot logically be considered as controlling variables at the next level of clustering. Furthermore, there is no *a priori* guarantee that the results of a hierarchical cluster analysis will consist of meaningful clusters. Clusters analysis is an aid to reasoning and a tool for discovery (Anderberg 1973). It is not an end in itself but rather a starting point for explanation. It is an analytic technique that should be used for illumination and not support, to paraphrase Colgan in the Prologue of this book.

Hierarchical cluster analysis is particularly well suited for ethological work. Ethologists seem to be prone to gathering multivariate data. The observation of a single individual for any length of time often generates data on a number of variables across a series of time units. This can be conceived of as a simple behaviors × time raw data matrix. If we replicate this for a number of individuals using the same variables for the same number of standard time units we have a behaviors × time × individuals raw data matrix. The individuals dimension can be collapsed by using some summary statistic (mean, total, etc.) and a behaviors × time raw data matrix results. In a similar fashion we can generate a behaviors × individuals raw data matrix. Either a behaviors × time or a behaviors × individuals raw data matrix is the starting point for hierarchical cluster analysis. Such data matrices are often the byproduct of the construction of a complete or partial ethogram. Diagrammatically such a raw data matrix could be illustrated as in Table 1. In Table 1 there are j behaviors and either i time units or i individuals. If possible, it is advisable that the raw data matrix of ij size be such that $i > j$.

It is assumed that the hierarchical cluster analysis of the raw data matrix will be performed by a computer. Even so, certain basic ideas central to hierarchical cluster analysis can be understood by doing simple one and two dimensional analyses by hand. This is done in Section 1. In that section we briefly examine the basic concepts and procedures of hierarchical cluster analysis. Section 2 is devoted to the similarity matrix. The raw data matrix is converted into a similarity matrix. There are a variety of techniques for doing this conversion, and many of them are discussed in Section 2. Diagrammatically the similarity matrix could be illustrated as in Table 2. Notice that we are concerned with the similarities between the j behaviors. The hierarchical cluster analysis is performed on the similarity matrix. Before this can be done,

Table 1. The Raw Data Matrix

Time or Individuals	Behaviors					
	1	2	3	4	⋯	j
1	X_{11}	X_{12}	X_{13}	X_{14}	⋯	X_{1j}
2	X_{21}	X_{22}	X_{23}	X_{24}	⋯	X_{2j}
3	X_{31}	X_{32}	X_{33}	X_{34}	⋯	X_{3j}
4	X_{41}	X_{42}	X_{43}	X_{44}	⋯	X_{4j}
5	X_{51}	X_{52}	X_{53}	X_{54}	⋯	X_{5j}
⋮	⋮	⋮	⋮	⋮		⋮
i	X_{i1}	X_{i2}	X_{i3}	X_{i4}	⋯	X_{ij}

Table 2. The Similarity Matrix

	Behaviors 1	2	3	4	⋯	j
1	S_{11}	S_{12}	S_{13}	S_{14}	⋯	S_{1j}
2		S_{22}	S_{23}	S_{24}	⋯	S_{2j}
3			S_{33}	S_{34}	⋯	S_{3j}
4				S_{44}	⋯	S_{4j}
⋮						⋮
j						S_{jj}

Behaviors

the user must select one of the many clustering algorithms that are available. Several of the basic algorithms are examined in Section 3. The algorithms are decision rules for defining clusters. The definition of what constitutes a cluster is difficult. The word *cluster* is surrounded by "formidable linguistic difficulties" (Williams 1971). For those who are interested in the concept of a cluster see Gasking (1960). In a very simple way a cluster could be thought of as "collections of points which are relatively close, but which are separated by empty regions of space from other clusters" (Sneath 1969). The clustering algorithm is a logical or mathematical statement that defines certain properties of a cluster and imposes them on the similarity matrix.

Section 4 deals with some applications of hierarchical cluster analysis to ethological data. We will see its use as a supplement to factor analysis, as a method for comparing groups of subjects, and as a way of depicting relationships between variables in large data sets. Section 5 briefly deals with the problems of the availability of computer programs.

1. EXAMPLES OF HIERARCHICAL CLUSTER ANALYSIS

Let us consider a method of discovering natural groups in the following eight numbers: 25, 14, 20, 9, 3, 21, 1, and 7. Since there is an inherent sequential ordering to numbers, then placing them in order would be the first step. The order is: 1, 3, 7, 9, 14, 20, 21, and 25. Several groupings might appear after this step. The numbers could be grouped in clusters of greater than and lesser than 10, or in clusters of odd and even numbers. Both of these arrangements employ a decision rule or algorithm that divides the set into two subsets that are rather arbitrary and

Table 3. The Ordering of Numbers on an Interval Scale

1	3	7	9	14	20	21	25

nonhierarchical. The eight numbers could be placed on an interval scale as in Table 3.

Using a visual-intuitive algorithm we might form clusters of: 1 and 3, 7 and 9, 20 and 21. Some might be tempted to join 20, 21, and 25. Generally, the number 14 causes problems. It could be left by itself as a single member cluster or jointed with 7 and 9 or with 20 and 21. Such visual-intuitive clustering is nonhierarchical and often suffers from the inconsistent application of an algorithm. The actual algorithm used was one that involved distances and involved the absolute value of these distances (e.g., $D_{13} = | X_{i1} + X_{i3} | = | 1 - 7 | = 6$). The visual-intuitive distance algorithm used also considered only adjacent numbers or nearest neighbors. This is the algorithm used in single-linkage hierarchical cluster analysis.

To do a single-linkage hierarchical cluster analysis, a matrix of absolute value distances is constructed. Since the numbers to be considered are sequentially arranged (1, 3, 7, . . . , 21, 25) and since only adjacent numbers are being considered for the single-linkage algorithm, a partial matrix of absolute value distances can be used as in Table 4.

This vector of absolute value distances (2, 4, 2, 5, 6, 1, 4) is a diagonal of the matrix that lies next to the major diagonal of the matrix (0, 0, . . . , 0, 0). This diagonal can be used to locate the smallest absolute value distances between adjacent numbers. This vector will be called the adjacent value diagonal ($S_{12}, S_{23}, S_{34}, \ldots, S_{(j-1)j}$).

The following steps can be used to perform a single-linkage hierarchi-

Table 4. The Partial Matrix of Absolute Value Distances

	1	3	7	9	14	20	21	25
1	0	2						
3		0	4					
7			0	2				
9				0	5			
14					0	6		
20						0	1	
21							0	4
25								0

cal cluster analysis. Before starting, it is well to remember that each number has two neighbors (except for the extreme numbers 1 and 25). As clusters are formed, a number may have a neighbor that is in a cluster and another neighbor that is not in a cluster, et cetera.

1. Locate the smallest value (or smallest remaining value) of D on the adjacent value diagonal of the matrix. Join in a cluster the two entities (numbers) represented by that value of D at a similarity scale value equal to D.
2. Examine the nearest neighbors of the entities *in* the cluster. These nearest neighbors are called candidates. Admit into the cluster the candidate with the smallest value of D if that value of D is less than the value of D for the candidate and its *other* nearest neighbor. If the candidate can be admitted into the cluster, then continue Step 2 with the expanded cluster until candidates cannot be admitted. If the candidate cannot be admitted, then return to Step 1.

The end product is a single-linkage hierarchical cluster structure illustrated as a dendrogram in Figure 2.

The single-linkage algorithm when applied to the D matrix gave a hierarchical structure quite similar to the visual-intuitive process that joined 1 and 3, 7 and 9, and 20 and 21 in a nonhierarchical fashion. (Clustering algorithms should give results that have some intuitive appeal.) Because it was a specifically stated and consistently applied algorithm it could solve the problem of where to place 14. Although 14 is well placed, its relation to entities like 1 and 3 might not be the most appealing to some when they look at Figure 2 for any length of time. It

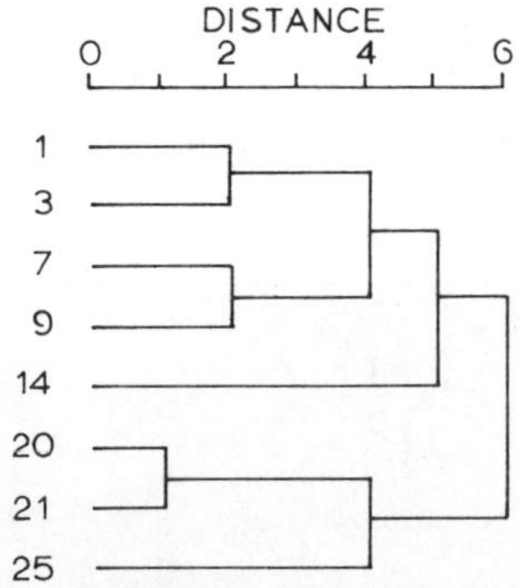

Figure 2. The dendrogram showing the results of the hierarchical cluster analysis (single-linkage) of the eight numbers in Table 3.

appears that 14 has similar relationships with cluster 1 and 3 and cluster 7 and 9. This is correct since at the $D = 4$ level of clustering these two clusters cease to be separate and are joined into a single cluster. But, it might be "nicer" to join 14 with 7 and 9 in some way that would not make it look as if 14 and 1 or 3 had some relationship.

A new algorithm could be generated and applied. The new algorithm should still join 1 and 3, 7 and 9, and 20 and 21 since that had such nice intuitive appeal but it should be stricter in the admission of candidates to clusters. The complete-linkage algorithm is a much stricter decision rule. To execute the complete-linkage algorithm the D matrix must be completed as in Table 5.

An examination of this expanded matrix reveals that the smallest value on the adjacent value diagonal is between 20 and 21 at $D = 1$. Clusters are initially formed by consulting the adjacent value diagonal. This diagonal will always contain the smallest values of $D > 0$ if the entities are sequentially ordered. The neighbors of 20 and 21 are considered and these are candidates 14 and 25. The candidate 14 is considered, but this time it is considered relative to its *farthest* neighbor in the 20 and 21 cluster. It is considered relative to 21, while 25 is considered relative to 20. These distances are 7 and 5 respectively. The diagonals of the matrix that lie to the right of the adjacent value diagonal contain the D values for the farthest distance to the nearest neighbor. The smallest D value in these diagonals is 5 indicating that 25 should be joined with 20 and 21 and not with anything else. Returning to the adjacent value diagonal 1 and 3, and 7 and 9 are formed into clusters. The entity 14 is again considered relative to 25 in cluster 20, 21, and 25 and relative to 7 in cluster 7 and 9. The D values are such that admission into 20, 21, and 25 is rejected. The entity 14 is considered for admission to 7 and 9 ($|7 - 14|$) and so is cluster 1 and 3 ($|9 - 1|$). This

Table 5. The Complete Matrix of Absolute Value Distances

	1	3	7	9	14	20	21	25
1	0	2	6	8	13	19	20	24
3		0	4	6	11	17	18	22
7			0	2	7	13	14	18
9				0	5	11	12	16
14					0	6	7	11
20						0	1	5
21							0	4
25								0

time 14 is admitted into 7 and 9 at $D = 7$. This process continues and the cluster structure emerges for the complete-linkage technique as shown in Figure 3.

Figure 2 for the single-linkage algorithm and Figure 3 for the complete-linkage algorithm should be compared. Both clustering algorithms give results that conform to the intital visual-intuitive approach. The natural order of the numerical sequence is preserved in both approaches. With the complete-linkage algorithm the entity 14 joins the 7 and 9 cluster directly and does not appear to have any direct relationship with 1 and 3. This is more appealing than the result of the single-linkage clustering. Also, if the single-linkage cluster structure were to be superimposed on the complete-linkage cluster structure, then one would see that clustering is *relatively* more compact using the complete-linkage algorithm. The initial clusters (20 and 21, 7 and 9, and 1 and 3) are formed at the same D level using both algorithms. With the complete-linkage technique the remaining additions to these clusters are strung out over greater distances making the initial clusters appear more compact. The complete-linkage algorithm tends to give clusters that look more like "good" clusters.

The previous example involving the application of the single-linkage and the complete-linkage algorithms to data illustrated some of the basic concepts of hierarchical cluster analysis. The illustraion was through a one dimensional problem. Hierarchical cluster analysis as a multivariate technique can be applied to N dimensional problems. To introduce the concept of dimensionality let us consider a two dimensional case.

This two dimensional problem was borrowed in part from Meisel (1972, page 149). Consider eight individuals (A through H) with two

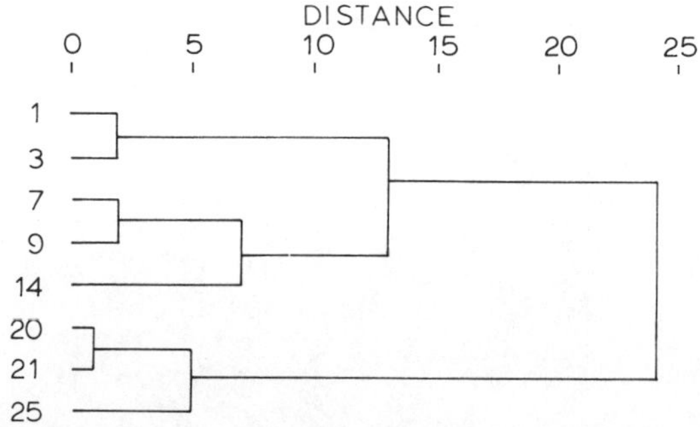

Figure 3. The dendrogram showing the results of the hierarchical cluster analysis (complete-linkage) of the eight numbers in Table 3.

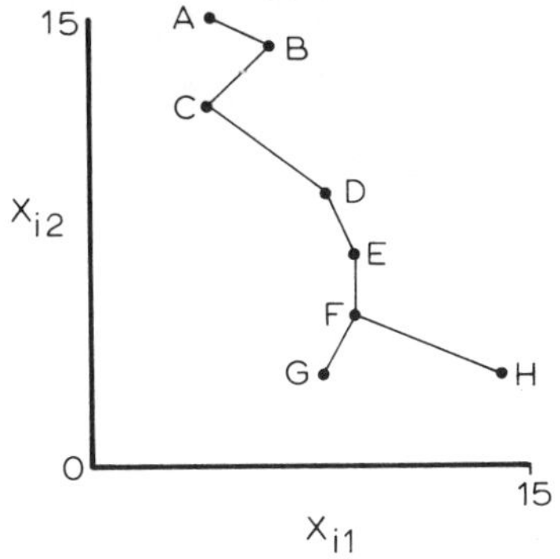

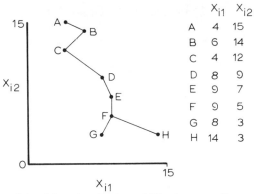

	X_{i1}	X_{i2}
A	4	15
B	6	14
C	4	12
D	8	9
E	9	7
F	9	5
G	8	3
H	14	3

Figure 4. The plot of the eight points (A through H) whose coordinates are shown on the right. The points are connected by lines (edges) that form a minimal spanning tree.

measures for each individual. These data can be arranged in a two dimensional space as shown in Figure 4.

The lines or edges connecting the points are referred to in graph theory as forming a "minimally connected graph" or a "minimal spanning tree." A minimal spanning tree is a connection of points by edges whose weight, as determined by the distance between the points, is a minimum (Andrews 1972; Meisel 1972; Zahn 1971). The edges of the graph in Figure 4 have weights that sum to 21.8 units. All other connections of points will have a sum that is greater than 21.8 units. Any sum of weights that is less than 21.8 units will involve a disconnected graph. Minimal connectivity is the criterion employed in the single-linkage clustering algorithm. A "maximally connected graph" is the opposite extreme of connectivity and is the criterion employed in the complete-linkage clustering algorithm. The two extremes can be conceptualized when one considers that there are $P - 1$ edges, whose sum is a minimum, for P points in a minimally connected graph and $P(P - 1)/2$ edges, whose sum is a maximum, for the same number of points in a maximally connected graph. Fewer than $P - 1$ edges in the graph will result in a mapping of points that are disconnected. One can achieve fewer than $P - 1$ edges by using an arbitrarily low cutoff or threshold for connectivity. To illustrate this principle, consider the D matrix in Table 6 for the eight points in Figure 4.

The distances in the D matrix are the lengths of the lines or edges connecting the points. There are 28 distances (zero distances are not counted) or $P(P - 1)/2$ distances for $P = 8$ points. If all of these values of D were used the result would be a maximally connected graph. If a

Table 6. The Distance Matrix for the Points in Figure 4

	A	B	C	D	E	F	G	H
A	0	2.2	3.0	7.2	9.4	11.2	12.7	15.6
B		0	2.8	5.4	7.6	9.5	11.2	11.5
C			0	5.0	7.1	8.6	9.9	13.5
D				0	2.2	4.1	6.0	8.5
E					0	2.0	4.1	6.4
F						0	2.2	5.4
G							0	6.0
H								0

distance threshold (d_t) of $d_t \leq 2.5$ was employed then a disconnected graph would result (Figure 5*A*). If $d_t \leq 3.5$, then the graph in Figure 5*B* would result. If $d_t \leq 6.0$, then the graph in figure 5*C* would result.

As one advances through increasing values of d_t the idea of a hierarchy emerges. At $d_t = 0$, the eight points exist but they are not connected. As d_t increases, progressively more and more points are connected until $P(P - 1)/2$ edges connect the P points. In the example used here, if $d_t = 15.6$, then the eight points would be connected by 28 lines.

The hierarchical organization of the eight pairs of numbers (eight points in a two dimensional space) used to demonstrate the graph theoretical approach can be used to illustrate hierarchical clustering in a manner similar to the original one dimensional example. Using the *D* matrix for these eight points a single-linkage cluster analysis can be performed. The cluster structure is shown in Figure 6.

If one were to drop perpendicular lines from the distance scale on the

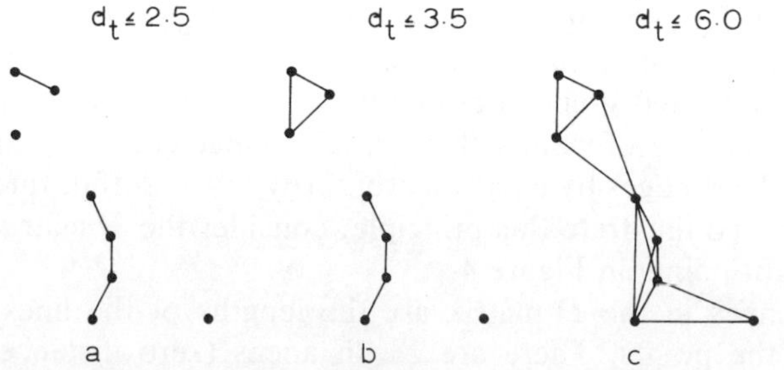

Figure 5. The effect of varying the distance threshold (d_t) on the connectivity of the eight points shown as a minimal spanning tree in Figure 4.

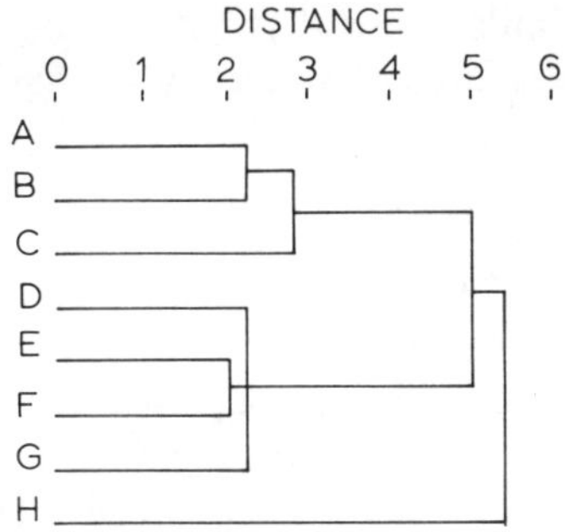

Figure 6. The dendrogram showing the results of the hierarchical cluster analysis using the single-linkage algorithm on the eight points shown as a minimal spanning tree in Figure 4. Note the relationship between the single-linkage clustering and the minimal spanning graphing.

single-linkage cluster analysis in Figure 6 at D values of 2.5, 3.5, and 6.0, then the relationship between the graph theoretical approach illustrated in Figure 5 *A*, *B* and *C* and the hierarchical cluster analytic approach of Figure 6 could be seen. For example, at $d_t = 2.5$ (Figure 5*A*) the points A and B are connected, and the points D, E, F, and G are connected. At $D = 2.5$ in the hierarchical cluster analysis (Figure 6) the clusters A and B, and D, E, F, and G exist.

Using some simple examples, one dimensional hierarchical cluster analysis and two dimensional hierarchical cluster analysis were illustrated. Two commonly used algorithms were demonstrated, single-linkage and complete-linkage. Also, some basic relationships between hierarchical cluster analysis and graph theory were explored. Little else will be said about graph theory. The real beauty and utility of hierarchical cluster analysis can be seen in its application to N dimensional problems. It can be used as a tool for discovering "strings or sheets of data wandering through N dimensional space. . . ." (Andrews 1972). Now we turn our attention to the general features of hierarchical cluster analysis by considering the methods of forming the similarity matrix and the properties of the algorithms.

2. THE SIMILARITY MATRIX

The raw data matrix may contain quantitative data that are in a uniform metric (e.g., frequencies only) or in a mixed metric (e.g., percentages, latencies, frequencies, lengths). The similarity matrix will contain measures of similarity or dissimilarity between the j behaviors. There are numerous measures of similarity or dissimilarity.

Distance

Return to the example of two dimensional hierarchical cluster analysis (Figure 4). Consider now that the column marked X_{i1} is a set of scores for a particular behavior and that X_{i2} is a set of scores of another behavior. In this example $j = 2$ and the pairs of scores can be arranged in a two dimensional space (j dimensional space). The first pair of scores (X_{11}, X_{12}) for individual A can be plotted at coordinates 4, 15 and the second pair of scores (X_{21}, X_{22}) for individual B can be plotted at coordinates 6, 14. The formula, $D_{AB} = \sqrt{[(X_{11} - X_{21})^2 + (X_{12} - X_{22})^2]}$, when applied to these coordinates yields a value of $D_{AB} = 2.2$. By calculating values of D for all possible pairs of scores in this example, a similarity matrix can be filled as was done in Table 6.

The value of D calculated in the above equation is for Euclidean distance. One can also use squared Euclidean distance D^2. Another quite similar distance measure is the "city-block" or "taxi cab" or "Manhattan" distance. If the line connecting two points (e.g., points A and B in Figure 4) is considered as the hypotenuse of a right triangle, then the sum of the lengths of the two arms of the triangle that join at the right angle is a measure of the distance between the two points. The "city-block" distance can be calculated by the formula, $|(X_{11} - X_{21})| + |(X_{12} - X_{22})|$. The "city-block" distance for the coordinates 4, 15 and 6, 14 is the value 3.

Edwards and Cavalli-Sforza (1964) have used a distance d^2 measure based on angular coefficients. Consider the points B and C in Figure 4 as being the ends of two lines that start at the intersection of the two axes. The angle between these two lines is θ. The equation for this distance measure is, $d^2_{BC} = 2 - 2\cos\theta$. For the points B and C, $\theta = 5°$, and $d^2_{BC} = 0.0076$.

The two dimensional examples above served as a handy example for calculating distances. The reader should recognize that the distances calculated were distances between individuals and not behaviors. The generalized similarity matrix in Table 2 is for behaviors. If our raw data matrix (Table 1) contains i individuals and j behaviors, then we will have j points in i dimensions. The distances between the j points will be the distances between the j behaviors. The distance between Behavior 2 (X_{i2}) and Behavior 4 (X_{i4}) is D_{24} and is represented in the similarity matrix as S_{24}. We want to find the distance between the columns of our raw data matrix to fill in the similarity matrix. For D_{24}, to get an estimate of S_{24}, we would solve the equation as follows:

$$\hat{S}_{24} = D_{24} = \sqrt{[(X_{12} - X_{14})^2 + (X_{22} - X_{24})^2 + \cdots + (X_{i2} - X_{i4})^2]}$$
$$= \sqrt{[\sum (X_{i2} - X_{i4})^2]}$$

Correlation r

The correlation between all possible pairs of behaviors may be the most useful measure of similarity for ethological applications of hierarchical cluster analysis. In most cases, the Pearson product moment correlation r is probably the easiest to use. In the similarity matrix, r_{24} is an estimate of S_{24} and is given by the formula

$$\hat{S}_{24} = r_{24} = \frac{\sum (X_{i2} - \bar{X}_{.2})(X_{i4} - \bar{X}_{.4})}{\sqrt{[\sum (X_{i2} - \bar{X}_{.2})^2 \sum (X_{i4} - \bar{X}_{.4})^2]}}$$

All of the symbols are as before. The symbol, $\bar{X}_{.2}$ refers to the mean of Behavior 2.

Something should be said about positive and negative values of r. An $r = -0.78$ and and $r = +0.78$ both indicate that the correlation of each set of variables accounts for the same percentage of the total variance (61%). In terms of similarity, an $r = 0$ indicates the minimum similarity and an $r = +1.0$ or an $r = -1.0$ both indicate the maximum similarity. From these arguments, the absolute value of r could be used as an index of similarity. The final decision should be left to the user and the intent of the hierarchical cluster analysis.

Association A

This material on association coefficients has been adapted from Sneath and Sokal (1973). Association coefficients are quantitative measures of agreement between pairs of columns in the raw data matrix. In this sense, association measures are similar to correlation coefficients as measures of similarity. Because of this relationship to correlation, the measures of association discussed here are only those measures that are useful for generating estimates of similarity from two-state (0 or 1) data. If each behavior was recorded as being either present or absent for each individual or in each time unit, then the raw data matrix would consist of j columns of 0, 1 data.

To fill the similarity matrix, it is necessary to generate a value of A_{ij} for each S_{ij} needed. Consider the two columns of data in Table 7 to have come from a raw data matrix like Table 1 and also that the column

Table 7. Two Columns of Raw Data and the Resulting Contingency Table

X_{i2}	X_{i4}
1	0
0	1
0	1
1	1
1	0
1	1
0	1
0	0
1	0
1	1
1	0
1	1

		X_{i4}	
		1	0
X_{i2}	1	$a = 4$	$b = 4$
	0	$c = 3$	$d = 1$

$m = a + d = 5$
$u = b + c = 7$
$n = m + u = 12$

on the left is for Behavior 2 (X_{i2}) and the column on the right is for Behavior 4 (X_{i4}). Thus, we will be calculating A_{24} as an estimate of S_{24}. Several measures of A_{ij} will be calculated to indicate the variety of association coefficients available. A contingency table is calculated that contains the following frequencies f: (a) 1, 1 matches, $f = 4$; (b) 1, 0 matches, $f = 4$; (c) 0, 1 matches, $f = 3$; and (d) 0, 0 matches, $f = 1$. The total number of matched pairs (1, 1 and 0, 0) is given by $m = a + d$. the total number of unmatched pairs (0, 1 and 1, 0) is given by $u = b + c$. The total number of pairs is given by $n = m + u$.

Jaccard's Association Coefficient $A(J)$

$$A(J)_{24} = \frac{a}{a + u} = 0.36$$

Dice's Association Coefficient $A(D)$

$$A(D)_{24} = \frac{2a}{2a + u} = 0.53$$

Sokal and Michener's Association Coefficient $A(S\&M)$

$$A(S\&M)_{24} = \frac{m}{n} = 0.42$$

Rogers and Tanimoto's Association Coefficient $A(R\&T)$

$$A(R\&T)_{24} = \frac{m}{m + 2u} = 0.26$$

Yule's Association Coefficient $A(Y)$

$$A(Y)_{24} = \frac{ad - bc}{ad + bc} = -0.50$$

Hamann's Association Coefficient $A(H)$

$$A(H)_{24} = \frac{m - u}{n} = -0.17$$

Product Moment Association Coefficient $A(PM\phi)$

$$A(PM\phi)_{24} = \frac{ad - bc}{\sqrt{[(a + b)(a + c)(c + d)(b + d)]}} = -0.24$$

Mutual Information *I(h; i)*

The reader should consult Orloci (1969) and Sneath and Sokal (1973) for a more complete treatment of information measures of similarity. Only mutual information, $I(h;\ i)$, will be considered here as a measure of similarity between two behaviors. (The symbol $I(h;\ i)$ is used by Orloci (1969) and by Sneath and Sokal (1973) to indicate mutual information and is used here for consistency.)

Return to the two behaviors, X_{i2} and X_{i4}, whose data were used to calculate the association coefficients (Table 7). These data will be used to calculate an estimate of similarity (S_{24}) based on mutual information, $I(h;\ i)$. The contingency table generated for the association data will be used here.

Mutual information, $I(h;\ i)$, is a function of: the total information in X_{i2} $[I(h)]$; the total information in X_{i4} $[I(i)]$; and the joint information $[I(h, i)]$. The functional relationship is shown below.

$$I(h;\ i) = I(h) + I(i) - I(h, i)$$

Where $I(h) = n \log_e n - (a + b) \log_e (a + b) - (c + d) \log_e (c + d)$

$$I(i) = n \log_e n - (a + c) \log_e (a + c) - (b + d) \log_e (b + d)$$

$$I(h, i) = n \log_e n - a \log_e a - b \log_e b - c \log_e c - d \log_e d$$

Solving the above equations, it was found that

$$I(h;\ i) = 7.638 + 8.151 - 15.433 = 0.356$$

This is then an estimate of similarity, $\hat{S}_{24}$. A value of $I(h;\ i)$ must be calculated for each S_{ij} needed.

Clear answers cannot be given to the question: "Which similarity measure should I use?" Certainly, if continuous data or multistate data were available, one would not reduce such data to two-state data just to employ an association measure. One could choose between distance and correlation estimates of similarity. The final choice by the user should be consistent with the kind of behaviors being studied and with the intended use of the resulting hierarchical cluster structure.

3. CLUSTERING ALGORITHMS

Following the choice of a similarity measure, the user must decide which clustering algorithm should be used. For a comprehensive survey of clustering algorithms the reader should consult Hartigan (1975). Clustering algorithms are decision rules that define the way that the clustering should occur. A few of the more common algorithms will be discussed.

Single-Linkage

This algorithm was used earlier in the illustrative examples. It goes by a variety of other names: nearest neighbor strategy (Lance and Williams 1967), minimum method (Johnson 1967), and the closest linkage method by others. All of the names describe the basic properties of the algorithm. Clusters are formed by establishing single links between the closest behaviors (highest correlation, association, or mutual information and smallest distance) in the i dimensional space. A cluster is formed between two unclustered behaviors if they are closer to each other than each is to any other behavior. A candidate behavior is admitted into an existing cluster if it is closer to its nearest neighbor in the cluster than to any other cluster or unclustered behavior. When a candidate is admitted into a cluster at some level of similarity, that level represents the minimum similarity that exists between the candidate and the existing cluster. The application of this algorithm produces "long straggly clusters" (Sneath and Sokal 1973) whose "chaining tendencies are notorious" (Lance and Williams 1967). Once a cluster is formed by this technique the first behaviors admitted and the last behavior admitted into the cluster are likely to be "relatively unalike" (Bailey 1967). The negative comments should not discourage the user. It is a fine algorithm provided its limitations are recognized.

Complete-Linkage

This algorithm was also used earlier in the illustrative examples. It is also called the farthest neighbor strategy (Lance and Williams 1967), the maximum method (Johnson 1967), and clustering by the compactedness criteria by others. Like single-linkage clustering, the two closest (most similar) behaviors in the *i* dimensional space are clustered. A candidate behavior is admitted into the existing cluster if it is closer to the most distant member of the cluster than it is to another unclustered behavior or the most distant member of another cluster. When a candidate is admitted into a cluster at some level of similarity, that level represents the maximum similarity that exists between the candidate and the cluster relative to other clusters and other behaviors. This method is the other extreme of the single-linkage method. It tends to produce either compact tight clusters or very few clusters. It is a very "stringent" algorithm (Bailey 1967) that admits candidates "with difficulty" (Sneath and Sokal 1973). If your goal is to get as many behaviors into as few large clusters as possible, then this algorithm should probably not be used. The single-linkage algorithm would be better. If on the other hand, you wanted a conservative clustering procedure with well defined clusters then this procedure would be excellent. You may obtain several single member clusters (unclustered behaviors) with the complete-linkage algorithm. Depending on your needs, this can be either a problem or a source of inspiration.

Averaging Methods

The averaging algorithms are somewhat intermediate on a scale of a lax cluster forming criterion (single-linkage) to a strict cluster forming criterion (complete-linkage). The averaging relates to the level at which a behavior is joined to an existing cluster or the level at which two existing clusters are joined. In this case, the level refers to the value on the similarity scale used to depict clustering. In the single-linkage technique the level was determined by the similarity between the candidate and its nearest neighbor in the relevant cluster. In the complete-linkage technique the level was determined by the similarity between the candidate and its farthest neighbor in the relevant cluster. The averaging techniques use a kind of average between the similarity values associated with the candidate and the relevant cluster. Two kinds of averages are used: arithmetic averages and centroids. Arithmetic averages are values and centroids are points. Centroids are points in a hyperspace and can be interpreted geometrically. In both the arithmetic average and the centroid techniques, the original similarity matrix is reformed by the formation of each new cluster. If two behaviors are united in a cluster, then an appropriate single value indicating the maximum similarity is substituted for them in a reformed similarity matrix. If the value of S_{12} indicates that behaviors 1 and 2 should be clustered, then similarity values are recalculated for all S_{ij} where $i = 1$ and $i = 2$ or where $j = 1$ and $j = 2$. Similarity values unaffected by the clustering of 1 and 2 are not recalculated (e.g., S_{34}). The recalculation after each clustering of the similarity matrix reduces the size of the matrix after each clustering.

The averaging techniques can involve weighting procedures. If a cluster of four behaviors exists, then it could be considered to have a weight of four. A candidate for admission into this cluster could be considered to have a weight of one. In weighted averaging algorithms, the candidate is given a weight equal to the total weight of the relevant cluster (candidate weight = 4, and cluster weight = 4). In unweighted averaging algorithms, the candidate has a weight of one and the relevant cluster has a weight equal to the number of members in the cluster. If the original cluster has a weight of four, then the new cluster that is forming will have a new total weight of five when the candidate is admitted. The original cluster determines four-fifths of the averaging.

The unweighted arithmetic average algorithms and the unweighted centroid algorithms tend to form new averages and centroids closer to the original averages and centroids. The weighted arithmetic average algorithms and the weighted centroid algorithms appear to be just a little stricter than the unweighted counterparts. While the averaging techniques are about midway between the extremes formed by single-linkage and complete-linkage algorithms, the weighted averaging techniques are a little closer to the complete-linkage algorithm with regard to the candidate admission criterion. On this point see Sneath and Sokal (1973, Figure 5–11, page 240). The averaging techniques consist of weighted and unweighted arithmetic average algorithms and weighted and unweighted centroid algorithms. They can be used with all of the previously discussed similarity measures.

It was stated previously that the clustering was intrinsic in Williams' (1971) organization of clustering strategies. The inherent structure of the data determines the resulting clustering structure. It should be obvious at this point that the choice of the algorithm is going to influence the cluster structure to some degree. The algorithms are extrinsic and must be determined in advance of the analysis. The algorithm does not determine the number of clusters that will form, and it does not determine the number of entities that will be contained in the clusters. The algorithm is an operational definition of a cluster in general terms. It is necessary because we lack a formal mathematical definition of a cluster. Once a cluster is operationally defined by selecting an algorithm or by writing a new one, then the process of clustering is intrinsic within the guidelines of the operational definition.

The choice of a clustering algorithm is very important. The algorithms discussed above are the most common ones. Again, the user should check Hartigan (1975) for a good description of a large variety of algorithms including variations on and combinations of the above algorithms. It is assumed that the reader of this chapter is more interested in hierarchically clustering some data than in exploring the intricate logic and mathematics of algorithms. The algorithms discussed in this chapter are some of the best general purpose algorithms. It is better to use an algorithm that you are familiar with than to use one of considerable complexity that only a mathematician can decipher. The purpose of hierarchical cluster analysis is to gain some increased insight about your data. You must understand what the algorithm is doing before the cluster structure can be illuminating. If you plan to publish the cluster structure, then you must precisely indicate the algorithm used. A dendrogram depicting the cluster structure without a precise indication of the algorithm is a useless series of lines.

4. APPLICATION AND EXAMPLES

The material in this section is designed to demonstrate some possible uses of hierarchical cluster analysis and it consists of uses that I have made of this analytic technique. The purpose is not to present new data to the reader; therefore the material on experimental procedures and observational techniques is very brief. In two instances, published correlation matrices have been reanalyzed using hierarchical cluster analysis. Correlation coefficients have been used as estimates of similarity in all cases. Therefore, similarity can be interpreted from that perspective in the dendrograms. The complete-linkage technique has been used as the clustering algorithm for all analyses.

Hinde (1970) discusses the use of factor analysis (see Chapter 8) as an aid in identifying common causal factors in behavior. If it is permissible to use factor analysis as an aid then it should be permissible to use hierarchical cluster analysis too. Neither should be used as the sole criterion. The logic is as follows: if the raw data matrix is behaviors × time, and if the proper size time units are selected, then many behaviors that are caused by a common factor should occur at approximately the same time (in or about the same time unit). Differential but fixed delays between the cause and the effects should be demonstrated in fixed sequences of effects. High correlations should exist between effects. While high correlations do not confirm causal factors, they do not deny them either. The same limitations apply here that apply to any sequence analysis. The results should be interpreted with extreme caution.

Hinde (1970) uses Wiepkema's (1961) classic study of the behavior of the bitterling *(Rhodeus amarus)* as an example of the application of factor analysis. Figure 7 shows the vector model from Wiepkema (1961) showing the arrangement of his three factors on the surface of a sphere. These factors were identified following a principal-axes factor analysis with a quadrimax rotation performed on a Spearman rank correlation matrix of the frequency that one behavior followed another behavior. Factor 1 was an aggressive factor and contained: CHS = chasing; HB = head butting; TU = turning beat; and JK = jerking. Factor 2 was a nonreproductive factor and contained: CHF = chafing; FF or FFL =

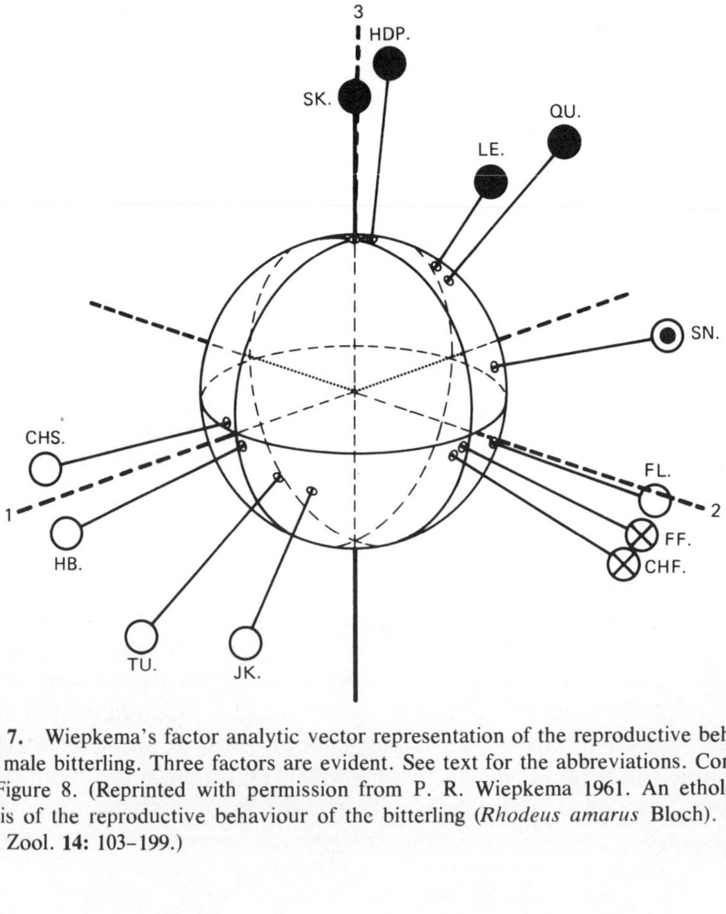

Figure 7. Wiepkema's factor analytic vector representation of the reproductive behavior of the male bitterling. Three factors are evident. See text for the abbreviations. Compare with Figure 8. (Reprinted with permission from P. R. Wiepkema 1961. An ethological analysis of the reproductive behaviour of the bitterling (*Rhodeus amarus* Bloch). Arch. Neerl. Zool. **14:** 103–199.)

finflickering; FL = fleeing; and SN = snapping. Factor 3 was a reproductive factor and contained: QU = quivering; LE = leading; HDP = head down posture; and SK = skimming. These three factors accounted for about 82 percent of the total common variance.

A hierarchical cluster analysis (complete-linkage) was performed on the published correlation matrix (Wiepkema 1961, Table IV, page 126). The resulting cluster structure is shown in Figure 8. The first cluster (top) corresponds nicely to Factor 1, the aggressive factor. The third cluster (bottom) corresponds nicely to Factor 2, the nonreproductive factor. Factor 3, the reproductive factor is poorly represented (middle) by a cluster (HDP and SK) that eventually joins Cluster 3 (bottom) and by a cluster (QU and LE) that eventually joins Cluster 1 (top).

The results of the hierarchical cluster analysis would indicate that the reproductive factor (Factor 3: QU, LE, HDP, and SK) contained behaviors, QU and LE, that were primarily related to aggressive factors and behaviors, HDP and SK, which were primarily related to nonreproductive factors. The actual factor loadings for QU raise some questions concerning its placement in Factor 3. The factor loading on Factor 1 was −0.67 and on Factor 3 was +0.64 yet QU was placed in Factor 3. The communality (h^2) was only .44 for LE indicating that three factors accounted for 44 percent of the common variance and left 56 percent of the common variance unexplained. It should be realized that the correspondence between one form of factor analysis and one

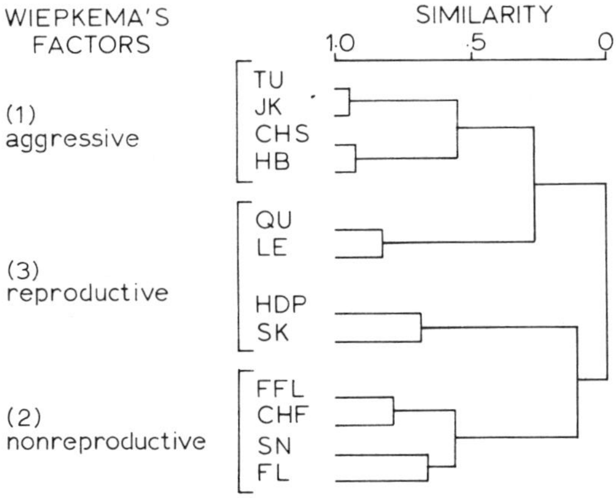

Figure 8. The dendrogram on the right is the result of a hierarchical cluster analysis (complete-linkage) of the reproductive behavior of the male bitterling. The original data is from Wiepkema (1961). On the left are the three factors identified by Wiepkema following a factor analysis of the same data. See text for abbreviations. Compare with Figure 7.

form of hierarchical cluster analysis was generally good. Together, these techniques should prove useful to ethologists if they are interpreted cautiously.

Hierarchical cluster analysis can also be useful for comparing patterns of behavior between experimental conditions. This can be done if the same kind of raw data matrix is available for each condition. Kleiman (1972) published two correlation matrices depicting maternal behavior in the green acouchi *(Myoprocta pratti)*. Both matrices contained the intercorrelations of eight maternal variables, one matrix for females with young in social groups and the other matrix for females with young not in social groups. Figure 9 shows the results of the hierarchical cluster

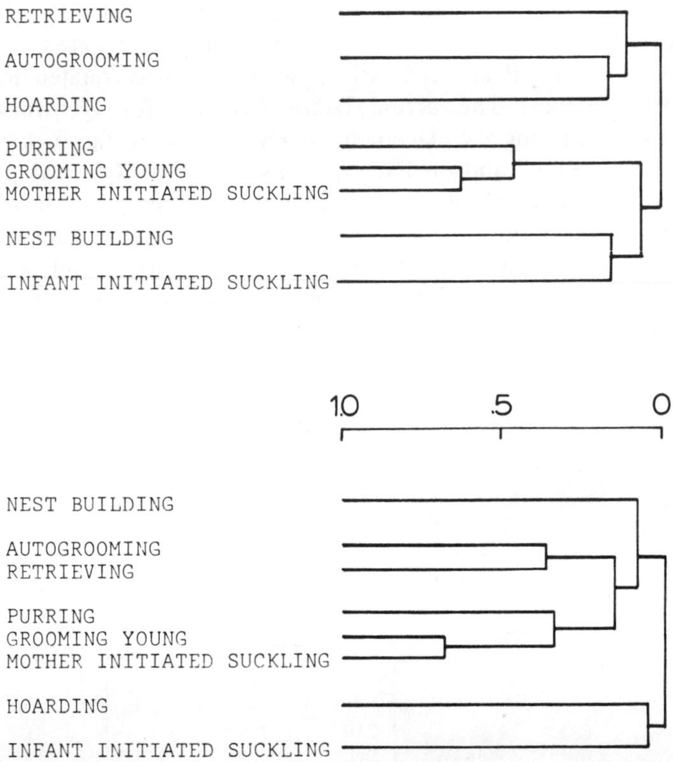

Figure 9. Dendrograms showing the results of hierarchical cluster analysis (complete-linkage) of Kleiman's (1972) data on the maternal behavior of the green acouchi. *Top:* the dendrogram showing the interrelationships between the behaviors of females and young in social groups. *Bottom:* the dendrogram showing the interrelationships between the behaviors of females and young not in social groups.

analysis (complete linkage) performed on each correlation matrix. The cluster: Purring, Grooming Young, and Mother Initiated Suckling remained as a fixed pattern of behavior under both conditions. This result was observed by Kleiman (1972) without the aid of cluster analysis.

If the same kinds of raw data matrices are available for several species then hierarchical cluster analysis can be used to locate common patterns of behavior shared by the various species. Figure 10 shows the results of doing just that on a number of behaviors presumably related to "activity" in two inbred strains of mice, *Mus musculus,* (*Top*: C57BL/6 and *Middle*: LP/J) and the Grasshopper mouse, *Onychomys leucogaster (Bottom).* In all cases the raw data matrix contained the frequencies of behaviors × time. Most of the behavior terms correspond to those in van Oortmerssen (1971) with the exception that grooming is divided into face and body grooming. The category Sniff Air was added to the above list. It is a behavior in the stationary quadruped posture and is characterized by a rapid, short excursion, lifting of the head often with the eyes closed or partially closed. The 12 behaviors applied equally well to *M. musculus* and to *O. leucogaster*. Ten males were in each of the three groups and were observed for 10 minutes each in a 20 gallon aquarium under low level red light. The two inbred strains of mice had cluster structures indicating low levels of clustering. The strains did not share any clusters in common. One would get the impression that "activity" was a loosely organized system of reasonably independent behaviors. For *O. leucogaster*, we see some tighter clusters. There is an appealing cluster of Shake, Scratch, Groom body, and Groom face.

Several of the behaviors in Figure 10 are followed by a minus sign. This indicates that there was a negative relationship between the behavior and the other members of the cluster. At times, this can be useful. For example, in the cluster structure for *O. leucogaster* (Figure 10, *Bottom*) the cluster Incomplete Upright, Push Dig, and Dig, the behavior Incomplete Upright has a negative relationship with Push Dig and Dig. This can be interpreted as: not being in the Incomplete Upright posture is positively related with the behaviors Push Digging and Digging.

The cluster structure in Figure 11 depicts the hierarchical cluster analysis (complete-linkage) of the maternal behavior of the Mongolian gerbil, *Meriones unguiculatus*. Briefly, the procedure was to record certain observations prior to removal of the litter, remove the litter, and replace the young one at a time recording the behavior of the female, and then to record certain observations when the young were all back in the nest. Data were gathered on eight litters for 25 days on 31 behaviors. Most of the behaviors were reasonably common measures of maternal

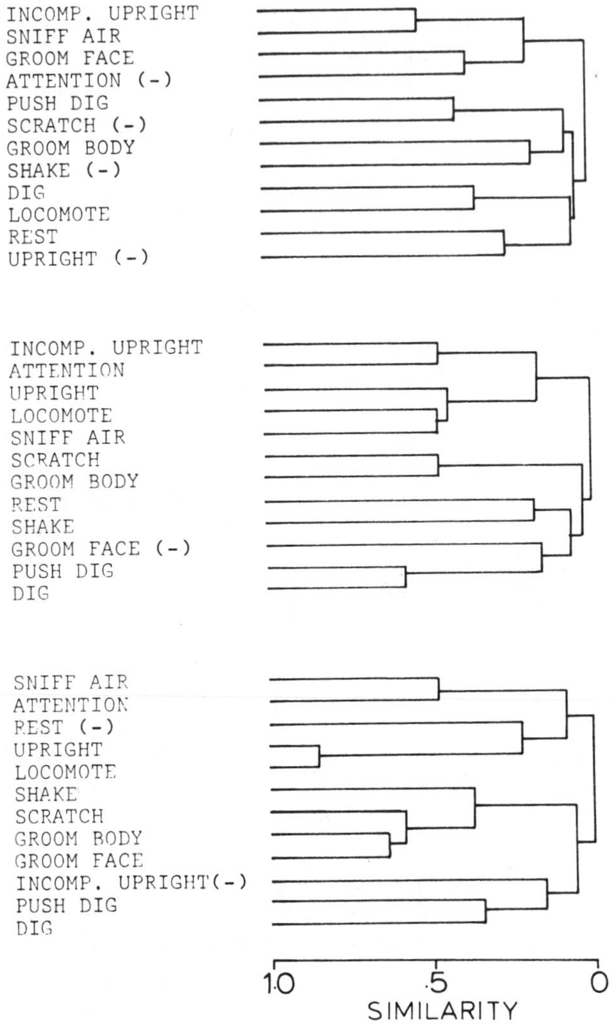

Figure 10. Three dendrograms depicting the hierarchical cluster analysis (complete-linkage) of "activity" data from three rodent forms: *Top: Mus musculus* C57BL/6; *Middle: M. musculus* LP/J; and *Bottom: Onychomys leucogaster*.

behavior and ontogeny. The measures of the ontogeny of the young were: age (confounded with days since parturition), weight, instantaneous percentage growth rate for weight (IPGR wt), mortality, rooting strength, and rate and percentage of ultrasonic vocalization. The use of the lower case "a" after a behavior indicates that the behavior occurred after the female made contact with a pup to be retrieved and before she made contact with a subsequent pup. The lower case "b" refers to the indicated behavior after all of the pups were back in the nest following retrieval. Retrieval Time (1) and (2) are retrieval times including maximum times for failure to retrieve (1) and excluding those maximum times (2). Retrieval Lat. is latency to retrieve. Disturbance and Defense are scaled reactions of the female to the removal of the cage top and then the removal of the young. Mother Power is calculated by: (weight of young × distance carried during retrieval)/Retrieval Time (2).

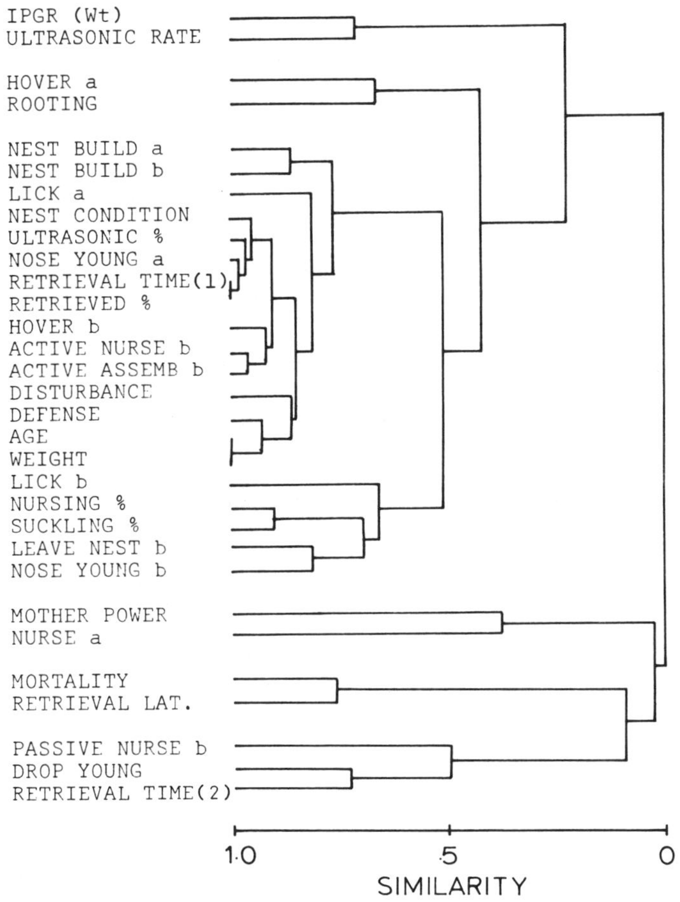

Figure 11. The dendrogram depicting the hierarchical cluster analysis (complete-linkage) of the maternal behavior of the Mongolian gerbil.

Six clusters emerge that are grouped into two superclusters. The two superclusters are independent and only unite at zero similarity. Of note is the absence of a separate cluster containing the developmental measures. This indicates that development and maternal care are inte-

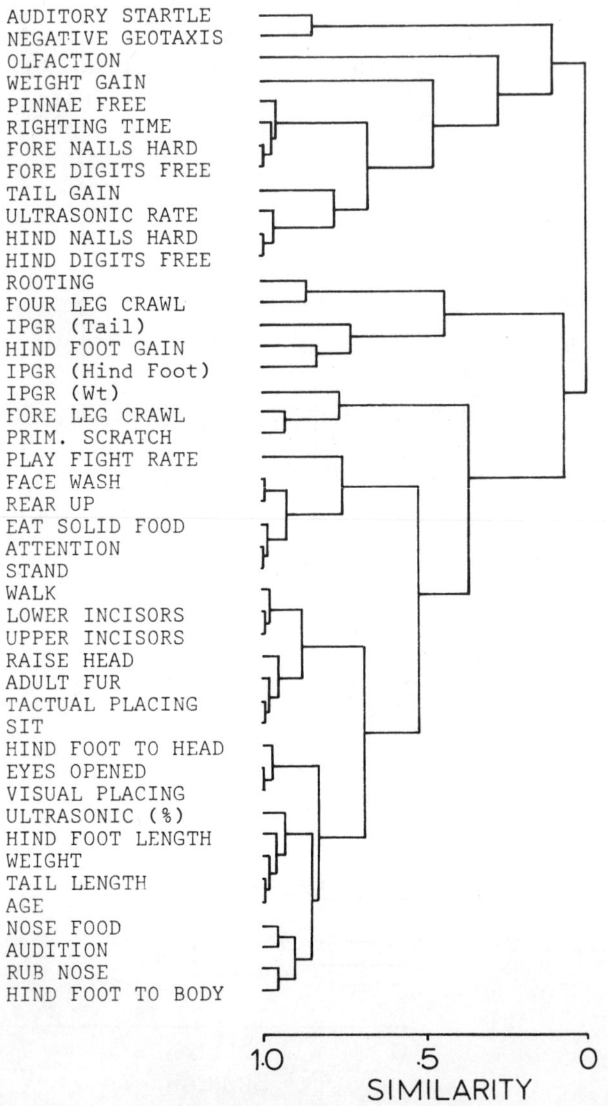

Figure 12. The dendrogram depicting the hierarchical cluster analysis (complete-linkage) of the behavior and morphological development of the Mongolian gerbil. (Redrawn from De Ghett 1972.)

grated. Also interesting is the evidence that maternal behavior is not a unitary concept.

Figure 12 illustrates the results of the application of hierarchical cluster analysis (complete-linkage) to the behavioral and morphological development of the Mongolian gerbil (De Ghett 1972). Forty-five measures of ontogeny were taken across 30 days of postnatal development. Eleven clusters emerged. The measures that cluster are those that have simiar developmental profiles. Many of these clusters are indicative of patterns of neural development.

Hinde (1970) urges those in the area of behavioral development not to forget that "development involves a nexus of causal relations" and cautions the researcher about doing and interpreting studies involving just a single variable. As the number of measures increases the number of pair relations increases rapidly. The number of pair relations is given by $N(N - 1)/2$. The 45 measures of development used in the study of the development of the Mongolian gerbil produced 990 relationships (correlations). Hierarchical cluster analysis provides a useful analytic tool for locating the best relationships according to the algorithm.

Lorenz (1950), in discussing the inductive methods used in ethology said:

> The development of any inductive natural science proceeds through three indispensable stages: the purely observational recording and describing of fact, the orderly arrangement of these facts in a system, and finally the quest for natural laws prevailing in the system.

Hierarchical cluster analysis aids in the system forming arrangement of data. It may even prove as a useful quantitative technique for identifying and defining systems. Fentress (1976) defines a system as "a definable set of elements . . . which are associated with one another by specified . . . rules." He then goes on to indicate that two or more systems can be distinghished on the basis of "relatively tight linkage between subcomponents within the system and relatively marked discontinuities between the subcomponents in different systems." A behavior system could be equated with a cluster and the specified rule that defines a system could be the clustering algorithm.

5. COMPUTER PROGRAMS

Hierarchical cluster analysis for multidimensional problems must be handled by a computer. Most of the computer programs appear to be unpublished and held in university computer centers. No doubt, this will change in the future and programs will become widely available. Anderberg's (1973) book contains many subroutines for hierarchical cluster analysis. Sneath and Sokal (1973) reference an IBM program for numerical taxonomy (IBM 704 taxonomy application Parts I & II — IBM 704 Program IB-CLF). Throughout this chapter I have tried to present the material in a way that demonstrates the logic of the techniques. A computer programmer should be able to read the material and write a suitable program for you.

CHAPTER

6

Multidimensional Contingency Table Analysis

PATRICK W. COLGAN and J. TERRY SMITH
Queen's University at Kingston

One of the first steps in a quantitative study of animal behavior is the observation of the frequencies with which various acts occur in a group of individuals, and the study of these data for patterns, regularities, and relationships. The methods of multidimensional contingency table analysis provide a systematic procedure for drawing valid conclusions from such frequency data. They are readily applicable to such problems as the structure of sequences of acts, the study of responses to stimuli, and the relationship of modes of behavior to other observable characteristics of the population.

When an ethologist classifies a group of individuals according to sex, stage of reproductive cycle, and whether or not a certain type of behavior was exhibited, the result is a contingency table, in this case three-dimensional, the cells of which contain the frequencies of the various category combinations of the variables "sex," "reproductive state," and "act." Panel 1 of Table 1 (Wiley 1976) shows how a typical three dimensional table can be displayed; the dimensions in this case are $2 \times 2 \times 3$.

More generally, when the individuals under study are classified according to a stratification structure on the population as well as by their behavioral characteristics, the resulting multidimensional contingency table summarizes a set of multivariate observations on the individuals. Important behavioral questions often can be related to the existence or absence of interactions among subsets of the classifying variables. The common statistical tools for problems involving several variables, for example regression analysis and analysis of variance, are not wholly satisfactory for two reasons: (1) the normal distribution on which they are based is a poor sampling model for frequency data; and (2) the usual application of these methods pays little heed to interactions among the variables, expecially those involving three or more variables. Another inadequate procedure frequently used is the reduction to all possible two-way contingency tables, a technique that, although consistent with an appropriate sampling model for the data, ignores or distorts possibly crucial information about the joint distribution of the variables. The methods of multidimensional contingency table analysis make it possible both to model the data appropriately and to determine those interactions that are important to a faithful summary of the data.

The methods described in this chapter grow out of the traditional analysis of two-way contingency tables. In the two-dimensional case, one computes expected cell frequencies predicted by the hypothesis of independence, and uses a chi-square statistic to test this hypothesis by comparing observed and expected cell frequencies. The use of more than two variables results in higher dimensional contingency tables and a variety of possible hypotheses describing relations among the variables.

Table 1. A 2 × 2 × 3 Contingency Table Based on the Social Behaviour of Carib Grackles (Wiley 1976)

Wing Elevation	Beak Elevation	Social Context: Near Male	Alone	Near Female
Panel 1 Observed Frequencies				
Low	Low	16	19	11
	High	30	10	4
High	Low	2	1	8
	High	7	3	4
Panel 2 Expected Frequencies under Model [13] [23]				
		15.05	17.58	10.56
		30.95	11.42	4.44
		2.95	2.42	8.44
		6.05	1.58	3.56
Panel 3 Standardized Residuals from Model [13] [23]				
		0.24	0.34	0.14
		−0.17	−0.42	−0.21
		−0.55	−0.91	−0.15
		0.38	1.13	0.24

These hypotheses are most easily studied through an associated set of linear models akin to those used in factorial analysis of variance, the terms of which relate to the interactions among various combinations of the variables. The fit of a particular model is tested by estimating expected cell frequencies in a way that conforms with that model, and by comparing these estimates with the observed cell frequencies by means of a chi-square statistic. Several different approaches to the analysis have been developed. This discussion will be based on a method developed in the comprehensive study by Bishop, Fienberg, and Holland (1975). Alternative approaches are described by Goodman (1970), Grizzle, Starmer, and Koch (1969), Ku and Kullback (1974), Nerlove and Press (1973), Bock (1974), and Cox (1972). Applications to ecology, geography, medicine, sociology, and politics have been discussed by Fienberg (1970), Altman (1975), Hamerton et al. (1977), Goodman (1972), and Gold (1976), respectively.

The notation to be used throughout the chapter can be illustrated in the case of a three-dimensional table of dimensions $I \times J \times K$. The observed frequency for a typical cell at level i of variable 1, level j of variable 2, and level k of variable 3 is denoted by x_{ijk}; the expected frequency for this cell is m_{ijk}. The replacement of a subscript by the symbol + denotes that the frequencies have been summed over that index; for example,

$$x_{i+k} = \sum_{j=1}^{J} x_{ijk}$$

denotes the marginal total for categories i and k of variables 1 and 3 respectively. The total frequency, x_{+++}, is denoted by N. The same notational conventions are used for cell probabilities p_{ijk}.

1. ASSUMPTIONS

The methods presented here are based on maximum likelihood estimation of expected cell frequencies in cases where the cell probabilities are given by one of the following distributions:

1. *Product Poisson.* All marginal totals are determined as an outcome of the sampling.
2. *Multinomial.* Here, only the total frequency N is determined prior to sampling.
3. *Product multinomial.* In this case a set of marginal totals is specified prior to sampling.

These distributions each correspond to a different sampling scheme. The product Poisson distribution arises when independent events are observed over a fixed period of time and classified by an n-way classification. In this case, the number of observations in any cell is Poisson distributed and all cells are mutually independent. An example of this sampling scheme is the classification of shark attacks on humans during 1976 according to month and severity of attack; it is clear that the total number of attacks as well as the marginal totals for months and levels of severity are determined as a result of the sampling. The multinomial distribution applies to the n-way classification of N randomly chosen individuals; here the number N is specified in advance, and the N individuals are selected without regard to the classifying variables. The product multinomial distribution is appropriate for the simultaneous classification of several groups of individuals when the numbers in the groups have been decided in advance. For example, if 40 individuals, with 10 selected from each of four species-by-sex subpopulations, are classified according to three levels of aggression, the result is a $2 \times 2 \times 3$ contingency table with the four species-by-sex margins x_{ij+} fixed by design. The correct distribution is the product of four independent trinomial distributions.

Because the statistics used for testing models are only asymptotically chi-square distributed, it is always assumed that the total frequency N is large enough so that the chi-square approximation is accurate to within reasonable limits. This problem is discussed below.

2. METHOD

In broad outline, the method has many points of analogy with regression analysis and factorial analysis of variance. As with these techniques, there is a process of working through a sequence of linear models to find the model with the smallest number of parameters that fits reasonably well. For each model considered, the parameters are estimated by maximum likelihood, and cell frequencies predicted by the model are calculated. The fit of the model is tested by the size of the discrepancies between the observed and predicted cell frequencies as summarized by a chi-square statistic.

Two-Way Tables

The details of the method are best illustrated using the familiar two-dimensional case. Suppose a table with I rows and J columns is used to determine whether or not two categorical variables are related in a population. The null hypothesis of independence specifies a model for the cell probabilities p_{ij} of the form

$$p_{ij} = p_{i+}p_{+j}$$

where $i = 1, \ldots, I$; $j = 1, \ldots, J$. The corresponding expression for the expected cell frequencies m_{ij} is

$$m_{ij} = Np_{i+}p_{+j}.$$

This is a multiplicative model which may be transformed into an additive model by taking logarithms; that is,

$$\log m_{ij} = \log N + \log p_{i+} + \log p_{+j}.$$

Introducing some new notation, we rewrite the above expression as

$$\log m_{ij} = u + u_{1(i)} + u_{2(j)}$$

(the subscripts 1 and 2 referring to the row and column variables, respectively). In other words, under the hypothesis of independence of row and column variables, the logarithm of the expected frequency for the (i, j) cell is a linear expression consisting of a general term, a term relating only to the marginal probability for row i, and a term relating only the marginal probability for column j. In the parlance of the analysis of variance, there is a grand mean term and two main effect terms, but no term for the interaction of the row and column variables. Such a model is called a *log-linear model.*

A more general log-linear model which includes a term for interaction might have the form

$$\log m_{ij} = u + u_{1(i)} + u_{2(j)} + u_{12(ij)} \tag{1}$$

where $i = 1, \ldots, I$; $j = 1, \ldots, J$. In this model, the "grand mean" term u is an average of the log-expected frequencies over all the cells, that is,

$$u = \frac{1}{IJ} \sum_{i=1}^{I} \sum_{j=1}^{J} \log m_{ij} \tag{2a}$$

while the "main effect" terms $u_{1(i)}$ and $u_{2(j)}$ and the "interaction " term $u_{12(ij)}$ are deviation terms, that is,

$$u_{1(i)} = \frac{1}{J} \sum_{j=1}^{J} (\log m_{ij} - u); \qquad i = 1, \ldots, I \tag{2b}$$

$$u_{2(j)} = \frac{1}{I} \sum_{i=1}^{I} (\log m_{ij} - u); \qquad j = 1, \ldots, J \tag{2c}$$

$$u_{12(ij)} = \log m_{ij} - u - u_{1(i)} - u_{2(j)};$$
$$i = 1, \ldots, I; j = 1, \ldots, J. \tag{2d}$$

As in the analysis of variance, these deviation terms sum to zero, that is, $u_{1(+)} = u_{2(+)} = u_{12(+j)} = u_{12(i+)} = 0$. We note that this model has IJ "free" parameters: one u, $(I - 1)$ of the $u_{1(i)}$, $(J - 1)$ of the $u_{2(j)}$, and $(I - 1) \times (J - 1)$ of the $u_{12(ij)}$, which is exactly the number of cells or data points.

The usual method of estimating these parameters by maximum likelihood is first to find the estimates of the expected cell frequencies, m_{ij}. For the multinomial sampling model, for which m_{ij} equals Np_{ij}, the likelihood can be written as

$$k \prod_{i=1}^{I} \prod_{j=1}^{J} (p_{ij})^{x_{ij}} = k \prod_{i=1}^{I} \prod_{j=1}^{J} \left(\frac{m_{ij}}{N}\right)^{x_{ij}} = k' \prod_{i=1}^{I} \prod_{j=1}^{J} (m_{ij})^{x_{ij}}$$

where k and k' are factors that do not depend on the parameters p_{ij} and m_{ij}. The expression on the right can be rewritten in terms of $\log m_{ij}$ as

$$k' \prod_{i=1}^{I} \prod_{j=1}^{J} \exp(x_{ij} \log m_{ij}) = k' \exp\left(\sum_{i=1}^{I} \sum_{j=1}^{J} x_{ij} \log m_{ij} \right) .$$

With the likelihood in this form it is easy to apply the appropriate theorems from mathematical statistics or alternatively to use a Lagrange multiplier for the constraint $m_{++} = x_{++} = N$ and show that the maximum likelihood estimates, $\hat{m}_{ij}$, of the expected cell frequencies are just the observed cell frequencies x_{ij}. The fit of the model in Expression 1 is therefore perfect, and we call this a *saturated model*. The estimates of the u-terms in this model can be obtained directly from the Equations 2a to 2d above. For example,

$$\hat{u} = \frac{1}{IJ} \sum_{i=1}^{I} \sum_{j=1}^{J} \log \hat{m}_{ij}$$

$$\hat{u}_{1(i)} = \frac{1}{J} \sum_{j=1}^{J} (\log \hat{m}_{ij} - \hat{u}); \qquad i = 1, \ldots, I.$$

The model of real interest is the one that specifies independence of the row and column variables. This model can now be regarded as a special case of the saturated model, obtained by setting the $u_{12(ij)}$ terms equal to zero. It is easily verified that the condition

$$u_{12(ij)} = 0; \qquad i = 1, \ldots, I; j = 1, \ldots, J$$

is mathematically equivalent to the independence condition for cell probabilities

$$p_{ij} = p_{i+}p_{+j}; \qquad i = 1, \ldots, I; j = 1, \ldots, J.$$

It can also be shown that the maximum likelihood estimates, $\hat{m}_{ij}$, for the expected cell frequencies under the restricted model

$$\log m_{ij} = u + u_{1(i)} + u_{2(j)} \tag{3}$$

are just the traditional ones given by

$$\hat{m}_{ij} = \frac{x_{i+}x_{+j}}{N}.$$

These estimates are completely specified by the model (Equation 3) and by the conditions that maximize the likelihood under the model, namely, $m_{i+} = x_{i+}$ and $m_{+j} = x_{+j}$, which require that the observed and expected marginal totals are the same.

In this framework, the hypothesis of independence is tested by testing the fit of the model (Equation 3) using either the Pearson statistic

$$X^2 = \sum_{i=1}^{I} \sum_{j=1}^{J} \frac{(x_{ij} - \hat{m}_{ij})^2}{\hat{m}_{ij}}$$

or the likelihood ratio statistic

$$G^2 = 2 \sum_{i=1}^{I} \sum_{j=1}^{J} x_{ij} \log \frac{x_{ij}}{\hat{m}_{ij}}$$

both of which have asymptotic chi-square distributions. In most examples where the asymptotic approximation is good, the two statistics will have similar values, and will therefore lead to the same conclusion. The likelihood ratio statistic has an advantage over the Pearson statistic for tables of three or more dimensions; it can be partitioned in a way that facilitates testing the conditional significance of individual terms in the model (more on this later).

The degrees of freedom, applicable to either statistic, for testing independence in a two-way table are well known to be $(I - 1)(J - 1)$. By referring to the identity

$$IJ = 1 + (I - 1) + (J - 1) + (I - 1)(J - 1)$$

it is easy to see that the number of degrees of freedom for testing the fit of the model (Equation 3) is also $(I - 1)(J - 1)$, which is obtained as the total number of cells IJ reduced by the number of "free" parameters estimated, $1 + (I - 1) + (J - 1)$, resulting from one u, $(I - 1)$ of the $u_{1(i)}$, and $(J - 1)$ of the $u_{2(j)}$. The same identity shows that if one estimates the parameters of the saturated model, there are no degrees of freedom left for testing the fit.

Higher Dimensional Tables

For tables of higher dimension, the principal steps of the method are essentially the same. For any particular model of interest the use of the maximum likelihood criterion results in a set of constraints that equate certain observed and expected marginal totals. These constraints yield estimates of the expected cell frequencies that either have explicit formulas in terms of the data or are obtained by an iterative numerical algorithm (see Darroch and Ratcliffe 1972). Approximate chi-square statistics are used to assess the fit of various models by comparing observed and expected frequencies. The whole process is easily automated for an electronic computer; moreover, many computing centres support one or more of the many package programs available. [The authors have found the routine C-TAB (Haberman 1973) to be satisfactory for general purposes.]

The main features that distinguish tables of three or more dimensions from those of two dimensions are the complexity of the linear models and the use of a hierarchy of models. We use the three-dimensional case to illustrate these ideas.

Models for Three-Way Tables

The saturated model for a three-dimensional table with I rows, J columns, and K faces is given by

$$\log m_{ijk} = u + u_{1(i)} + u_{2(j)} + u_{3(k)} + u_{12(ij)} + u_{13(ik)} + u_{23(jk)} + u_{123(ijk)}$$

where $i = 1, \ldots, I$; $j = 1, \ldots, J$; $k = 1, \ldots, K$. Here m_{ijk} is the expected frequency in the (i, j, k) cell. As for the two-way case, the u terms are deviation or "effect" terms, so that

$$u = \frac{1}{IJK} \sum\sum\sum \log m_{ijk}$$

and

$$u_{1(+)} = u_{2(+)} = u_{3(+)} = u_{12(+j)} = u_{12(i+)} = \cdots = u_{123(ij+)} = 0.$$

For example, $u_{12(ij)}$ represents the contribution to the log-expected frequency of the (i, j, k) cell by the interaction between variables 1 and 2.

The possible submodels of the saturated model are given in Table 2 along with the corresponding degrees of freedom for testing fit and convenient labels that relate to the interpretations of the models. Models that eliminate any of the main effect terms, u_1, u_2, or u_3, usually apply to a very restricted set of problems since they impose equal probabilities for categories of variables whose terms do not appear in the model. Our major concern is thus with models that contain all the main effect terms.

The highest-order u-terms containing each variable uniquely characterize a particular model in the hierarchy of Table 2. For example, the model

$$\log m_{ijk} = u + u_{1(i)} + u_{2(j)} + u_{3(k)} + u_{12(ij)}$$

Table 2. Hierarchical Models for an $I \times J \times K$ Table

Expression for $\log m_{ijk}$		Description	Degrees of Freedom
$u + u_{1(i)} + u_{2(j)} + u_{3(k)}$	(1)	complete independence	$(I-1)(J-1) + (I-1)(K-1) + (J-1)(K-1) + (I-1)(J-1)(K-1)$
$(1) + u_{12(ij)}$		independence of 3 from 1 and 2	$(I-1)(K-1) + (J-1)(K-1) + (I-1)(J-1)(K-1)$
$(1) + u_{13(ik)}$		independence of 2 from 1 and 3	$(I-1)(J-1) + (J-1)(K-1) + (I-1)(J-1)(K-1)$
$(1) + u_{23(jk)}$		independence of 1 from 2 and 3	$(I-1)(J-1) + (I-1)(K-1) + (I-1)(J-1)(K-1)$
$(1) + u_{12(ij)} + u_{13(ik)}$		conditional independence of 2 and 3 given 1	$(J-1)(K-1) + (I-1)(J-1)(K-1)$
$(1) + u_{12(ij)} + u_{23(jk)}$		conditional independence of 1 and 3 given 2	$(I-1)(K-1) + (I-1)(J-1)(K-1)$
$(1) + u_{13(ik)} + u_{23(jk)}$		conditional independence of 1 and 2 given 3	$(I-1)(J-1) + (I-1)(J-1)(K-1)$
$(1) + u_{12(ij)} + u_{13(ik)} + u_{23(jk)}$	(8)	no three-factor interaction	$(I-1)(J-1)(K-1)$
$(8) + u_{123(ijk)}$		saturated model	0

may be identified by the terms u_{12} and u_3. Thus, a convenient shorthand notation for this model is [12] [3]. This notation has further meaning in signifying that, under this model, the maximum likelihood estimates of the expected cell frequencies m_{ijk} are required to satisfy the following constraints on the marginal totals:

$$m_{ij+} = x_{ij+} \qquad \text{and} \qquad m_{++k} = x_{++k}.$$

The marginal tables $\{x_{ij+}\}$ and $\{x_{+k}\}$ are referred to as the *fitted marginal totals* for this model. (Indeed, these marginal tables are known to be sufficient statistics for the parameters of the model, and are sometimes called *sufficient configurations*.) As another example, the model of conditional independence of variables 2 and 3 given variable 1 can be represented by [12] [13].

For calculating degrees of freedom we use the identity

$$\begin{aligned} IJK = 1 + (I-1) + (J-1) + (K-1) &+ \\ (I-1)(J-1) + (J-1)(K-1) + (I-1)(K-1) &+ \\ (I-1)(J-1)(K-1). & \end{aligned}$$

For a $3 \times 5 \times 8$ table, the identity would be

$$120 = 1 + 2 + 4 + 7 + 8 + 14 + 28 + 56$$

so that the degrees of freedom for the complete independence model

$$\log m_{ijk} = u + u_{1(i)} + u_{2(j)} + u_{3(k)}$$

are computed either as $120 - (1 + 2 + 4 + 7) = 120 - 14 = 106$ or as $8 + 14 + 28 + 56 = 106$. The degrees of freedom calculated in this way need to be adjusted if any of the expected cell frequencies are estimated to be zero. This will usually occur because zero frequencies along an entire row or column of the observed table causes one or more cells of the fitted marginal totals to be zero (or because logical zeros have been used, as discussed shortly). The required adjustment is to decrease the nominal degrees of freedom, obtained as above, by an adjustment A of the form

$$A = T - M$$

with T equal to the number of zeros in the expected table for the given model, and M equal to the sum of the corrected number of zeros in each marginal total fitted under that model (the correction being due to the number of zeros, if any, in the marginal totals of the fitted marginal totals). In a loose sense, the corrected degrees of freedom may be regarded as equal to the difference between the effective number of cells with informative data and the effective number of parameters that can be estimated. In rare cases, zeros occur among the estimated cell frequencies for certain models because of pathologies in the pattern of the data. Bishop, Fienberg, and Holland (1975, p. 70) give an example of such a case where the above adjustment results in negative degrees of freedom, which is obviously incorrect.

Strategies for Choosing Models

There are many strategies for deciding on a particular model as the best one to represent the data. One reasonable approach is to start at the bottom of Table 2 and proceed through a nested sequence of models, each candidate for fit obtained from the previous one by deleting one or more terms. The process terminates when the chi-square statistic is significant for the first time at one of the usual significance levels. If the computer program being used provides estimated standard errors for the u-terms in the model, a rough guideline for deciding which terms to delete is the significance of the u-terms regarded as approximate normal variates.

An important rule in exploring a sequence of models is the following:

Hierarchy rule. *A term may not be deleted from a model before all higher order terms containing that combination of variables are deleted.*

For example, a model of the form

$$\log m_{ijk} = u + u_{1(i)} + u_{2(j)} + u_{3(k)} + u_{123(ijk)}$$

is *not* a proper model since the terms $u_{12(ij)}$, $u_{13(ik)}$, and $u_{23(jk)}$ have been dropped before the term $u_{123(ijk)}$. This rule, a consequence of the estimation method, explains why Table 2 does not contain all possible combinations of the eight different u-terms. Another useful maxim is that once a term has been dropped, it may not be reintroduced into a later model in the sequence. This enables straightforward interpretation of the significance probability measuring the fit of the model finally adopted, and lessens the tendency toward "overfitting" the data.

Further information about the importance of an individual term is the conditional significance of that term given the presence of all the other terms in the current model. The appropriate statistic is the difference of the likelihood ratio chi-square statistics for two models that differ only by the term in question; this difference statistic has an approximate chi-square distribution with degrees of freedom obtained by subtraction in the same way. In other words, if $G^2(1)$ and $G^2(2)$ are the likelihood ratio chi-square statistics for two models for a three-way table, model 1 which includes the term $u_{12(ij)}$ and model 2 which is obtained from model 1 by deleting $u_{12(ij)}$, and if $G^2(1)$ and $G^2(2)$ have k_1 and k_2 d.f., respectively, then the conditional significance of the term $u_{12(ij)}$ given all the terms in model 2 is measured by the approximate chi-square difference statistic $[G^2(2) - G^2(1)]$ based on $(k_2 - k_1)$ d.f. It is important to note that this procedure is appropriate only for likelihood ratio statistics because they have the property of being partitionable into components that relate to individual terms in the model. Pearson chi-square statistics do not have this partitioning property and thus should not be used to calculate difference statistics for individual terms.

Another strategy for choosing the best model is to start at the top of Table 2 and proceed through a sequence of models obtained by adding successive terms until the first model is encountered that fits reasonably well. The decision about which term to add next may be based on a study of the *residuals* for the current model; a residual is obtained for each cell by subtracting the estimate of the expected frequency under the model from the observed frequency. Standardized residuals, obtained by dividing residuals by the square roots of the estimated expected frequencies, may be regarded roughly as standard normal variates (though they are *not* independent), and those of unusually large magnitude may be singled out for attention. Certain sign patterns among the significant residuals are the result of interactions between variables that are not accounted for by the model, and hence an indication that a particular interaction term needs to be included. As with the previous strategy, the conditional significance of individual terms can be used to rank candidates for addition to the model. For example, starting with the complete independence model of Table 2, one might compute the difference statistic for each of the terms $u_{12(ij)}$, $u_{13(ik)}$, and $u_{23(jk)}$ to determine which is the most significant conditional on the terms u, $u_{1(i)}$, $u_{2(j)}$, and $u_{3(k)}$. The term with the most significant difference statistic would then be added to the complete independence model, and the fit of the resulting model would be tested.

Special Techniques

It is often necessary to deal with *n*-way classifications in which certain category combinations are logically impossible or, at the very least, unobservable. As Slater (1973) has discussed for the analysis of act sequences using dyad tables in which rows and columns give preceding and succeeding act, respectively, the occurrence of one type of act may logically rule out the occurrence of another type of act. For example, if food and drinking cups are separated by some distance, feeding cannot follow drinking without locomotion intervening; hence, the (feeding, drinking) cell of the table will never contain observations. It may also be impossible to observe a sequence of two acts when the succeeding act is the same as the preceding act.

These two types of impossibilities require that some cells of the table will be constrained to have zero expected frequency; such cells are said to contain *logical zeros*. (Note that a cell with an observed frequency of zero, a sampling zero, need not have a logical zero.) Some of the available computer routines can be made to take account of logical zeros in the calculation of maximum likelihood estimates of parameters. Because logical zeros necessarily appear as zeros in the table of estimated expected frequencies, the adjustment rule must be applied in obtaining the correct degrees of freedom. Pathological cases arise here as well; indeed the example mentioned above applies equally well when the estimated zeros are logical zeros. There are certain kinds of tables for which the handling of logical zeros is not as straightforward as described here. One such case is the *separable table*, which is literally split into two or more disconnected pieces by logical zeros. Chapter 5 of Bishop, Fienberg, and Holland (1975) gives a detailed account of the problem of logical zeros.

For some applications of multidimensional contingency table analysis, it is desirable to regard one of the variables as a response and the others as factors, even though the analysis as presented here treats the variables symmetrically. By choosing an appropriate model and interpreting it in special ways, one can focus on the contributions of the various factors to the response regarded as a dependent variable. Suppose that in a three-way table variable 3 is to be regarded as the response, and variables 1 and 2 as the factors, and suppose that the best-fitting model is determined to be

$$\log m_{ijk} = u + u_{1(i)} + u_{2(j)} + u_{3(k)} + u_{12(ij)} + u_{13(ik)}. \tag{4}$$

We conclude from this result that variable 3 depends upon variable 1 ($u_{13} \neq 0$) but not on variable 2 ($u_{23} = 0$). If variable 3 is a dichotomous variable, that is, $K = 2$, we can derive a *logit model* corresponding to the best-fitting log-linear model which gives an explicit quantitative relation between variable 3 and the factor variables, 1 and 2. To obtain the logit model for Expression 4, we simply subtract the two versions of Expression 4, one for $k = 1$ and the other for $k = 2$. It is clear that terms not involving k will cancel and we are left with

$$\log\left(\frac{m_{ij1}}{m_{ij2}}\right) = u_{3(1)} - u_{3(2)} + u_{13(i1)} - u_{13(i2)}.$$

If we let $p_{ij} = p_{ij1}$, $v = u_{3(1)} - u_{3(2)}$, and $v_{1(i)} = u_{13(i1)} - u_{13(i2)}$, and recall that $m_{ijk} = Np_{ijk}$, we can rewrite the result as

$$\log \frac{p_{ij}}{1 - p_{ij}} = v + v_{1(i)}.$$

This gives, for the (i, j) combination of variables 1 and 2, the logit or log-odds for category 1 over category 2 of variable 3 as a linear function of terms depending only on variables 1 and 2; in fact, in this case the logit will change only over categories of variable 1. When *estimated* u-terms are used to compute this function, one obtains a numerical log-odds value for each (i, j) combination which can be displayed either in tabular or graphical form as a convenient device for focusing on the estimated factor-response relationship.

In an application with variables designated as response and factors, such as described above, it will often be the case that the marginal frequencies for the factor variables are fixed by the sampling design. For example, suppose that the factor variables, 1 and 2, are sex and species, and variable 3 is strength of response to some stimulus. Usually, a specified number of animals will be chosen from each sex-species category (i, j), yielding IJ different samples of sizes N_{ij}. These samples are each classified according to strength of response to yield the $I \times J \times K$ table. When such a table is analyzed using the methods described here, it is important that all log-linear models to be tested for fit contain the terms corresponding to the marginal totals fixed by the sampling design. In our example, a model that does not include the u_{12} term is invalid since omission of that term is equivalent to requiring the marginal frequencies, x_{ij+}, for the factor variables to be estimated rather than remaining fixed at their sampling values, N_{ij} .

Another application involving margins that are fixed by sampling is the comparison of several populations. One may wish, for example, to compare three different hybrids of sunfish (variable 1) on the basis of their response to different interlopers (variable 2) at different stages of the reproductive cycle (variable 3). In this case, the marginal totals x_{i++} are determined by the number of individuals N_i of each hybrid used in the experiment. Any of the models in Table 2 is appropriate, since each contains the u-term corresponding to the fixed margins, $u_{1(i)}$. The interpretation of the best-fitting model, however, can be made to bear upon the problem of comparing the populations. If, for example, the best-fitting model for the data to compare hybrids of sunfish turns out to be

$$\log m_{ijk} = u + u_{1(i)} + u_{2(j)} + u_{3(k)} + u_{23(jk)}$$

it can be concluded that the interaction between interloper and reproductive state is independent of hybrid, or, in other words, that the hybrids are not significantly different with regard to the way type of interloper interacts with reproductive state.

The analysis of a multidimensional contingency table should not stop with selection of a best-fitting model. A useful technique for the next stage is an analysis of residuals. In terms of the expected frequencies predicted by the best-fitting model, one computes, for each cell, the standardized residual $(O - E)/\sqrt{E}$, where O and E represent observed and expected frequencies respectively. As indicated earlier, these standardized residuals may be regarded as approximate standard normal variates and those of unusually large magnitude may be singled out for further study. For example, a single cell in the table may be found to be largely responsible for a particular interaction term. A more refined, though still informal, method of detecting a large residual is to compare the magnitude of the standardized residuals with the square root of the upper 5 percent point of the appropriate chi-square distribution (degrees of freedom for the fitted model) divided by the number of cells.

3. EXAMPLES

The first example involves an analysis of the songspread display of the male carib grackle *(Quiscalus lugubris)* (Wiley, 1975). For each of 115 independent occurrences of this display, Wiley rated the Beak Elevation (two levels), Wing Elevation (two levels), and Social Context (three levels: whether a male, female, or no bird was within one meter). The data were thus cast in a $2 \times 2 \times 3$ contingency table (Table 1). The question of interest was whether these data showed an interaction between Beak Elevation and Wing Elevation. That is, letting x_{ijk} be the number of occurrences of songspread display with Wing Elevation i, Beak Elevation j, and Social Context k, does an adequate model include a term for the interaction of factors 1 and 2? Table 3 shows the results of fitting the full set of models. [The procedure used by Wiley, originally proposed by Kullback (1959), is shown to be invalid in Bishop et al. (1975, pp. 362–363); Wiley's conclusions, however, are correct.] It is clear from the significance probabilities in Table 3 that an appropriate model for the data is the following:

$$\log m_{ijk} = u + u_{1(i)} + u_{2(j)} + u_{3(k)} + u_{13(ik)} + u_{23(jk)}.$$

Note that there is no $u_{12(ij)}$ term, and that the difference χ^2 for this term is only $3.10 - 1.02 = 2.08$ based on $3 - 2 = 1$ d.f. ($P > .100$).

Before interpreting this result, let us consider how we might have

Table 3. Fitted Models for Wiley (1975) Data

Model	LRχ^2	d.f.	P
[1] [2] [3]	25.84	7	0.0005
[12] [3]	25.44	6	0.0003
[13] [2]	15.91	5	0.007
[23] [1]	13.03	5	0.023
[12] [13]	15.51	4	0.004
[12] [23]	12.64	4	0.013
[13] [23]	3.10	3	0.376
[12] [13] [23]	1.02	2	0.601

arrived at the same model without generating the whole of Table 3. We proceed through the hierarchy of models by examining the significance of estimated effects at each stage, starting with all parameters included and dropping those that appear insignificant. The model of no three-factor interaction,

$$\log m_{ijk} = u + u_{1(i)} + u_{2(j)} + u_{3(k)} + u_{12(ij)} + u_{12(ik)} + u_{23(jk)}$$

has a likelihood ratio χ^2 of 1.02 with 2 d.f.; so this model certainly fits the data. Which terms are superfluous? Using a program that supplies standard errors of estimated effect terms, we examine the standardized values of the three two-factor effects, computed in terms of this model. They are:

$$\frac{\hat{u}_{12(11)}}{\text{S.E.}} = 1.382$$

$$\frac{\hat{u}_{13(11)}}{\text{S.E.}} = 1.409, \quad \frac{\hat{u}_{13(12)}}{\text{S.E.}} = 1.637, \quad \frac{\hat{u}_{13(13)}}{\text{S.E.}} = -3.269$$

$$\frac{\hat{u}_{23(11)}}{\text{S.E.}} = -2.659, \quad \frac{\hat{u}_{23(12)}}{\text{S.E.}} = 0.351, \quad \frac{\hat{u}_{23(13)}}{\text{S.E.}} = 2.303.$$

(Note that the value for only one level of a two-level factor need be given, since estimated effects must add to zero across levels, and standard errors will be the same for both levels in this case.) Regarding the standardized estimates as approximate standard normal variables, we can single out $\hat{u}_{13(13)}$, $\hat{u}_{23(11)}$, and $\hat{u}_{23(13)}$ as significant. From this we conclude that the model that includes the terms u_{13} and u_{23} but omits u_{12} will probably provide a good fit. Can another term be dropped? To check this, we fit the model

$$\log m_{ijk} = u + u_{1(i)} + u_{2(j)} + u_{3(k)} + u_{13(ik)} + u_{23(jk)}$$

($\chi^2 = 3.10$, d.f. = 3, $P = .376$), and examine the significance of the $u_{13(ik)}$ and $u_{23(jk)}$ terms. This time, the standardized values for $\hat{u}_{13(ik)}$ are 1.018, 1.708, and −2.929, while those for $\hat{u}_{23(jk)}$ are −2.632, 0.581, and 1.869. The two values with magnitude greater than 2.6 confirm that u_{13} and u_{23} are important terms and cannot be dropped. The final model is thus the same one chosen by examining Table 3.

The final model is that of conditional independence of variables 1 and 2 given the level of variable 3. In other words, Beak Elevation and Wing Elevation fluctuate independently for a given level of Social Context. This is equivalent to saying that they interact only in that they are both dependent on Social Context. Wiley considers the implications of this conditional independence in terms of information transfer in the communicative systems.

The final interpretation of a chosen model should always be accompanied by examination of the expected frequencies and the residuals. Panels 2 and 3 of Table 1 show expected frequencies and standardized residuals for the model [13] [23]. The magnitudes of the expected frequencies are only marginally acceptable according to the guidelines given by Cochran (1954) (see section 4) since four of them, or one-third, are smaller than 4. This indicates that the sample size may not be large enough for acceptable accuracy of the chi-square approximation to the distribution of the goodness-of-fit statistic. Thus, significance probabilities in Table 3 should not be interpreted too literally, but should be used rather to assess the *relative* goodness-of-fit of the various models. In spite of this caveat, the model [13] [23] remains the best choice for these data. The standardized residuals do not provide strong information since they are all rather small, although the pattern of their signs indicates that the u_{12} term has been dropped from the model.

A second example illustrates the use of a contingency table analysis in producing useful graphical displays, and indicates how individual cells in the table may be tested for fit by means of logical zeros. The research involves aggressive responses in reproductively active male pumpkinseed sunfish *(Lepomis gibbosus)* (Colgan and Gross 1977). The data, a 4 × 3 × 4 contingency table shown in Table 4, were derived by classifying a total of 8508 independent acts by 49 fish according to internal state, variable 1 (as reflected by the reproductive Period: Nesting, Spawning, Brooding, and Vacating), external stimulus, variable 2 (a stimulus Dummy pumpkinseed sunfish in a Subordinate, Normal, or Aggressive posture), and Response type, variable 3 (Retreat, Approach, Opercular Spread, and Bite). Preliminary analysis suggested that individual differences among fish were small and that the internal and external factors were concurrent (Houston and McFarland 1976) and therefore possibly

Table 4 A 4 × 3 × 4 Contingency Table Based on the Aggressive Behavior of Pumpkinseed Sunfish. Periods: Ne: Nesting; Sp: Spawning; Br: Brooding; Va: Vacating. Dummies: S: Subordinate; N: Normal; A: Aggressive.

No. Days (d_i)	Period	Dummy	Response				No. Fish (s_j)
			Retreat	Approach	Opercular Spread	Bite	
		S	4	4	7	1	16
2	Ne	N	17	53	77	2	49
		A	6	28	33	1	15
		S	92	178	270	41	16
3	Sp	N	476	749	1041	419	49
		A	185	269	333	203	15
		S	83	153	264	58	16
4	Br	N	442	684	1034	322	49
		A	156	233	301	148	15
		S	4	9	12	1	16
1	Va	N	12	31	24	8	49
		A	6	15	13	6	15

additive. Variable 3 was regarded as a response to the two factors, variables 1 and 2.

Table 5 shows the results of fitting the full set of hierarchical log-linear models. The only model giving a clearly good fit is the model of no three-factor interaction [12] [13] [23]. Of the rather ill-fitting models with only two two-factor terms, the best model is the one omitting a Period × Dummy term, u_{12}

$$\log m_{ijk} = u + u_{1(i)} + u_{2(j)} + u_{3(k)} + u_{13(ik)} + u_{23(jk)}.$$

The goodness-of-fit chi-square for this model is 44.89 with 24 d.f. (P = 0.006), and the difference chi-square measuring the importance of the u_{12} term has a value of 44.89 − 17.08 = 27.81 (d.f. = 24 − 18 = 6, $P <$ 0.001). It should be noted, however, that the goodness-of-fit chi-square for this model is substantially less significant than those for the other models that omit a single two-factor term.

We shall study the above model in greater depth, in view of its theoretical appeal, to see how it predicts the additivity of variables 1 and 2, and how the observed data depart from these predictions. We shall see also how the model can be used to scale the Period effect on a

Table 5. Fitted Models for the Table 4 Data

Model	LRχ^2	d.f.	P
Panel 1 All Cells			
[1] [2] [3]	203.78	39	8×10^{-10}
[12] [3]	179.63	33	2×10^{-10}
[13] [2]	127.99	30	3×10^{-10}
[23] [1]	120.67	33	6×10^{-10}
[12] [13]	103.85	24	$< 10^{-10}$
[12] [23]	96.53	27	1×10^{-9}
[13] [23]	44.89	24	0.006
[12] [13] [23]	17.08	18	0.517
Panel 2 Excluding the (1, 1, 2) Cell			
[12] [13]	94.11	23	2×10^{-9}
[12] [23]	86.29	26	2×10^{-8}
[13] [23]	19.70	23	0.660
[12] [13] [23]	6.93	17	0.984

motivational axis. The following graphical technique is a standard device in factorial analysis of variance.

With the factor Response held fixed at, say, level 2 (Approach), the above model gives

$$\log m_{ij2} = u + u_{1(i)} + u_{2(j)} + u_{3(2)} + u_{13(i2)} + u_{23(j2)}.$$

The terms can be regrouped and relabeled, with

$$w = u + u_{3(2)}$$

$$w_{1(i)} = u_{1(i)} + u_{13(i2)}$$

$$w_{2(j)} = u_{2(j)} + u_{23(j2)}$$

giving the alternative expression

$$\log m_{ij2} = w + w_{1(i)} + w_{2(j)}.$$

Since the number of fish tested with each dummy varied, as did the number of days each period lasts, it is useful to work with a "per fish per day" expected frequency of acts

$$\dot{m}_{ij2} = \frac{m_{ij2}}{d_i s_j}$$

where d_i is the number of days in Period i, and s_j is the number of fish presented with Dummy j. Thus

$$\log \hat{\hat{m}}_{ij2} = \log m_{ij2} - \log d_i - \log s_j$$
$$= w + (w_{1(i)} - \log d_i) + (w_{2(j)} - \log s_j).$$

If we now use the estimated effect terms and expected cell frequencies to plot the points $(\hat{w}_{1(i)} - \log d_i, \log \hat{\hat{m}}_{ij2})$ for each of the 12 (i, j) combinations, joining points at the same level of variable 2, we obtain the solid lines in Figure 1. Table 6 gives the estimated cell frequencies for this model; values of d_i and s_j are given in Table 4. The point $(\hat{w}_{1(1)} - \log d_1, \log \hat{\hat{m}}_{112})$, for example, is computed as

$$\hat{w}_{1(1)} = \hat{u}_{1(1)} + \hat{u}_{13(12)} = -1.675 + 0.489 = -1.186$$

and

$$\log \hat{\hat{m}}_{112} = \log \hat{m}_{112} - \log d_1 - \log s_1$$
$$= \log 12.15 - \log 2 - \log 16$$
$$= -0.968.$$

The graph in Figure 1 is analogous to the sort used to study interaction

Table 6. Expected Frequencies for the Model [13] [23] Fitted to the Table 4 Data

Period	Dummy	Response			
		Retreat	Approach	Opercular Spread	Bite
Ne	S	3.33	12.15	18.98	0.33
	N	17.24	53.59	74.68	2.48
	A	6.43	19.25	23.34	1.18
Sp	S	92.92	171.00	266.69	55.34
	N	480.84	754.09	1049.40	411.50
	A	179.24	270.91	327.93	196.16
Br	S	84.03	152.98	259.39	44.07
	N	434.87	674.64	1020.70	327.71
	A	162.10	242.37	318.96	156.22
Va	S	2.71	7.86	7.95	1.25
	N	14.05	34.68	31.28	9.31
	A	5.24	12.46	9.77	4.44

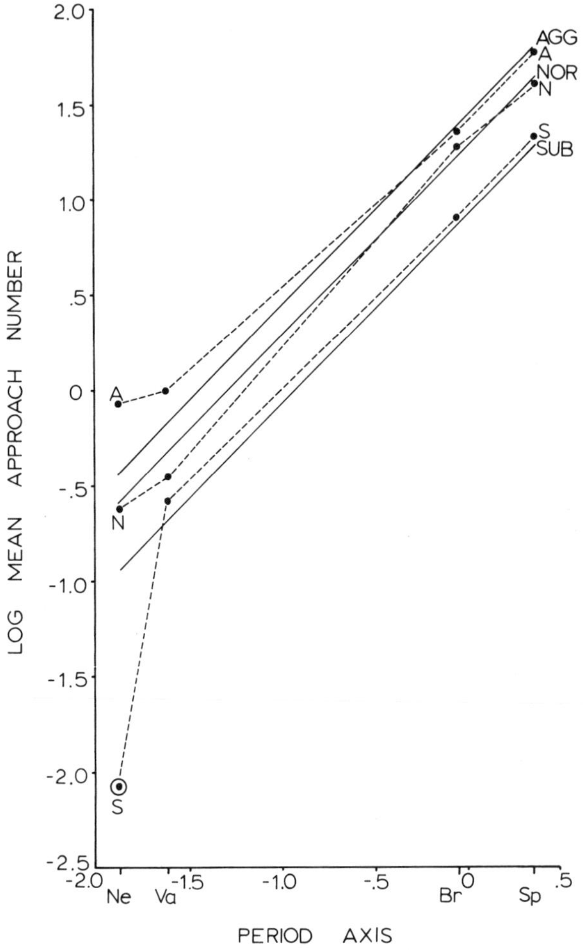

Figure 1. Observed and expected log mean numbers of approaches. Periods: Ne: Nesting; Sp: Spawning; Br: Brooding; Va: Vacating. Dummies: A: Aggressive; N: Normal; S: Subordinate.

in the analysis of variance. The relative magnitudes of the per-day Period effects are indicated by the spacing of points long the abscissa, which can therefore be regarded as a motivational axis; the vertical spacing of the lines measures the Dummy effects. The additivity of Period and Dummy effects, as required by the fitted model, is reflected in the parallelism of the lines.

An indication of the additivity of the data is given by a plot of the points ($\hat{w}_{1(i)} - \log d_i$, $\log \dot{x}_{ij2}$) on the same graph (Figure 1), where $\dot{x}_{ij2}$ =

$x_{ij2}/d_i s_j$; these points are labeled by the symbols S, N, and A for the three levels of variable 2. The deviations of these points from the lines predicted by the model reflect the poor fit of the model and the apparent nonadditivity of Period and Dummy effects. There is clearly a large departure from additivity for the cell (1, 1, 2) as indicated by the point that is circled: the Approach response elicited by the Subordinate dummy is much less, relative to that of the other two dummies, during the Nesting period than during the three later periods. This conclusion is substantiated by a study of the standardized residuals, $(x_{ijk} - \hat{m}_{ijk})/\sqrt{\hat{m}_{ijk}}$, shown in Table 7: among all those cells at level 2 of variable 3, the (1, 1, 2) cell with a standardized residual of −2.34 can be readily identified as indicating a poor fit.

At this stage, it is useful to ask to what extent the (1, 1, 2) cell is responsible for the poor fit of model [13] [23]. To answer this, we assign a logical zero to the cell and recompute the chi-square statistic for the model. The second panel of Table 5 shows the results of fitting the last four models with the expected frequency for cell (1, 1, 2) constrained to be zero. We see that again the models [12] [13] and [12] [23] fit very badly, while now the model [13] [23] fits very well. The difference χ^2 with a value of (44.89 − 19.70) = 23.19, based on 1 d.f., is highly

Table 7. Standardized Residuals for the Model [13] [23] Fitted to the Table 4 Data

Period	Dummy	Response			
		Retreat	Approach	Opercular Spread	Bite
	S	0.37	−2.34	−2.75	1.15
Ne	N	−0.06	−0.08	0.27	−0.31
	A	−0.17	1.99	2.00	−0.17
	S	−0.10	0.54	0.20	−1.93
Sp	N	−0.22	−0.19	−0.26	0.37
	A	0.43	−0.12	0.28	0.49
	S	−0.11	0	0.29	2.10
Br	N	0.34	0.36	0.42	−0.32
	A	−0.48	−0.60	−1.01	−0.66
	S	0.78	0.41	1.44	−0.23
Va	N	−0.55	−0.62	−1.30	−0.43
	A	0.33	0.72	1.03	0.74

significant, which indicates that, under the hypothesis that the model [13] [23] fits the data, the cell (1, 1, 2) is decidedly an outlier. Thus, Period and Dummy effects may be regarded as additive, given the level of Response, provided one is prepared to ignore the Approach response to the Subordinate dummy during the Nesting period.

A third example concerns the use of contingency table methods for the analysis of sequence data, illustrating several of the difficulties commonly encountered with this application. This example also serves to demonstrate the study of individual differences and the general handling of logical zeros. The research (Ballantyne and Colgan, in press) examined sequences of responses by male sunfish in a social test situation. The data consist of four 10 × 10 tables, one for each of four individual fish, giving frequencies of all possible consecutive pairs of 10 different response types; the resulting 4 × 10 × 10 table which classifies a total of 939 distinct preceding-succeeding act dyads into 400 cells is shown in Table 8.

We assume that the response type exhibited at a particular time depends only on the previous response type and not on earlier response types; this implies that the successive acts of each fish form a first-order Markov chain. Although sequential dependencies exist among responses that are not adjacent, the major interaction is with the immediately previous response type; unfortunately, more distant relations can be established only with very large data sets. We assume further that the transition probabilities are stationary. Under these conditions the meth-

Table 8. Frequencies of Response Dyads for Four Male Pumpkinseed Sunfish

Fish	Preceding Act	Following Act									
		G	Ne	OS	C	La	Nu	BM	Cir	TM	PSP
1	G	0	64	0	0	1	0	0	0	23	0
	Ne	13	0	36	3	4	0	10	1	1	0
	OS	10	0	0	3	4	0	7	0	0	0
	C	6	1	2	0	0	0	1	0	0	0
	La	8	0	1	1	0	0	0	1	0	0
	Nu	0	0	0	0	0	0	0	0	0	0
	BM	14	0	0	1	0	0	0	0	0	4
	Cir	1	0	0	0	1	0	0	0	0	0
	TM	23	2	0	0	0	0	0	0	0	0
	PSP	4	0	0	0	0	0	0	0	0	0
2	G	0	55	0	1	0	0	0	0	22	0
	Ne	10	0	33	9	9	1	1	0	0	0

Table 8. *(Continued)*

Fish	Preceding Act	Following Act									
		G	Ne	OS	C	La	Nu	BM	Cir	TM	PSP
	OS	16	5	0	8	11	0	3	0	1	0
	C	12	2	3	0	1	0	3	0	0	0
	La	9	0	6	4	0	0	1	1	0	0
	Nu	1	0	0	0	0	0	0	0	0	0
	BM	4	3	1	0	0	0	0	0	0	0
	Cir	0	0	2	0	0	0	0	0	0	0
	TM	16	0	0	0	0	0	0	0	0	0
	PSP	0	0	0	0	0	0	0	0	0	0
3	G	0	59	1	2	0	0	1	0	5	0
	Ne	23	0	25	7	4	0	9	0	0	0
	OS	14	8	0	5	4	0	1	3	0	0
	C	7	3	2	0	0	0	0	0	0	0
	La	3	1	2	0	0	1	0	3	0	0
	Nu	0	0	0	0	0	0	1	0	0	0
	BM	9	0	1	0	0	0	0	0	0	2
	Cir	2	1	0	0	2	0	0	0	1	0
	TM	5	0	1	0	0	0	0	0	0	0
	PSP	2	0	0	0	0	0	0	0	0	0
4	G	0	55	0	1	0	0	0	0	17	0
	Ne	20	0	28	10	0	0	2	0	0	0
	OS	11	2	0	4	2	0	9	0	1	0
	C	10	0	3	0	1	0	4	0	0	0
	La	2	0	1	1	0	0	0	0	0	0
	Nu	0	0	0	0	0	0	0	0	0	0
	BM	5	3	1	2	1	0	0	0	0	0
	Cir	0	0	0	0	0	0	0	0	0	0
	TM	16	2	0	0	0	0	0	0	0	0
	PSP	0	0	0	0	0	0	0	0	0	0

Key to symbols: G = Gap
Ne = Nears
OS = Opercular spread
C = Chase
La = Lateral display
Nu = Nudge
BM = Biting movements
Cir = Circle
TM = Territorial maintenance
PSP = Pharyngeal sound production

ods of Section 2 may be applied to the dyad frequencies as though they were governed by four independent multinomial distributions, one for each fish (see, for example, Billingsley 1961).

Since it was not possible, for all response types, to observe the occurrence of a given response type following itself, the expected frequencies on the diagonal of each 10×10 table are taken to be logically zero; that is, $m_{ikk} = 0$ for $i = 1, 2, 3, 4$; $k = 1, 2, \ldots, 10$.

In the analysis of these data, variable 1 represents Fish, variable 2 is Preceding Act, and variable 3 is Succeeding Act; hence x_{ijk} denotes the observed number of act dyads for fish i in which response type j was followed by response type k, with $I = 4$, and $J = K = 10$. A question of major interest is whether the response patterns differ among the four fish. Homogeneity across fish would be suggested by the good fit of either of the models [12] [23] or [13] [23], or any of their submodels. The results of fitting several models are shown in Table 9.

The presence of zeros among the expected frequencies for several of the models requires special treatment of the degrees of freedom. We note that the numbers of zeros in each of the observed marginal totals are as follows: 5 in x_{ij+}, 5 in x_{i+k}, 50 in x_{+jk}, and 0 in each of x_{++k}, x_{+j+}, and x_{i++}. (The logical zeros account for 40 out of 50 zeros in x_{+jk}.) Thus the calculations for the models of Table 9 are as follows:

$$
\begin{array}{lll}
[12]\,[13]\,[23] & : 243 - 220 + (50 + 5 + 5) & = 83 \\
[12]\,[13] & : 324 - 89 + (5 + 5) & = 245 \\
[12]\,[23] & : 270 - 211 + (50 + 5) & = 114 \\
[13]\,[23] & : 270 - 209 + (50 + 5) & = 116 \\
[1]\,[23] & : 297 - 200 + 50 & = 147 \\
[1]\,[2]\,[3] & : 378 - 40 & = 338
\end{array}
$$

Ostensibly, the results in Table 9 indicate that none of the models below the saturated model fits the data. Because the apparent require-

Table 9. Fitted Models for the Data in Table 8

Model	LRχ^2	d.f.	P	No. Iterations	Max. Deviation
[1] [2] [3]	1076.00	338	0	7	0.004
[1] [23]	253.06	147	1×10^{-7}	1	2×10^{-9}
[13] [23]	192.64	116	1×10^{-5}	1	1×10^{-10}
[12] [23]	192.00	114	7×10^{-6}	1	9×10^{-10}
[12] [13]	946.46	245	0	7	0.005
[12] [13] [23]	125.46	83	0.002	26	0.076

ment of the three-factor interaction term u_{123} to explain the variation in the data conflicts with the biological intuition that the between-fish variation is not highly significant, we are led to look for the causes of this result. In the first place, the tables of expected frequencies for all the fitted models contain a high proportion of small values: for example, for the [23] model, only 52 of the 360 observable cells contain expected frequencies of 5 or greater. This indicates that the total sample size ($N =$ 939) is not great enough to provide an accurate approximation by the chi-square distribution, and in effect the significance probabilities in Table 9 are not sufficiently reliable to be used for inference. Second, the sample size problem is compounded by the pattern of observations in the table. Upon closer inspection of Table 8, we note that the four faces are each highly "skewed," that is, there are some large frequencies, and many that are either very small or zero. This pattern frequently results in the poor fit of all but the saturated model, and often causes difficulties with the convergence of the iterative estimation procedure. The latter effect shows up in the need for 26 iterations to reduce the maximum change in expected frequencies between iterations to a value of 0.076 (the convergence criterion was a maximum deviation of 0.10).

The problem of skewed tables is quite likely to arise in the study of sequence data by these methods, because Markov chains in which the states are linearly "chained" will necessarily exhibit many unidirectional transitions between pairs of "adjacent" states and a large number of transition pairs which are very rarely observed. That is, x_{ijk} is often much greater or smaller than x_{jik}, and many of the x_{ijk} are zero or close to zero. The Markov chain in the present example is clearly a case in point. One remedy that is recommended for tables exhibiting such "unevenness" is the separate study of portions of the table that are more homogeneous. This will never completely solve the problem for tables such as this one, since deletion of infrequent *response types* will not necessarily eliminate infrequent *transition pairs*.

Another possible remedy is the lumping together of similar response types. Several such recombinations were attempted, but all attempts failed to produce a table for which the no-three-factor model would both converge and fit. (In practice, one should investigate the validity of maintaining the Markov model for a lumped chained; see, for example, Thomas and Barr 1977).

Given that there are difficulties with the formal application of the method to these data, which are typical of those obtained in sequence studies, are there any conclusions that may be drawn with some reasonable degree of support? In this study the investigators were interested in studying the transition pattern in more detail, and so wished to combine the data for the four fish, regarding them in effect as four replications of a single fish. Although the chi-square statistics for the models [12] [23] and [13] [23] are highly significant, this cannot be taken as proving that the fish are different because of the above-mentioned difficulties with the data. Acting on external evidence, the investigators chose to regard the individual fish as not significantly different, and summed the table over fish to obtain a 10 × 10 table of one-step transition frequencies. Standardized residuals were obtained by fitting the model [1] [2] to this table and the magnitudes of the positive standardized residuals were used to identify the most important transitions (regarding the standardized residuals as approximate standard normal variates and using a cut-off value of +1.96). The results are summarized in the kinematic graph (see Chapter 10) of Figure 2 where

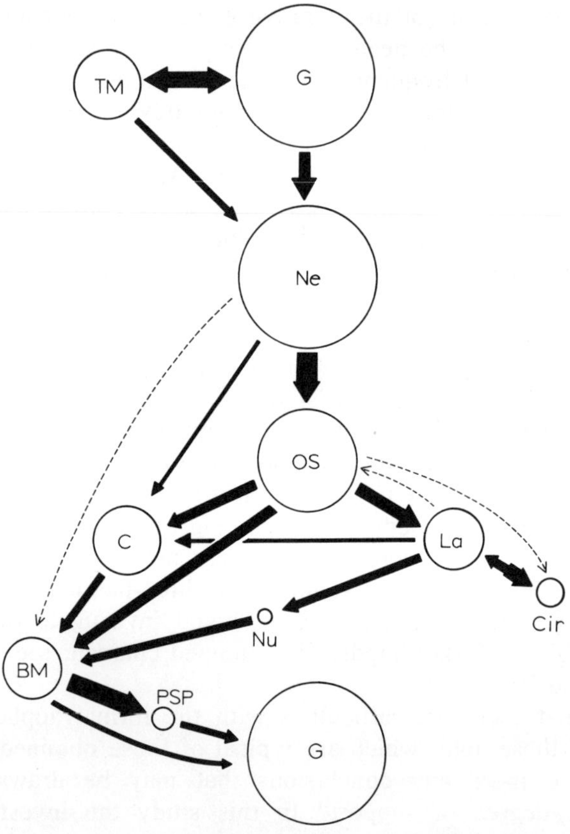

Figure 2. Dyadic sequences in pumpkinseed sunfish. See Table 8 for key to symbols.

the total frequency of each act is indicated by the size of the circle and the magnitude of large positive standardized residuals by the thickness of the arrow. (The dotted arrows represent smaller positive standardized residuals. Note that the response type "Gap" is shown twice in the figure.) The analysis has thus resulted in a highly useful graphical summary of the data despite difficulties that prevented traditional sorts of inferences.

4. INTERPRETATION AND LIMITATIONS

The examples show the manner in which multidimensional contingency table analysis can be used to detect structure in cross-classified nominal data. Difficulties that may arise fall into two main categories: failure of the assumptions, and problems with sample size. Problems with assumptions usually concern the independence of the individuals (or events) being classified, since the three kinds of sampling models cover most examples of cross-classified data. Because there is no simple way of testing the independence assumption, the results of the analysis must be viewed with scepticism whenever there is a suspicion of lack of independence.

The problem of sample size is one that occurs frequently in this kind of analysis. It has two facets that are ultimately related. The first is that the statistics for testing the fit of models are only *asymptotically* chi-square distributed, and hence, a minimum sample size (total frequency, N) is required for the validity of the significance probabilities.

One indication of inadequate sample size is an excess of small expected cell frequencies. The rules given by Cochran (1954) apply to two-dimensional tables but are useful for higher dimensional tables as rough indicators of possible difficulties. These rules may be paraphrased as follows: If fewer than 20 percent of the cells have expected frequencies below 5, expected frequencies as small as 1 are permissible. If most of the expected frequencies are 5 or below, any expectations less than 2 may be danger signals. In practice, one should not be too conservative about sample size whenever the analysis is of an exploratory nature to detect structure and is not intended for drawing hard and fast conclusions.

At the other end of the sample size scale, an analysis can be affected by too many observations as well; the result here is that most models will fail to fit even though they give a reasonable description of the data. The reason for this is that, if a large enough sample is taken, any hypothesis will be rejected, no matter how minute are the discrepancies from it. A good rule of thumb is to take as a baseline the significance probability of the model with the highest-order interaction term deleted. Then, in moving from an ill- to well-fitting model, the jump in significance probability should be a substantial fraction of that baseline value.

The other facet of the sample size problem relates to the large number of observations required to analyze a table even of moderate size. At worst, the number of parameters to be estimated is equal to the number of cells in the table; this number can be large for a table of three or more dimensions with four or more categories per variable. For example, three variables with four levels each could require the estimation of 64 parameters. By contrast, a regression analysis with three independent variables is not likely to involve more than 7 parameters (using two-variable interactions at most). Hence, the contingency table analysis is expensive in observations, and for this reason may be unsuitable for certain kinds of problems. As a general rule, a table with fewer than two observations per cell on average is likely to present difficulties with the analysis. Such difficulties are exacerbated when the observed frequencies are skewed or uneven, such as those encountered in the study of highly chained sequence data.

If more than three variables are involved, the interpretation of fitted models may be difficult. In general, one can give clear interpretations to models with interaction terms involving no more than two variables. For this reason, if two models fit the data more or less equally well, the model with the fewer higher order interactions is to be preferred.

Finally, it should be noted that if the levels of one or more variables can be ordered in a natural way, then the ordering information will be ignored in a contingency table analysis. Alternative techniques should be used for such data such as orthogonal polynomials (Goodman, 1971). When some of the variables are continuous (measured on a ratio or an interval scale), the temptation is to group the continuous variables, losing valuable information, or to use regression or analysis of variance methods even though the response variable may be categorical. These alternatives are usually inappropriate and should be rejected in favor of techniques that mix continuous and discrete variables, such as the method discussed by Nerlove and Press (1973).

In spite of occasional difficulties, the methods discussed in this chapter should be used whenever possible for the analysis of multidimensional contingency table data. They are strongly recommended as useful and defensible alternatives to the inadequate techniques they replace, namely, the reduction to all possible two-way tables, or the forcing of categorical data through the techniques derived from normal theory such as regression analysis and analysis of variance.

CHAPTER
7

Multidimensional Scaling

IAN SPENCE
University of Western Ontario

The originally projected title of this chapter was "Nonmetric Multidimensional Scaling," and had it so remained it would have included only a discussion of that branch of scaling which has been referred to as "nonmetric." This seemed to me to be an unfortunate restriction, since, while there would be no difficulty in devoting the whole chapter to procedures that could honestly be described as nonmetric, it would be a pity to exclude a variety of exciting and useful techniques that were directly inspired by the nonmetric upheaval of the early 1960s. As we shall see, that some of these data analytic procedures are not nonmetric is truly a rather unimportant consideration.

To describe every multidimensional scaling technique that has been evolved would be an immense, not to say tedious, task, even if one were to restrict the scope to nonmetric procedures. Consequently I do not attempt a feat of such Augean proportions, but instead concentrate on what I feel are the basic principles underlying most multidimensional scaling procedures. Naturally I describe some approaches more thoroughly than others, but whenever possible I point the reader to sources where he can satisfy his appetite if I have succeeded in merely tickling his palate. Some more experienced readers may find my choice of emphasis not to their liking. Nevertheless, I hope that the menu will be sufficiently varied that few will not find something to appeal to their interests, and further that even those familiar with this branch of data analysis will find something new in the chapter.

There are few areas of scientific enquiry that would seem to me to be more suited to the application of multidimensional scaling techniques than ethology. Much of the discipline is devoted to discovering some order in data sets that describe relations among entities. Often it is desired to find some simple structure or explanation that will succinctly describe these observed relations, since the raw observations may not themselves be susceptible to immediate understanding. Even experienced workers, after pondering the complexities of the data, may not, by working solely at an intuitive level, be able to do full justice to their hard won data. Data reduction procedures can be most useful in helping to clarify things by producing compact representations that do as little violence as possible to the original data.

Consider the preceding-following matrix in Table 1. This is taken from Blurton Jones (1968). However it is not presented in exactly its original form. The original data were observed frequencies of the occurrence of column behavior j after the occurrence of row behavior i in the great tit *(Parus major)*, and these are shown in the lower rows of Table 1. These frequency counts are, in some loose sense, a measure of how "close" behavior j is to behavior i. If the count is high then one behavior follows

Table 1. Preceding-Following Behaviors Matrix for *Parus Major* (Blurton Jones, 1968).

	1	2	3	4	5	6	7	8	9	10	11	12	13
1. Attack	4	17	16	11	10	13	11	0	6	0	0	9	4
	2	6	4	5	4	27	3	0	3	0	0	2	3
2. Head-down	26	0	5	14	4	13	2	8	5	0	0	5	18
	20	0	2	11	3	44	1	8	4	0	0	2	24
3. Horizontal	25	3	0	12	13	11	3	2	10	8	0	4	9
	13	1	0	6	6	26	1	1	6	1	0	1	8
4. Head-up	5	9	8	8	14	15	5	4	13	0	2	5	12
	4	5	3	6	9	51	2	4	11	0	1	2	15
5. Wings-out	22	13	10	5	2	10	2	7	7	0	0	2	19
	20	9	5	5	2	40	1	8	7	0	0	1	29
6. Feeding	2	5	18	13	11	3	3	5	13	8	1	16	1
	6	10	23	33	24	38	4	14	36	5	1	20	5
7. Incomplete feeding	4	10	15	4	4	13	7	22	0	0	12	8	0
	1	2	2	1	1	15	1	7	0	0	2	1	0
8. Hopping around	1	10	0	4	2	4	46	0	3	6	11	11	3
	1	5	0	3	1	14	19	0	2	1	5	4	3
9. Hopping away	0	4	6	9	5	1	8	4	1	6	31	15	10
	0	2	2	6	3	4	3	3	1	1	13	5	11
10. Crest-raising	0	0	0	6	7	3	0	11	17	0	30	13	12
	0	0	0	1	1	2	0	2	3	0	3	1	3
11. Fluffing	0	4	5	6	3	3	0	23	13	35	0	6	3
	0	1	1	2	1	4	0	10	5	3	0	1	2
12. Looking around	5	0	5	0	3	6	12	12	11	30	8	0	9
	2	0	1	0	1	11	3	6	5	3	2	0	6
13. Hopping towards	5	25	12	8	21	4	2	2	2	7	5	6	0
	13	54	18	24	55	46	4	8	7	5	10	9	2

Note: Raw frequencies in lower rows.

another with high reliability, and the two are clearly related in some way. If the count is low, then, all other things being equal, the two behaviors are not closely related—the extreme situation being when a particular behavior is never followed by another. However, all things are not generally equal in a situation like this: for example, in absolute terms, we simply do not observe as many instances of crest-raising as of feeding. Consequently, if feeding follows fluffing four times and crest-raising follows fluffing three times, is it the case that both of these behaviors bear about the same relationship to fluffing? Probably not, even in the restricted framework established here.

What we are really interested in is the *proportion* of times behavior *j* follows *i*. It is a simple matter to calculate this using the iterative proportional fitting procedure (IPFP) of Deming and Stephan (1940) which is described fully in Bishop and others (1975) and also in Chapter 6 of this volume. A good nontechnical exposition and example is to be found in Fienberg (1971). Without going into detail, the procedure iteratively adjusts the matrix entries such that the row and column marginals are equal, and this is done in such a way as to preserve the interaction structure of the table. If the marginals are set arbitrarily to be equal to 100, the entries in the upper rows of the table can be interpreted as percentages. So now we know that feeding follows fluffing 3 percent of the time, whereas crest-raising follows fluffing 35 percent of the time. I believe that this method of adjusting preceding-following matrices can be useful in many situations. (The anthropologist A. K. Romney was the first person to advocate the use of this technique in connection with multidimensional scaling.)

The data matrix is of course still asymmetric—of which more later—but let us suppose that we ignore the asymmetry. Provisionally, it will be assumed that the observed differences between (*i*, *j*) and (*j*, *i*) pairs can be attributed to random fluctuation and not to any systematic effects. In this case, however, the assumption is unlikely to be true.

Now examine Figure 1. At the moment it is not particularly important to understand the details of the construction process. It will be noted that behaviors that tend to follow each other with high frequency are situated close together in the space whereas behaviors that are unrelated are far apart in the space. Thus we have constructed a simple spatial model of the observed coöccurrences where distances in the model can be identified with strength of relationship between the behaviors.

Figure 1 strongly suggests that the behaviors can be divided into three different groups, with the behaviors more similar within groups than between groups. The reader is invited to see if he can visually detect these groupings before reading the following quotation from Blurton Jones:

> . . . the movements studied can be arranged into a number of partially discrete groups . . . Together with [attacking] go head-down, head-up, horizontal, wings-out . . . As another group, hopping away, crest-raising, looking around and [fluffing] belong together. Pecking food objects and incomplete pecking movements, and perhaps hopping around, form another group (Blurton Jones 1968, p. 100).

Blurton Jones goes on to speculate that behaviors within each of the groups may have certain causal factors in common, these being inde-

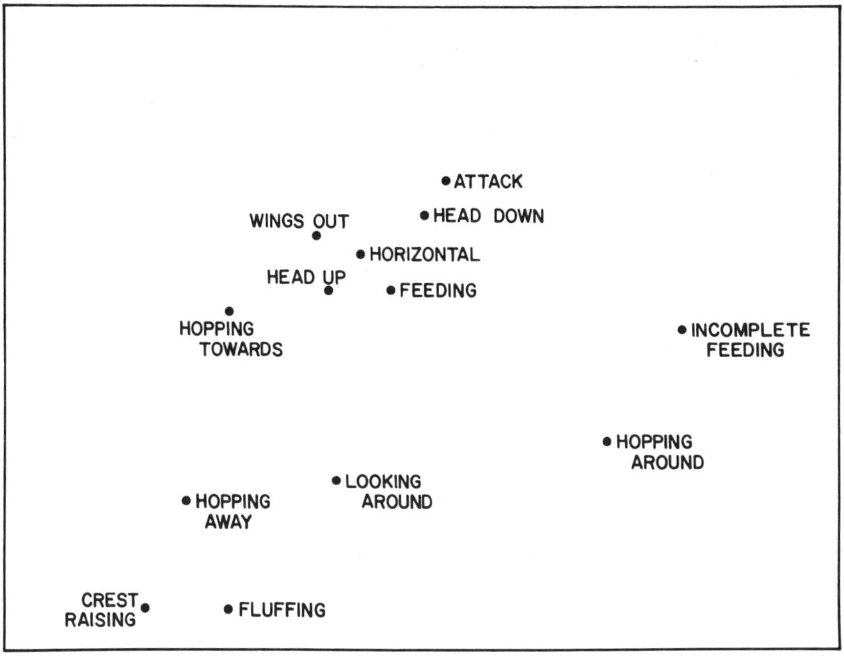

Figure 1. A geometrical representation of the Blurton Jones data.

pendent across groups. It is important to note that he arrived at this interpretation by a quite different, and rather more laborious, statistical process; multidimensional scaling was not used. Instead a variety of correlational and nonparametric statistical procedures were employed. It is not my intention to belittle the techniques in Blurton Jones' study. On the contrary, it is even probable that, for his purposes, a more detailed picture was obtained by these methods. Notwithstanding, I submit that a geometrical representation of the data may be informative: certainly the same general interpretation results independent of the approach employed to analyze the data. It is perhaps possible that other more detailed hypotheses may be suggested by the two dimensional plot. For example, it seems possible that attack behaviors might often be preceeded by a particular sequence of other behaviors such as wings out—horizontal—head-down—attack. This of course could be tested by examining the original observational record.

The origins of procedures that allow the construction of point configurations based on a knowledge of distance information alone are quite old. For centuries mapmakers have been faced with this problem, albeit in only two—or sometimes three—dimensions, and they were reasonably successful in finding solutions. However, as Tobler (1977) recounts, it was less than a century ago that sophisticated mathematical techniques, including iterative processes, began to appear in surveying and cartography. Geometers also had interests clearly germane to the problem, and one finds that a substantial amount of relevant theory had been worked out decades ago (e.g., Blumenthal 1953).

Work on the problem of quantifying similarity in psychology has a long history with its roots in psychophysics, but it was not until 1938 that Richardson attempted to perform what we would now call multidimensional scaling. This was made possible by the work of two mathematicians, Young and Householder (1938), who provided the necessary theorems for the solution to Richardson's problem. Torgerson (1952) revived interest in this procedure and also contributed several developments of his own. From that point on a number of people have improved and refined these techniques, which have come to be known as traditional metric scaling procedures.

During the 1950s, and even in one case as early as 1941, three workers were absorbed by problems relevant to many of the techniques and issues discussed in this chapter. These independent contributions were not to come together until the 1960s, and in retrospect it seems somewhat surprising that there was not more cross-fertilization at an earlier date. Guttman, since the early 1940s, had been developing methods for the geometric and vector representation of qualitative and ordinal data. However, only recently have his contributions become well known. Coombs (1964), in his remarkable "Theory of Data," developed during the 1950s, was concerned with data collection procedures and particularly with the properties of nonmetric data. He and his collaborators suggested a number of methods for analyzing data of this kind. However, many of them were not readily translatable to the language of the computer, and consequently his work has come to be valued more for its conceptual than its practical insights. Almost at the same time, Shepard (1962), as a direct result of his work on stimulus and response generalization in animals (Shepard 1958), was to suggest a method of multidimensional scaling that does not require prior knowledge of the function that relates the distances and the dissimilarities. By a stroke of good fortune, Shepard was then working at Bell Telephone Laboratories and elicited the collaboration and interest of the mathematician J. B. Kruskal. Thus in 1964 the first really practical and numerically stable multidimensional scaling algorithm was published, and the program was made widely available. The response of the scientific community was nothing short of astonishing—clearly geometrical models had considerable intuitive appeal for many researchers! In the last 12 years, literally thousands of empirical applications of the procedure have been published, and their range has spanned virtually all disciplines.

But not only applied researchers were interested: many others of a more statistical and mathematical bent were seduced by the data analytic possibilities of the new technique. A variety of generalizations and improvements of the basic algorithm have been spawned, and in addition much work has been done on examining the behavior of the algorithms in a wide range of situations. Some of this work is reviewed in subsequent sections of this chapter.

1. ASSUMPTIONS

The assumptions involved in most forms of multidimensional scaling are rather simple, and hence may be summarized quite briefly. The consequences of not satisfying these assumptions are not often easy to evaluate. However in Sections 2 and 4 there is some discussion of what is known in such cases.

Common to many varieties of multidimensional scaling is the assumption that the objects may be properly represented by points in a metric, usually Euclidean, space. Therefore, in the geometrical representation produced by these scaling procedures the usual metric distance axioms must be satisfied. The data, however, often need not satisfy these axioms.

Metric Scaling

In the simplest case, it is assumed that the interobject dissimilarities are known without error and that they are indeed metric distances; then, as will be seen, it is a simple matter to recover the point coordinates of the configuration. This ideal state of affairs will seldom, if ever, be obtained! Nonetheless, identical procedures may be used to estimate the coordinates in the presence of error, and even if the interobject dissimilarities are some linear transformation of the "true" distances, it is not difficult to modify the computing procedures to provide a reasonable solution. The mathematical conditions necessary for the appropriate application of metric methods have been explored by Young and Householder (1938), Torgerson (1958), and Gower (1966), among others. When the relationships between the dissimilarities can be assumed known only in terms of their order, from smallest to largest, then metric scaling is not strictly applicable, and a nonmetric alternative should be considered.

Individual Differences Metric Scaling

This form of scaling was introduced in psychology where it was assumed that dissimilarity data obtained from a number of subjects could be represented by postulating a common perceptual space for all subjects, which was then simply expanded or contracted along particular axes to produce the individual spaces. Individual differences among subjects could be accounted for by defining a set of idiosyncratic weights for each individual that would reflect the importance attached by that subject to the dimensions of the common space. Consequently, this model allows substantial differences among the individual data matrices; differences, moreover, that are not simply attributable to error. Apart from the introduction of individual weights to reflect the saliences of the dimensions, the model and other assumptions are the same as for metric scaling.

It turns out that this weighted Euclidean model may be usefully employed in situations other than the psychological. For example, the several data matrices could contain dissimilarities collected at various points in time, and the dimension weights would then give some indication of the change in relative importance of the dimensions over time. However, independent of the substantive context, the model is only appropriate if one can sensibly make the assumption that the systematic variation from data matrix to data matrix can be adequately described in terms of the shrinking or expansion of the axes of a common multidimensional space. An example below makes this notion clearer.

Nonmetric Scaling

Frequently the assumption that the data (or some linear transformation thereof) can be viewed as metric distances is not even approximately satisfied. This is usually because one cannot be confident of the metric properties of the response measure. For example, in Table 1, is it reasonable to assume that an entry of 24 percent represents twice the amount of association between a pair of behaviors as 12 percent? If such an assumption cannot be made, the methods of metric scaling may not yield very good results—although, in most empirical situations, metric procedures seem to do quite well, even if the metric assumption is not met. Nevertheless, if one is not sure of the measurement status of the data it is usually quite reasonable to assume that 24 percent merely represents a *greater degree* of association than 12 percent, and likewise for all other pairs of entries in a matrix like Table 1. Such comparisons induce an order on the set of dissimilarities, and an experimenter may feel more comfortable with this ordinal level of measurement. The goal then would be to construct a configuration whose interpoint distances were in an order that was as close as possible to the observed ordering of the dissimilarities. Shepard (1962) and Kruskal (1964) were among the first to provide practical computer algorithms for this problem and to demonstrate that the ordinal constraints were usually strong enough to allow a metric solution.

ALSCAL

ALSCAL (Takane, Young, and de Leeuw 1977) is a relatively new computer program that has the ability to deal with any of the above cases, and also several more. Although a computer program, and not a model like those described in the preceding sections, ALSCAL *implements* a very general model, and consequently it was felt worthwhile to devote a separate section to this procedure. The level of measurement in the data may be categorical, ordinal, interval, or ratio. The distance function employed is very general and includes as special cases both the simple and the weighted Euclidean models. It is even possible to deal with asymmetries in the input data matrices. A more detailed description is given later in this chapter.

Maximum Likelihood and Bayesian Estimation

Ramsay (1977) has recently proposed that if certain assumptions are made regarding the error distribution of the distances, the well known statistical techniques of maximum likelihood and Bayesian estimation can be utilized in constructing multidimensional scaling solutions. In addition to the usual benefits of being able to examine the plausibility of the error model, maximum likelihood procedures allow formal hypothesis testing based on asymptotic chi-square theory. Further, an interesting feature of the companion Bayesian approach is the ability to construct credibility regions—analogous to the confidence regions of classical statistics. Hence, some idea of size of the uncertainty regions within which the points are located can be obtained.

Ramsay's models are basically metric and encompass both the unreplicated Euclidean case and also the replicated weighted Euclidean model. Certain kinds of nonlinear transformation are allowed, providing they come from a power family ($y = x^p$). The discontinuous, stepwise transformations used by Kruskal and Guttman in nonmetric scaling are not permitted. Finally, in contrast to most metric procedures, there is a missing data capability.

2. METHOD

Metric Scaling

The problem of metric scaling is given the distances **D** how are the coordinates of the n points **X** to be found? Assume that the m-dimensional configuration **X** has its origin at the centroid of the points, that is

$$\sum_{i=1}^{n} x_{ia} = 0 \qquad \text{for } a = 1, 2, \ldots, m \tag{1}$$

The scalar products referred to the origin are defined as

$$b_{ij} = \sum_{a=1}^{m} x_{ia} x_{ja} \tag{2}$$

or in matrix terms,

$$\mathbf{B} = \mathbf{XX}' \tag{3}$$

Hence if **B** were known the problem would be a simple one of matrix decomposition (cf. principal components analysis, Chapter 8). Somehow **B** must be constructed using our knowledge of the distances. The relation between a distance and the coordinates is

$$\begin{aligned} d_{ij}^2 &= \sum_{a=1}^{m} (x_{ia} - x_{ja})^2 \\ &= \sum_{a=1}^{m} x_{ia}^2 + \sum_{a=1}^{m} x_{ja}^2 - 2 \sum_{a=1}^{m} x_{ia} x_{ja} \\ &= d_i^2 + d_j^2 - 2b_{ij} \end{aligned} \tag{4}$$

where d_i^2 is the squared distance of the ith point from the origin and b_{ij} is a scalar product, as defined in Equation 2. Consider the following average:

$$d_{\cdot j}^2 = \frac{1}{n}\sum_{i=1}^{n} d_{ij}^2 = \frac{1}{n}\left(\sum_{i=1}^{n} d_i^2 + \sum_{i=1}^{n} d_j^2 - 2\sum_{i=1}^{n} b_{ij}\right)$$

$$= d_{\cdot}^2 + d_j^2 - \frac{2}{n}\sum_{i=1}^{n} b_{ij}$$

$$= d_{\cdot}^2 + d_j^2 - \frac{2}{n}\sum_{i=1}^{n}\sum_{a=1}^{m} x_{ia}x_{ja}$$

$$= d_{\cdot}^2 + d_j^2 - \frac{2}{n}\sum_{a=1}^{m} x_{ja}\sum_{i=1}^{n} x_{ia}$$

$$= d_{\cdot}^2 + d_j^2 \qquad \text{(the last term is zero because of Equation 1)}$$

Similarly, it is easy to show that

$$d_{i\cdot}^2 = d_{\cdot}^2 + d_i^2 \quad \text{and} \quad d_{\cdot\cdot}^2 = 2d_{\cdot}^2$$

Using the above, after rearranging Equation 4, we get

$$b_{ij} = -\tfrac{1}{2}(d_{ij}^2 - d_i^2 - d_j^2)$$

$$= -\tfrac{1}{2}[d_{ij}^2 - (d_i^2 + d_{\cdot}^2) - (d_j^2 + d_{\cdot}^2) + 2d_{\cdot}^2]$$

$$= -\tfrac{1}{2}(d_{ij}^2 - d_{i\cdot}^2 - d_{\cdot j}^2 + d_{\cdot\cdot}^2) \tag{5}$$

or in matrix notation

$$\mathbf{B} = -\tfrac{1}{2}\mathbf{M}(\mathbf{D}{*}\mathbf{D})\mathbf{M} \tag{6}$$

where **D*D** is the Hadamard (elementwise) product and contains squared distances. **M** is a mean centering matrix

$$\mathbf{M} = (\mathbf{I} - \frac{1}{n}\mathbf{U})$$

where **I** is the identity matrix and **U** is a matrix of unities. We now have an expression for **B** entirely in terms of the (squared) distances, which are known. The nonzero eigenvalues of **B** are the elements of $\mathbf{\Lambda} = \text{diag}(\lambda_1, \ldots, \lambda_n)$ and their number is equal to both the rank of **B** and the dimensionality of the space. The eigenvectors **X** are scaled such that $\mathbf{X}'\mathbf{X} = \mathbf{\Lambda}$.

The major names associated with the above development are Young and Householder (1938), Torgerson (1958), and recently Gower (1966). Other related discussion may be found in Messick and Abelson (1956) and Schönemann (1970).

When, as is usually the case, the distances are not known exactly, the above theory can be used to provide a least squares solution. Suppose that $\mathbf{\Delta}$, a matrix of distance estimates, is known, and it is assumed that $\delta_{ij} = d_{ij} + e_{ij}$. We define a matrix of estimated scalar products, analogous to Equation 6, as

$$\hat{\mathbf{B}} = -\tfrac{1}{2}\mathbf{M}(\mathbf{\Delta}*\mathbf{\Delta})\mathbf{M} \tag{7}$$

and consider the following residual matrix

$$\boldsymbol{\xi} = \hat{\mathbf{B}} - \mathbf{XX}' \tag{8}$$

Hence the least squares loss function, which provides a measure of the badness of fit, can be written

$$\theta = \mathrm{tr}(\boldsymbol{\xi}'\boldsymbol{\xi}) = \mathrm{tr}[(\hat{\mathbf{B}} - \mathbf{XX}')'(\hat{\mathbf{B}} - \mathbf{XX}')] \tag{9}$$

noting that the loss function is defined in terms of scalar products, and not distances. It can be shown, by matrix differentiation for example, that θ has a minimum with $\mathbf{X}$ the set of the first m eigenvectors of $\mathbf{B}$, scaled as before. Note that $\mathbf{X}$ is unique only up to (distance preserving) Euclidean transformations: the configuration may be translated to a new origin, or subjected to a rigid rotation without distorting the distances.

In practice, $\mathbf{\Delta}$ may be related to the "true" distances in a slightly more complicated fashion

$$\delta_{ij} = a + bd_{ij} + e_{ij} \qquad (b > 0,\ a \neq 0)$$

The errors cause no real problem, as we have seen, since a least squares fit to the estimated scalar products may be obtained. The constant b simply represents a change of scale and is seldom of any importance. However, the addition of a in the linear transformation may cause the estimated scalar product matrix $\mathbf{B}$ to be negative definite, and consequently a Euclidean representation will be impossible, some coordinates having imaginary values. This is the problem of the "additive constant" (Torgerson 1958) and is usually dealt with by adding an arbitrary quantity c to the δ_{ij} prior to the construction of $\mathbf{B}$: this value is chosen sufficiently large to ensure positive semidefiniteness of $\mathbf{B}$. Although somewhat simple-minded, this procedure generally works quite well. However Cooper's (1972) solution is much more elegant, if computationally more involved.

Individual Differences Metric Scaling

The above may be generalized quite easily to the weighted Euclidean model, which has been discussed by Carroll and Chang (1970), Bloxom

(1968), Horan (1969), Tucker (1972), Schönemann (1972), Harshman (1970), and Krane (1976), among others. Here, instead of having only one $\boldsymbol{\Delta}$ matrix, we have several—$\boldsymbol{\Delta}_1, \boldsymbol{\Delta}_2, \ldots, \boldsymbol{\Delta}_N$ Consequently N estimated scalar product matrices may be computed, each of which could be subjected to individual decomposition, as above. However, generally a more economical representation is desired, and the usual choice is

$$\hat{\mathbf{B}}_k = \mathbf{X}\mathbf{W}_k{}^2\mathbf{X}' \tag{10}$$

where $\mathbf{W}_k = \text{diag}(w_{k1}, w_{k2}, \ldots, w_{kn})$, a diagonal matrix of weights corresponding to the kth matrix and $\mathbf{X}$ is common to all of the decompositions. It also turns out that this representation, in addition to being parsimonious, has the important property of dimensional uniqueness.

In order to obtain an intuitive appreciation of the model, consider the following example: given $\mathbf{D}_1$, $\mathbf{D}_2$, and $\mathbf{D}_3$, a set of errorless distance matrices, the individual solutions are

$$\mathbf{B}_k = \mathbf{X}_k\mathbf{X}_k{}' \qquad (k = 1, 2, 3)$$

Suppose that the configurations turned out to be those shown in the top half of Figure 2. The similarity between these individual spaces is

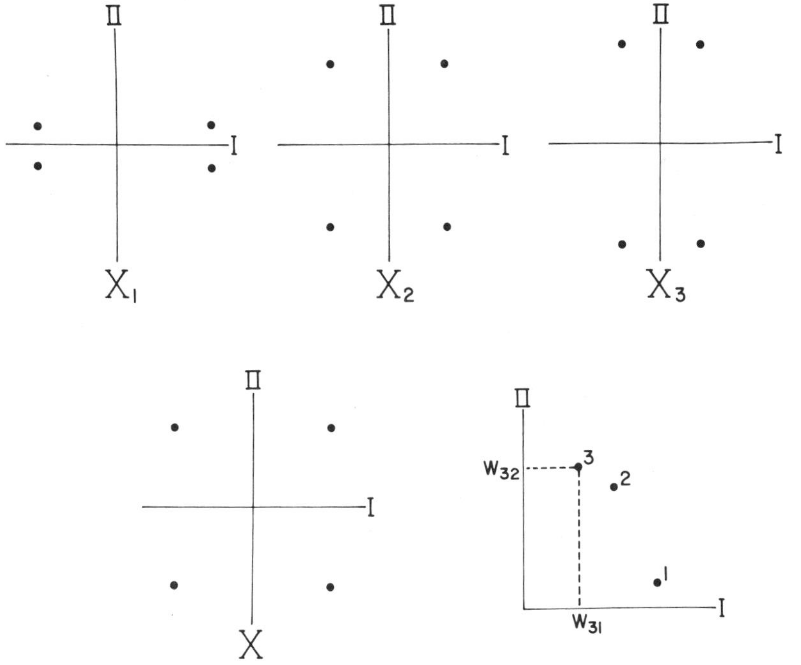

Figure 2. An illustration of the weighted Euclidean model.

obvious. In fact, each could be produced from any other by a differential stretching or contraction of one or both axes. Indeed all three can be considered to be a rescaling of a fourth configuration **X**. If **X** is as shown in the lower half of the figure, and if w_{31} and w_{32} have values similar to those indicated, then

$$\mathbf{X}_3 = \mathbf{X}\mathbf{W}_3 = \begin{bmatrix} x_{11} & x_{12} \\ x_{21} & x_{22} \\ x_{31} & x_{32} \\ x_{41} & x_{42} \end{bmatrix} \begin{bmatrix} w_{31} & 0 \\ 0 & w_{32} \end{bmatrix}$$

$$= \begin{bmatrix} w_{31}x_{11} & w_{32}x_{12} \\ w_{31}x_{21} & w_{32}x_{22} \\ w_{31}x_{31} & w_{32}x_{32} \\ w_{31}x_{41} & w_{32}x_{42} \end{bmatrix}$$

Hence $\mathbf{X}_3$ and **X** are related by the diagonal matrix $\mathbf{W}_3$ which contains the weights required to rescale differentially the axes of **X** to produce $\mathbf{X}_3$. In general

$$\mathbf{X}_k = \mathbf{X}\mathbf{W}_k$$

Consequently, the interpoint distances in the kth matrix are of the form

$$d_{ijk}^2 = \sum_{a=1}^{m} (x_{iak} - x_{jak})^2$$

$$= \sum_{a=1}^{m} w_{ka}^2 (x_{ia} - x_{ja})^2$$

and have been referred to as elliptical distances. Furthermore, in terms of scalar products

$$\mathbf{B}_k = \mathbf{X}_k X_k' = \mathbf{X}\mathbf{W}_k{}^2\mathbf{X}'$$

The space spanned by the columns of **X** is frequently referred to as the group stimulus, or common, space. $\mathbf{W}_k$ is a weight matrix and $\mathbf{X}_k$ spans the individual, or private, space. At this point it should be noted that some authors choose to write the above equations using the square roots of the weights (e.g., Carroll and Chang 1970). This is not an important distinction, so long as one is aware of the convention that has been adopted—indeed later in this chapter the weights are defined in such a fashion.

From the point of view of economy this model is attractive since the data can be represented in terms of a single configuration and a set of weights for each individual data set. However, a more important virtue of this model is the fact that it is not rotationally indeterminate provided the data fulfill certain minimal conditions of variation across individuals. Given the following loss function

$$\theta = \sum_{k=1}^{N} \text{tr}[(\mathbf{B}_k - \mathbf{X}\mathbf{W}_k{}^2\mathbf{X}')'(\mathbf{B}_k - \mathbf{X}\mathbf{W}_k{}^2\mathbf{X}')]$$

there is usually only one **X** and one set of $\mathbf{W}_k$ which will minimize θ—for further details regarding the necessary conditions see Harshman (1970, 1972) and Kruskal (1976). Any rotation of **X** will result in a poorer fit. In psychology, much has been made of this property, and it has been assumed that the columns of **X** are in fact aligned with the common psychological dimensions being used by the subjects. The subjects' individual spaces differ only in terms of the importance or salience they attach to these dimensions. Thus, in the example, subject 1 weights dimension I much more than II; subject 2 weights the dimensions approximately equally; and subject 3 attaches more importance to dimension II. Some good empirical examples of this feature are to be found in Carroll and Chang (1970) and Carroll and Wish (1974).

The problem of finding **X** and **W** is not as easily solved as in the traditional metric case. Although a variety of possibilities have been tried, the one described here is currently the most widely used and is historically the oldest. Recall the above decomposition with the estimated scalar products substituted

$$\hat{\mathbf{B}}_k \cong \mathbf{X}\mathbf{W}_k{}^2\mathbf{X}'$$

or

$$\hat{b}_{ijk} \cong \sum_{a=1}^{m} w_{ka}^2 x_{ia} x_{ja} \qquad (11)$$

Carroll and Chang (1970) have viewed this as essentially a regression problem: if the coordinates were known the weights could be obtained by ordinary least squares multiple regression, with the known $x_{ia}x_{ja}$ products as the values of the independent variables, and the $\hat{b}_{ijk}$ as the values of the dependent variable. Hence values for the weights that minimize the standard least squares loss function conditional on some fixed **X** may be found. The process may be repeated with the weights fixed and the coordinates to be estimated, and the whole cycle iterated until convergence. This procedure has come to be known as an alternating least squares strategy and involves, at each stage, a conditional least squares minimization.

The description given in Carroll and Chang (1970) is more refined, and a little more general, but its essential features are as described here.

Starting values for the iterations can be, and usually are, simply a set of pseudorandom numbers. In practical situations this works reasonably well, although frequently at the expense of considerable computing time. A number of alternative approaches, which are probably superior, have been devised by Ramsay (1977), de Leeuw and Pruzansky (1975), and Krane (1976). The last author uses a somewhat different rationale, more in the tradition of Bloxom (1968) than Carroll and Chang (1970), and employs a quasi-Newton optimization technique to minimize a least squares objective function which includes the set of additive constants among its parameters. As yet, the utility of his approach has been insufficiently tested.

Nonmetric Scaling

As previously mentioned, the approach taken in nonmetric scaling does not require the assumption of a linear relation between the dissimilarities and the distances: all that is demanded is an ordinal relation. Thus, when the configuration is constructed the following order isomorphy requirement should be satisfied as well as possible

$$d_{ij} \leqq d_{kl} \text{ iff } \delta_{ij} \leqq \delta_{kl} \tag{12}$$

where **D** contains the distances of the solution, **Δ** the dissimilarities, and "iff" means "if and only if." Ignoring the ambiguity of the equalities and the presence of errors, for the moment, this means that a configuration should be found whose distances satisfy

$$d_{ij} = f(\delta_{ij}) \tag{13}$$

where f is a nondecreasing function. In practice this requirement cannot usually be satisfied exactly because of several factors, the most important of which is measurement error, and hence an approximate correspondence must be tolerated. This is precisely analogous to the situation in ordinary regression: what set of bivariate data falls exactly on the line?

The fitting of this model is done differently than in the metric case. Instead of working with scalar products the distances are fit directly the the data—a much more natural approach, and one that is intuitively more satisfying to most people, although this does give rise to some technical difficulties. Obviously, a configuration whose distances will minimize the following least squares function,

$$L = \sum_{i,j}^{n} (d_{ij} - \delta_{ij})^2 \tag{14}$$

would imply a linear relation between the dissimilarities and the distances, and hence the procedure would be metric. However a modification of L to allow a monotonic relation may be used

$$S = \sum_{i,j}^{n} [d_{ij} - f(\delta_{ij})]^2 \tag{15}$$

where f is the nondecreasing function introduced above. The problem is now more complicated than in the metric case: not only must a suitable **X** be found, but also the monotonic transformation which relates the data and the distances. Thus the meaning of the title of Shepard's pioneering paper becomes apparent—"Multidimensional Scaling with an Unknown Distance Function."

A variety of algorithms have been devised to perform this task. For the record, I will acknowledge the better known of these: M-D-SCAL (Kruskal 1964), SSA-I (Lingoes and Roskam 1973), POLYCON (Young 1972), TORSCA (Young 1968), MINISSA-I (Roskam 1969), EMD (McGee 1966), and KYST (Kruskal, Young, and Seery 1973). All of these, plus several others, attempt to provide a solution to the above problem. Although there are considerable differences among these algorithms, each contains a basic core that iteratively performs the following steps.

1. Given a configuration **X**, calculate **D**, the matrix of interpoint distances.
2. Compute a set of values, $\mathbf{D}^t$, which are some transformation of the distances in **D** having the same rank order as the dissimilarities in $\mathbf{\Delta}$. The $\mathbf{D}^t$ will not, in general, satisfy the metric distance axioms.
3. Compute a new **X** and hence a new **D** such that the d_{ij} are a "good" approximation to the d_{ij}^t.
4. Go to 1.

Thus the procedure alternates between satisfying the ordering requirement (in Step 2) and the distance requirement (in Step 3).

Within this general framework, a number of options and refinements are possible. The best known and also one of the first algorithms is due to Kruskal (1964) and very similar procedures have been used by Young (1968) and McGee (1966). It is worth describing some of the details of these programs in relation to the above general scheme.

1. The first **X** is chosen arbitrarily, or in some fashion that makes use of the raw data. Young (1968), for example, uses a modification of traditional metric scaling to provide an initial configuration.

2. The set of d^t_{ij} is chosen to minimize the discrepancy between the current d_{ij} and the d^t_{ij} subject to the ordinal constraints of the data. The following conditional least squares problem is solved: find the d^t_{ij} that minimize $\Sigma^n_{i,j}\ (d_{ij} - d^t_{ij})^2$ for fixed d_{ij} subject to $d^t_{ij} \leqq d^t_{kl}$ iff $\delta_{ij} \leqq \delta_{kl}$. This is a problem in isotonic regression (Kruskal 1964b; Barlow et al. 1972) and the best fitting values d^t_{ij} are usually denoted $\hat{d}_{ij}$. The solution is obtained by forming blocks of minimal size of the d_{ij} such that the blocks means are strictly increasing. Kruskal and Young give complete and clear descriptions.

3. Use a gradient method to alter **X** such that a suitably normalized version of the following objective function

$$S = \sum_{i,j}^{n} (d_{ij} - \hat{d}_{ij})^2$$

is improved. Note that the $\hat{d}_{ij}$ are fixed at this stage. A simple method of conceptualizing this gradient process is due to Gleason (1967). Imagine a line connecting any two of the points i and j. At the moment we only need a single subscript for each since their position relative to each other on the line is all that matters. Now consider the following scheme for improving the position of i:

$$\begin{aligned} x_i^{\text{new}} &= x_i + \left(1 - \frac{\hat{d}_{ij}}{d_{ij}}\right)(x_j - x_i) \\ &= x_i + \left(\frac{d_{ij} - \hat{d}_{ij}}{d_{ij}}\right) d_{ij} \\ &= x_i + (d_{ij} - \hat{d}_{ij}) \end{aligned} \tag{16}$$

If the current d_{ij} is smaller than the current $\hat{d}_{ij}$ then $x_{\text{i}}^{\text{new}}$ will become smaller, hence increasing d_{ij}; the converse is also true. Considering all pairs of points simultaneously (necessitating a second subscript) leads to the analogous iterative relation

$$x_{ia}^{\text{new}} = x_{ia} + \alpha \sum_{j=1}^{n} \left(1 - \frac{\hat{d}_{ij}}{d_{ij}}\right)(x_{ja} - x_{ia}) \tag{17}$$

The weighting factor α may be varied, either automatically by the program or by the user, depending on how much it is desired to emphasize the discrepancies $d_{ij} - \hat{d}_{ij}$.

4. Compute the new **D** and go to 2.

The Guttman-Lingoes-Roskam approach differs mainly in how the d^t_{ij} are selected.

1. The first **X** is constructed by a method that uses the rank order information in the data.
2. The d^t_{ij} chosen are simply a permutation of the d_{ij} and are denoted by d^*_{ij}. The permutation is selected subject to the constraint

$$d^*_{ij} \leqq d^*_{kl} \quad \text{iff} \quad \delta_{ij} \leqq \delta_{kl}$$

Obviously the d^*_{ij} have the same scale as the d_{ij}, being the same values numerically, but in a different order.
3. A process, similar to that in Equation 17, is used to improve **X**, the major difference being that this is usually iterated for about five cycles with the same d^*_{ij}.
4. Compute the new **D** and go to 2.

The treatment of ties is similar in both approaches, even though the terminology used differs. Two basic situations may be distinguished.

Situation I

$$\delta_{ij} = \delta_{kl} \text{ implies } d^t_{ij} \leqq d^t_{kl} \text{ or } d^t_{ij} > d^t_{kl}$$

Thus tied data may be untied, reflecting a belief that the tie is the result of lack of resolution in measurement.

The order of the d^t values after the tie is broken is the same as the current order of d_{ij} and d_{kl}. Kruskal calls this the Primary Approach; Guttman and Lingoes (e.g. Guttman 1968) use the term semistrong monotonicity.

Situation II

$$\delta_{ij} = \delta_{kl} \text{ implies } d^t_{ij} = d^t_{kl}$$

This second constraint implies that observed ties in the data are not simply the result of insufficient resolution, and hence, if dissimilarities are equal, the corresponding distances should be too. The terms used by Kruskal and Guttman-Lingoes are, respectively, the Secondary Approach and strong monotonicity.

Since the above may not be very easy to follow, I have constructed a small example which the interested reader can follow through in order to clarify some of these ideas. I have assumed that $n = 7$ and that we are at about the midway point in the iterative process. The 21 (= 7 × 6/2) dissimilarities have been converted to ranks, and the distances are those constructed in the previous iteration. The problem is how to find the d^t_{ij}, and the solutions are given in the right-hand columns of Table 2. The

Table 2. Example to Illustrate Different Methods of Computing the d^t_{ij}

		$\hat{d}_{ij}$		d^*_{ij}	
δ_{ij}	d_{ij}	I	II	I	II
1	4	2.4	2.5	1.0	1.0
2	1	2.4	2.5	2.0	2.0
3	3	2.4	2.5	2.0	2.0
4	2	2.4	2.5	3.0	3.0
6	2	2.4	4.7	4.0	4.3
6	4	4.0	4.7	4.0	4.3
6	8	6.3	4.7	5.0	4.3
8	6	6.3	6.0	6.0	6.0
9.5	5	6.3	6.5	6.0	6.0
9.5	8	7.0	6.5	6.0	6.0
11.5	6	7.0	7.1	6.0	6.5
11.5	10	7.2	7.1	7.0	6.5
13	7	7.2	7.1	7.0	7.0
14	6	7.2	7.1	8.0	8.0
15	7	7.2	7.1	8.0	8.0
16	6	7.2	7.1	8.0	8.0
17	8	8.0	7.1	10.0	10.0
18.5	10	10.0	11.0	10.0	11.0
18.5	12	12.0	11.0	12.0	11.0
20.5	13	13.0	13.5	13.0	13.5
20.5	14	14.0	13.5	14.0	13.5
$S_1 = \sum (d_{ij} - d^t_{ij})^2 / \sum d^2_{ij}$		0.160	0.203	0.218	0.243

Note: The distances corresponding to tied dissimilarities have been arranged in ascending order. This is irrelevant if Approach II is adopted, but is mandatory if Approach I is employed.

index of fit is defined as follows:

$$S_1 = \frac{\sum (d_{ij} - d^t_{ij})^2}{\sum d^2_{ij}}$$

Figure 3 presents these results graphically—although we are not fitting a function, but rather a set of points, the computed d^t_{ij} have been connected by straight line segments to improve the presentation. Such plots are known as Shepard diagrams and are part of the output of many scaling programs. At the end of the iterative process, one can see at a glance the approximate form of the previously unknown distance function, and also how well the data are reproduced by the distances of the final configuration. The closer the d_{ij} are to the d^t_{ij}, the better the fit, and vice versa. If the fit is perfect all $d_{ij} = d^t_{ij}$, and the set of points would form a never decreasing sequence as one moved to the right in the diagram. These diagrams are very useful for detecting outliers, in much the same sense as in ordinary regression: one can readily detect which distances have not been fitted well, by observing the magnitude of the deviations, measured in a horizontal direction, from the d^t_{ij}. Informal residual examination can be of great help in deciding on the appropriateness of the model or the number of dimensions to retain; unusual residuals can signal degeneracy, or, more mundanely, transcriptional

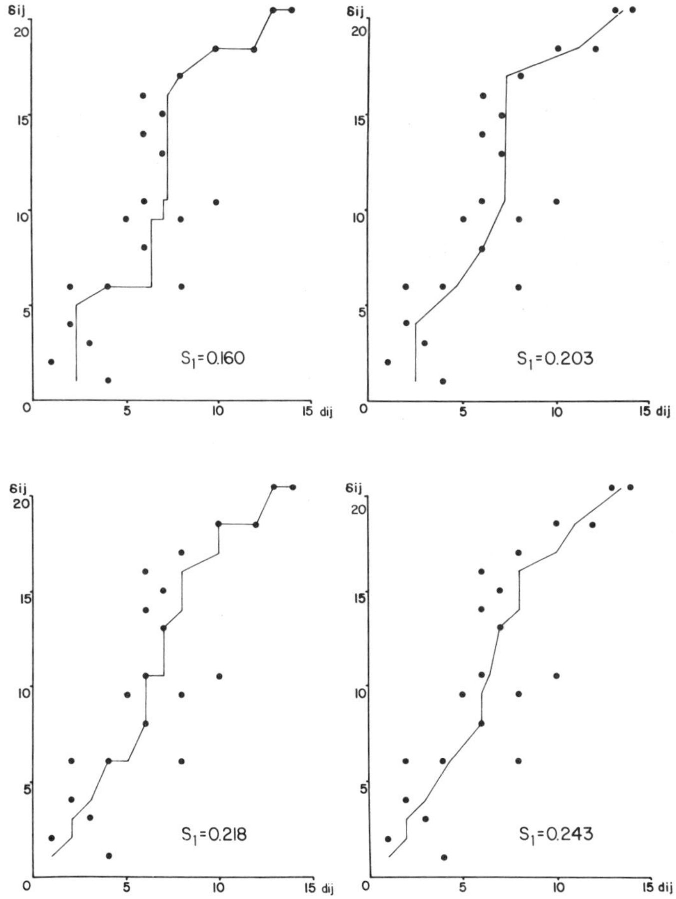

Figure 3. An example of four different monotone regressions.

and other technical errors. As close attention should be devoted to the Shepard diagram as is lavished on the configuration plots from a scaling analysis!

We have not, until now, formally considered the question of goodness (or badness) of fit. There are a large variety of possible measures that may be used. For a discussion of these the reader should consult Roskam (1969), Lingoes and Roskam (1973), or Spence (1970). In this chapter, it will merely be mentioned that there is not a great deal to choose among them, and consequently one may as well stick with the oldest and best known measure

$$S_1 = \frac{\sum_{i,j}^{n} (d_{ij} - d_{ij}^t)^2}{\sum_{i,j}^{n} d_{ij}^2} \tag{18}$$

This is simply a normalized residual sum of squares. When the d_{ij}^t are replaced by $\hat{d}_{ij}$, the measure $\hat{S}_1$ is known as Kruskal's Stress Formula One, and when the d_{ij}^* are used, the value $S_1^* = 2\phi_1$, or twice the value of the Guttman-Lingoes Normalized Phi coefficient. The denominator in the above could be replaced by a variance-like quantity $\Sigma_{i,j}^n (d_{ij} - d_{..})^2$, but in conventional nonmetric scaling situations, there does not seem to be any great advantage in this. Furthermore, since the behavior of $\hat{S}_1$ is very well known, via Monte Carlo studies as well as empirical applications, it would seem sensible to use the more familiar measure.

Earlier in this section, a rather intuitive iterative relation was introduced. In practice, most algorithms utilize a specific objective function, like $\hat{S}_1$, and employ numerical optimization techniques such as the method of gradients to find successive configurations that will yield progressively better values of the objective function. All such multivariable optimization techniques suffer from one potentially serious drawback: the iterations may not converge to the best possible solution. This problem is further discussed below. It may also be mentioned that the optimization process does not require all (i, j) pairs, since the summations may be performed over whichever pairs are present. Thus nonmetric scaling procedures can easily handle missing data—provided, of course, that not too many dissimilarities are missing.

ALSCAL

ALSCAL is the acronym used by Takane, Young, and de Leeuw (1977) for a very recent multidimensional scaling program. This is an ambitious effort to produce a data analysis procedure that will handle any of the preceding models. Actually ALSCAL is capable of doing much more, and is claimed to be more rapid, efficient, and robust than some more specialized programs. Consequently, the reader might be excused for thinking that the program name comes from "*al*l *scal*ing." In fact, the acronym indicates, in general terms, the numerical method used to find the solution—*a*lternating *l*east squares *scal*ing. ALSCAL will compute solutions for the unweighted and weighted Euclidean models, both under metric and nonmetric assumptions, and will handle missing data—something that metric procedures typically cannot do. The data matrices may be replicated or unreplicated and can be symmetric or asymmetric. Indeed ALSCAL is the first widely available algorithm to use a formal model for asymmetric data that does not regard the observed asymmetries as merely the product of error. (I know of only two other models that have been developed, one by Waldo Tobler, a geographer at the University of Michigan, and the other by Richard Harshman at U.C.L.A. However, so far neither has been published, nor are the programs yet generally available.)

The ALSCAL model for distances is defined as

$$d_{ijk}^2 = \sum_{a=1}^{m} v_{ia} w_{ka} (x_{ia} - x_{ja})^2$$

implying,

$$d_{jik}^2 = \sum_{a=1}^{m} v_{ja} w_{ka} (x_{ia} - x_{ja})^2 \qquad (19)$$

Nonnegativity constraints are usually imposed on the weights v_{ia} and w_{ka}. Note that the model is similar to the weighted Euclidean model, but in addition to weights for replications w_{ka} there are also weights for the points v_{ia}. It should also be observed that d_{ijk} is not necessarily equal to d_{jik} because v_{ia} and v_{ja} are not constrained to be equal. The weighted Euclidean model is obtained from the above by setting all $v_{ia} = 1$, and the simple Euclidean model results when in addition the $w_{ka} = 1$. The unweighted asymmetric scaling model is obtained by setting only the w_{ka} equal to one. The significance of the weights for the points is discussed in the example below.

The (unnormalized) objective function that is minimized by ALSCAL is

$$\phi = \sum_{k=1}^{N} \phi_k = \sum_{k=1}^{N} \sum_{i,j}^{n} [d_{ijk}^2 - f_{ik}(\delta_{ijk}^2)]^2 \qquad (20)$$

where d^2_{ijk} is the squared Euclidean distance of the coordinate representation and δ^2_{ijk} is the squared value of the observed dissimilarity between i and j on the kth replication. The transformation f_{ik} is permitted to be category preserving, order preserving, or to be an interval or ratio transformation, at the option of the user. Furthermore, the user is allowed to decide whether this transformation should be allowed to differ over rows i of the matrices, and over replications k. Consequently the ALSCAL model is a very flexible one. However, this flexibility is bought at the expense of conceptual simplicity and an increase in computational complexity.

Interested readers will find a complete description of the algorithm in Takane, Young, and de Leeuw (1977). The basic idea is quite simple although the execution is rather complicated: the program alternates between estimating the v_{ia}, with the other parameters fixed, then the w_{ka} likewise, and finally the x_{ia}, with all weights fixed. Prior to each cycle of these conditional least squares subproblems the best fitting transformations $f_{ik}(\delta^2_{ijk})$ are found. The whole process is iterated several times. The algorithm has the advantage that the objective function will never increase, and this fact, plus some mild regularity assumptions, guarantees convergence, although not necessarily to a stationary point. Empirical evidence seems to support the claim that this algorithm, like INDSCAL, is not seriously troubled by suboptimal solution problems (except possibly when the data are categorical).

Maximum Likelihood and Bayesian Estimation

All of the procedures described thus are are essentially descriptive statistical methods. Although more involved technically than the data reduction procedures learned by every beginning student of applied statistics, there is really little conceptual difference between the calculation of a sample mean and a simple Euclidean metric scaling solution. Both represent a convenient and frequently useful condensation of the data, and both are implicitly based on least squares estimation techniques. There is, however, no hypothesis testing involved.

Ramsay's models allow for the testing of certain hypotheses and, in addition, offer other benefits not shared by older procedures. For example, he points out that previous metric procedures operate on scalar products that are calculated from the squares of the original dissimilarities, and hence contain more error than the dissimilarities themselves: thus the older analysis is statistically inefficient. Furthermore, since most metric scaling models employ a least squares criterion based on scalar products there is the implicit and probably unjustified assumption that the scalar products are independently and normally distributed about their population values. Consequently Ramsay deals with the distances directly, avoiding the intermediary scalar products—this is not unlike the approach taken in nonmetric scaling.

There are, however, distinguishing features in his approach: the most important of these is that the irregular monotone transformations of Kruskal and Guttman are not permitted. Only smooth transformations from a power family may be used. This is done to avoid using up the large number of degrees of freedom required in fitting $\hat{d}_{ij}$ or d^*_{ij}. Incidentally, it has been known for some time that the computation of these discontinuous steplike functions entails the loss of degrees of freedom, but to my knowledge nobody knows how to calculate just how many degrees of freedom are lost. In practical terms the consequences can be clearly seen—nonmetric procedures become quite unstable when applied to matrices of small order or with many missing elements. The reason being that there are insufficient degrees of freedom available to allow the estimation of the coordinates *and* the transformation. In Ramsay (1977), since only power transformations are permitted, only one degree of freedom is lost when the exponent is estimated.

The distance model employed is virtually identical to that in individual differences metric scaling. Specifically,

$$d_{ijk} = \left[\sum_{a=1}^{m} w_{ka}(x_{ia} - x_{ja})^2 \right]^{1/2}$$

This specializes to the unweighted Euclidean model when all w_{ka} are equal to unity. Each observed dissimilarity, δ_{ijk}, is assumed to be a sample from a probability distribution with expectation d_{ijk} and constant error variance σ^2. This can be written

$$\delta_{ijk} \sim f(\delta_{ijk} \mid d_{ij}, \sigma^2)$$

where f is a probability density function. The distance d_{ijk} is related to d_{ij} by the weighted Euclidean distance function given above. Although a variety of possibilities may be considered when f is chosen, Ramsay feels that the "best all round candidate" is the lognormal distribution. This error distribution has the attractive properties of (a) restricting the range of the δ_{ijk} to the nonnegative reals, and (b) having a standard deviation that increases with d_{ij}. These features are often found in empirical data sets. Hence

$$f(\delta_{ijk} \mid d_{ij}, \sigma^2) = (2\pi)^{-1/2}(\sigma\delta_{ijk})^{-1} \exp\left[\frac{-\log^2(\delta_{ijk} \mid d_{ij})}{2\sigma^2}\right]$$

Of course, any other plausible f may be used. The next step is to define the likelihood function, which under independence is

$$L = \prod_{i,j,k} f(\delta_{ijk} \mid d_{ij}, \sigma^2)$$

where the product is taken over all (i, j, k) triples that are available—hence missing data present no difficulty. The maximization of the logarithm of L is, as usual, more convenient, the log likelihood being

$$\log L = \sum_i \sum_j \sum_k \log f(\delta_{ijk} \mid d_{ij}, \sigma^2)$$

Finding the configuration, weights, and the exponent of the power transformation is a problem in numerical optimization. Either a Newton-Raphson method may be used, or a gradient like procedure that has been suggested by Ramsay (1975). Although this is, in principle at least, a standard problem, the details are quite involved and can be found in Ramsay (1977).

A number of tests based on the magnitude of the log likelihood function are possible. One of the most interesting is a test for the appropriate number of dimensions: it turns out that the quantity

$$X^2 = -2(\log L_{m-1} - \log L_m)$$

has an asymptotic χ^2 distribution, and consequently the improvement in fit obtained by going from $(m - 1)$ to m dimensions may be statistically evaluated.

Ramsay shows that his algorithm is as rapid as INDSCAL, if not more so, and also demonstrates that it is capable of recovering a configuration and set of weights more accurately than INDSCAL, when synthetic examples are analyzed.

Since the development of a likelihood function is a keystone of a Bayesian estimation procedure, it is not difficult for Ramsay to extend the above approach. Using indifference priors, he shows how the posterior densities may be obtained and describes how they may be used to construct credibility regions for the estimates obtained during the scaling. An example of this feature is to be found below.

3. EXAMPLES

The following illustrations will give the reader some idea of what can be done with the scaling techniques described above. It is not an easy task to find examples that can be briefly presented without losing too much after editing, or without becoming incomprehensible. Consequently, I have done some analyses myself using data that are available in the literature, rather than attempt to condense published applications (such as Golani 1973; Holloway and Jardine 1968; Miller 1975; Morgan et al. 1976). As a result not all of my examples are necessarily optimal as illustrations, either from a substantive or methodological point of view, but I hope they will be sufficient to convey the sense of what can be achieved by the use of multidimensional scaling.

Metric Analysis

Actually the results of one metric scaling have already been presented: the introductory example at the beginning of the chapter was analyzed using metric procedures. Hence a second example is really not necessary. However, it is worth pointing out that I also analyzed the Blurton Jones data using a nonmetric procedure and obtained virtually the same configuration. This will normally be the case, and only when the relation between the dissimilarities and the distances is markedly nonlinear will there be much difference in the solutions obtained.

The other common situation where a difference between metric and nonmetric solutions may be considerable is when the number of objects is small. As noted in Section 4, Interpretation and Limitations, nonmetric scaling is performed at the user's peril if the order of the matrix is small. In such cases, metric procedures are to be preferred. However, after a metric analysis is completed, it is a good idea to plot the distances against the dissimilarities, or the residuals against the distances in order to see whether some nonlinear transformation of the raw data would be beneficial.

An INDSCAL Application

Robertson (1976) has recently used the weighted Euclidean model to examine individual differences in sexual discrimination by Siamese fighting fish (*Betta splendens* Regan). In one of her experiments, 12 males were individually presented with 10 different stimuli—live male and female *B. splendens* and eight dummy *B. splendens*. Four of the dummies were intended to depict males and four were females. Each of these four pairs of stylized fish was intended to portray respectively lifelike, aggressive, submissive, and reproductive fish. Robertson recorded 14 behaviors, such as time spent near the stimulus, time spent with raised opercula, number of approaches made, and number of bites to the stimulus. Using these data, she computed, for each individual *B.*

splendens male, a Spearman rank correlation matrix whose entries were taken as a measure of overall similarity of behaviors observed when the fish was presented with stimulus i and, at another time, stimulus j.

The twelve 10×10 data matrices were analyzed by INDSCAL, and the group stimulus space for two of the dimensions obtained is shown in Figure 4. The corresponding subject weight space is shown in Figure 5.

Provisional interpretation of the dimensions is fairly straightforward: it seems that two criteria that may be used to discriminate are (a) whether the target is male or female, and (b) whether or not the target is a displaying male. Robertson suggests that the basis for the simple male-female discrimination is fin length and the presence or absence of body pattern. Whether or not a stimulus is perceived as a displaying male depends on the darkness of the body coloration and the length of fins. It is clear that these two dimensions are not independent and, in fact, Robertson did obtain some subject weights that were negative but small—an outcome that is often associated with correlated dimensions. In Figure 5 these small negative weights have been set to zero.

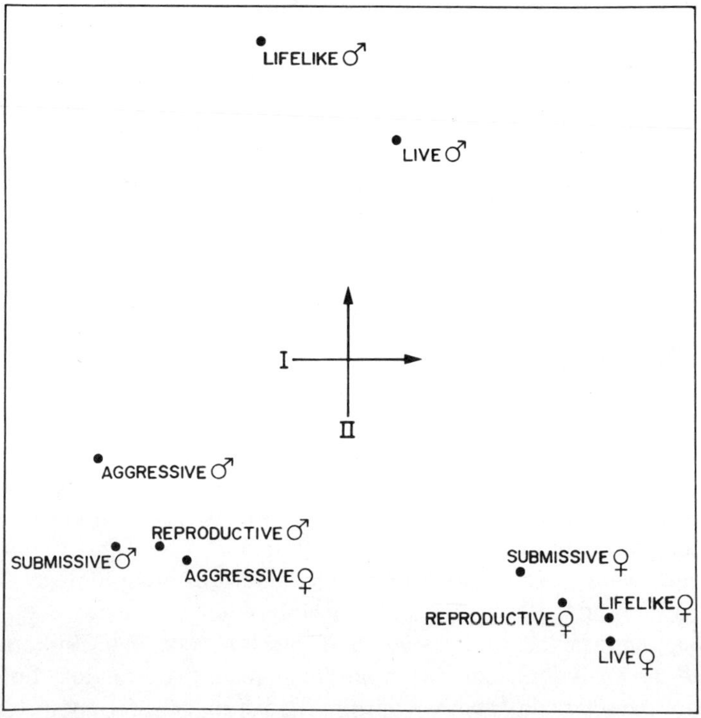

Figure 4. Two INDSCAL dimensions from Robertson (1976).

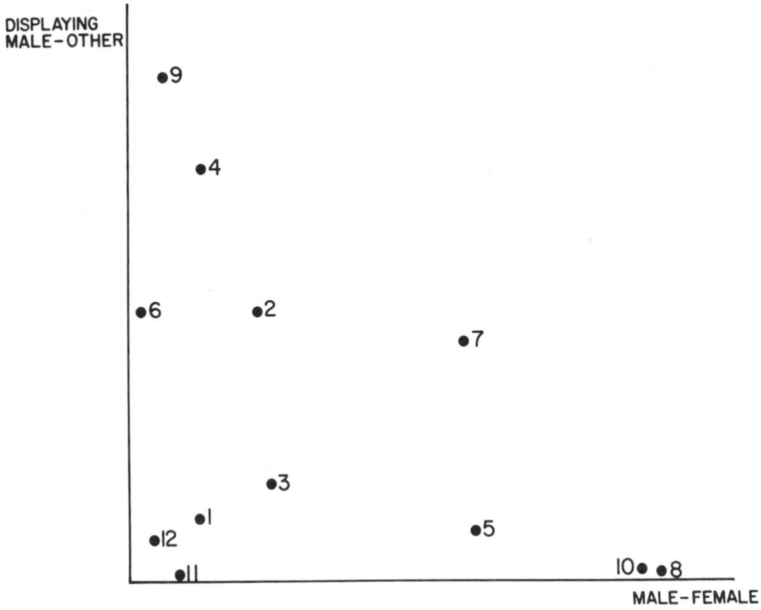

Figure 5. The weight space from Robertson (1976).

The subject space shows three or four groups: those using fin length and body pattern (5, 10, and 8), those using fin length and darkness of coloration (2, 6, 4, and 9), a single fish using both (7), and four fish using neither criterion (1, 3, 11, and 12). These last four may have been using different criteria, or indeed their behavior may not have changed much in the presence of the different target stimuli. Consequently their particular individual similarity matrices are not well described by the INDSCAL solution.

Robertson (1976) gives a more extended interpretation relating the scaling solution to territorial and reproductive aspects of the behavior of *B. splendens*. The foregoing discussion only briefly summarizes the major features of the scaling analysis.

A Nonmetric Analysis

In order to illustrate some of the features of nonmetric algorithms a synthetic example is used. This also shows, in miniature form, how a Monte Carlo experiment (see Chapter 11) is performed. First a configuration was constructed: often this is randomly determined in a Monte

Carlo experiment, but in this instance a cross consisting of 13 points was used. Each arm of the cross was one unit long. The dissimilarities were computed from

$$\delta_{ij} = (d_{ij}^{e})^3$$

where d_{ij} is the errorless distance between points i and j and d_{ij}^{e} is an error perturbed value obtained from d_{ij} by sampling from a lognormal error distribution based on d_{ij}. Specifically log d_{ij}^{e} is a sample from a normal distribution with mean equal to log d_{ij} and constant variance σ^2. The value of σ was arbitrarily set equal to 0.1. This error model is discussed in Ramsay (1977). In addition, a rather extreme power transformation has been applied to the error containing distances—namely the cube of the d_{ij}^{e}. Also, one of the larger dissimilarities δ_{17} has been divided by 10 to simulate a keypunching error—a misplaced decimal point, or improper alignment in the assigned field.

The TORSCA (Young 1968) program was used to analyze the dissimilarity matrix in two dimensions, and the configuration is shown in Figure 6. As can be seen, the cross has been fairly well recovered. The form of the nonlinear transformation used to distort the original distances is clearly apparent in Figure 7, as is the increase in variance of the residuals with increasing distance—a consequence of lognormal errors. The indicated outlier is actually d_{17}, and this would normally indicate some kind of problem with this data point such as the technical error which was simulated.

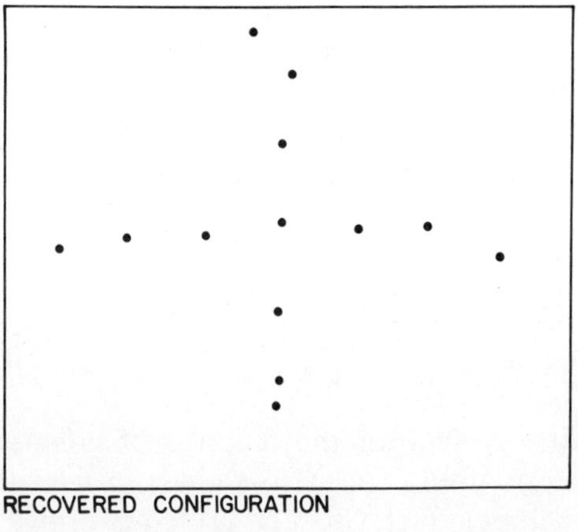

Figure 6. The two-dimensional solution for the synthetic data example.

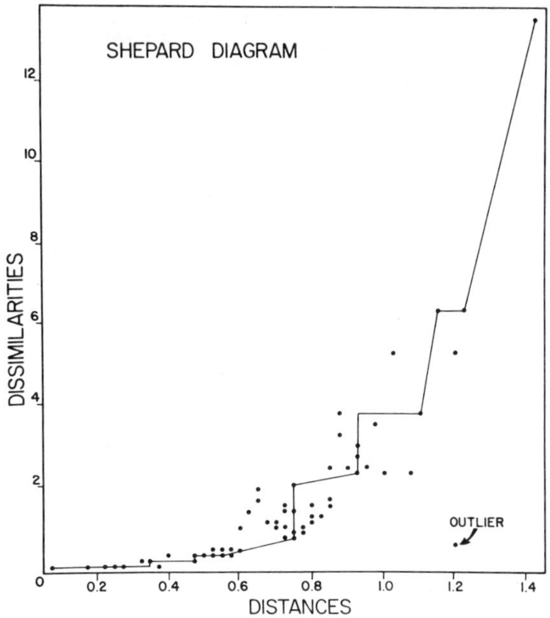

Figure 7. The relationship between the dissimilarities and the recovered distances for the synthetic data example.

Two ALSCAL Analyses

We shall deal with the first of these rather quickly since this example uses ALSCAL in the weighted Euclidean mode. The novel feature is that instead of being interval or ratio the data are categorical—an even weaker level of measurement than ordinal!

The data concern sleeping groups formed by a troop of vervet monkeys on seven different nights and are drawn from Struhsaker (1967) and Cohen (1970). The nights chosen were a subset of the reported data—4 February, 24 February, 31 March, 6 April, 30 April, 11 May, and 1 June, all in 1964. Each evening the monkeys divided into sleeping groups that spent the night separated from one another by at least one break in the tree canopy. The binary matrix in Table 3 records the grouping of 30 April: a one indicates that row monkey *i* was in the same group as column monkey *j*, and a zero means that they were in separate groups on that evening. The six other data matrices (not shown) are similar, but reflect different groupings. Table 4 gives a brief description of the 16 animals involved. There is a small problem in that II left the

Table 3. Vervet Monkey Data for 30 April 1964.

	I	II	III	IV	V	VI	VII	VIII	IX	X	XI	XII	XIII	XIV	XV	XVI
I	—	0	0	0	0	0	0	0	0	0	1	1	0	0	1	0
II	0	—	0	0	0	0	0	0	0	0	0	0	0	0	0	0
III	0	0	—	0	0	0	0	0	0	0	0	0	0	0	0	0
IV	0	0	0	—	0	0	0	0	1	1	0	0	1	1	0	0
V	0	0	0	0	—	1	1	1	0	0	0	0	1	0	0	1
VI	0	0	0	0	1	—	1	1	0	0	0	0	1	0	0	1
VII	0	0	0	0	1	1	—	1	0	0	0	0	1	0	0	1
VIII	0	0	0	0	1	1	1	—	0	0	0	0	1	0	0	1
IX	0	0	0	1	0	0	0	0	—	1	0	0	1	1	0	0
X	0	0	0	1	0	0	0	0	1	—	0	0	1	1	0	0
XI	1	0	0	0	0	0	0	0	0	0	—	1	0	0	0	1
XII	1	0	0	0	0	0	0	0	0	0	1	—	0	0	0	1
XIII	0	0	0	1	0	0	0	0	1	1	0	0	—	1	0	0
XIV	0	0	0	1	0	0	0	0	1	1	0	0	1	—	0	0
XV	1	0	0	0	0	0	0	0	0	0	1	1	0	0	—	0
XVI	0	0	0	0	1	1	1	1	0	0	0	0	1	0	0	—

Table 4. Descriptions of the Individual Vervet Monkeys

I	Adult male
II	Older adult male
III	Adult male
IV	Adult female
V	Juvenile male
VI	Adult female
VII	Young juvenile female
VIII	Young juvenile female
IX	Young juvenile female
X	Juvenile female
XI	Subadult female
XII	Adult female
XIII	Two young indistinguishable juvenile males
XIV	Infant male (son of IV)
XV	Infant female (daughter of XII)
XVI	Infant male (son of VI)

group in mid-April, however this is simply dealt with by having zeros in row 2 and column 2 of the relevant matrices. Also XIII represents two young indistinguishable males—they have been treated as the same animal.

Probably three or four dimensions are needed to account for these data, however only the first two dimensions, and the weights, are displayed in Figure 8. In the common space it is possible to see the stable groupings that have formed, independent of the variations observed from night to night. For example, it is clear that VI, VII, VIII, and XVI form a strong group—this is a mother-infant pair, plus two young juvenile females. The members of the other mother-infant pairs, (XII, XV) and (IV, XIV), are also located close to each other in the space. The adult male III is identified as being something of an outcast, and indeed, inspection of the original data shows that he frequently slept alone, or switched from one group to another. The two young juvenile females IX and X seem to have formed a close liaison, but they are not strongly connected to any other group. Adult males I and II have not formed any strong bonds.

The dimensions of this space are not susceptible to easy naming, even if the first four are considered. What they represent is the change in grouping across nights. Thus the weight space can be used to determine on which nights similar groupings occurred. Figure 9 is a plot of the nights in the first two dimensions of the weight space. On the first, sixth, and seventh nights the sleeping groups were similar; and on the third and

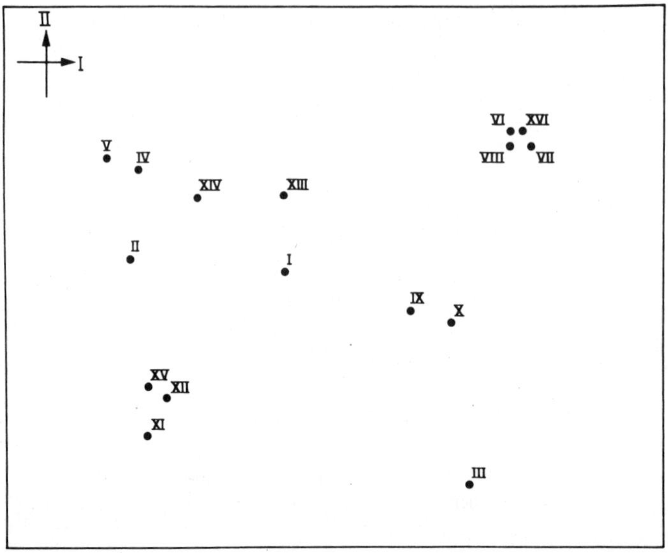

Figure 8. The first two dimensions of an ALSCAL analysis of the vervet monkey data.

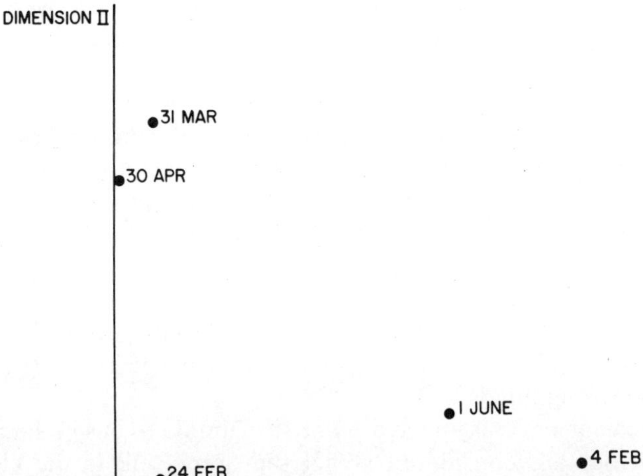

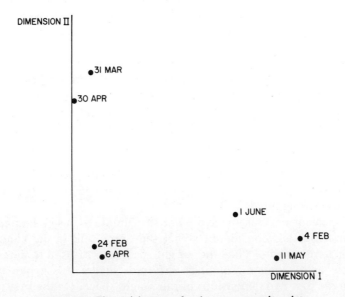

Figure 9. The weight space for the vervet monkey data.

fifth nights other different groupings were formed. The groupings of the second and fourth nights are not well accounted for by this two dimensional representation: the addition of a third, and possibly a fourth, dimension is required.

The foregoing summary interpretation, and a possible extension based on more dimensions and including data matrices from more nights, might possibly be integrated with a discussion of seasonal variation in reproductive behavior. This would be a rather large task, however, and will not be attempted here.

The second ALSCAL analysis is of the Blurton Jones data, and the asymmetric option has been used. The space of behaviors and the weight space are shown in Figure 10. Remember that the solution space is not rotationally invariant so the dimensions and the weights may be directly interpretable. I have attempted such a tentative description.

The first dimension could be labeled Approach-Withdraw and the second dimension Undirected-Directed. These are, of course, merely subjective descriptions. However, behaviors on the left of the configura-

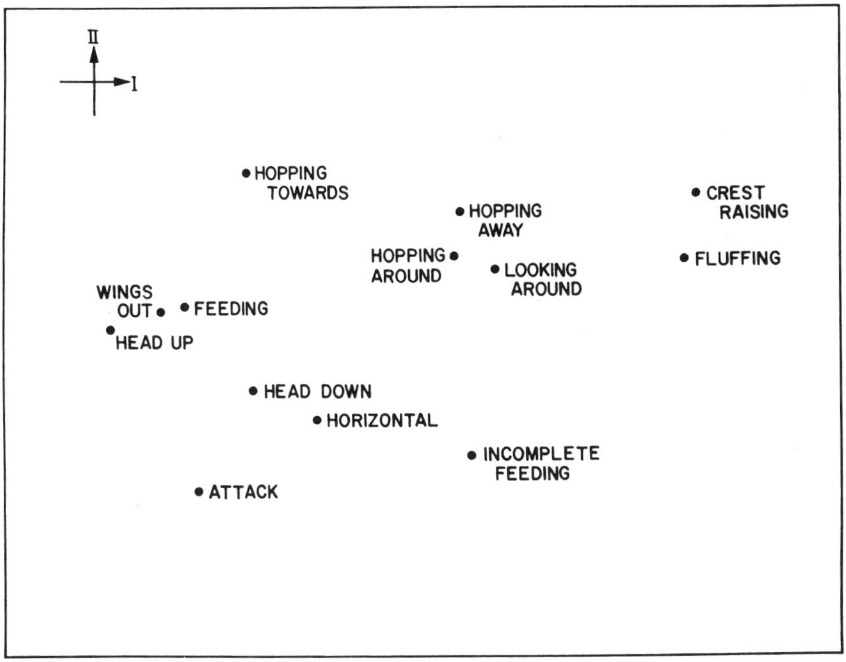

Figure 10. The first two dimensions of the asymmetric ALSCAL analysis of the Blurton Jones data.

tion all seem to involve an approach tendency whereas those on the right do not, or involve active withdrawal. Likewise, the vertical dimension has the more "focused" or directed behaviors at the bottom. It will be noted that this configuration is different in detail from that in Figure 1, but the same general arrangement persists (ignoring the reflection and rotation). The differences result because of two changes in the analysis procedure: (a) the complete asymmetric matrix has been used, and (b) the measurement level has been assumed to be ordinal.

The weight space is given in Figure 11. It can be seen that only six behaviors have large weights on the dimensions. Wings out, attack, and head down are weighted heavily on the Approach-Withdraw dimension, and crest raising, fluffing, and incomplete feeding weight the Undirected-Directed dimension more strongly. The other seven behaviors have small weights that do not strongly favor either dimension.

Two idiosyncratic spaces are shown in Figure 12 for the attack and crest-raising behaviors. The first weights the Approach-Withdraw dimension heavily, whereas the second weights the Undirected-Directed dimension. The distances in these spaces reflect the relative likelihood of other behaviors following attack and crest raising respectively. Thus

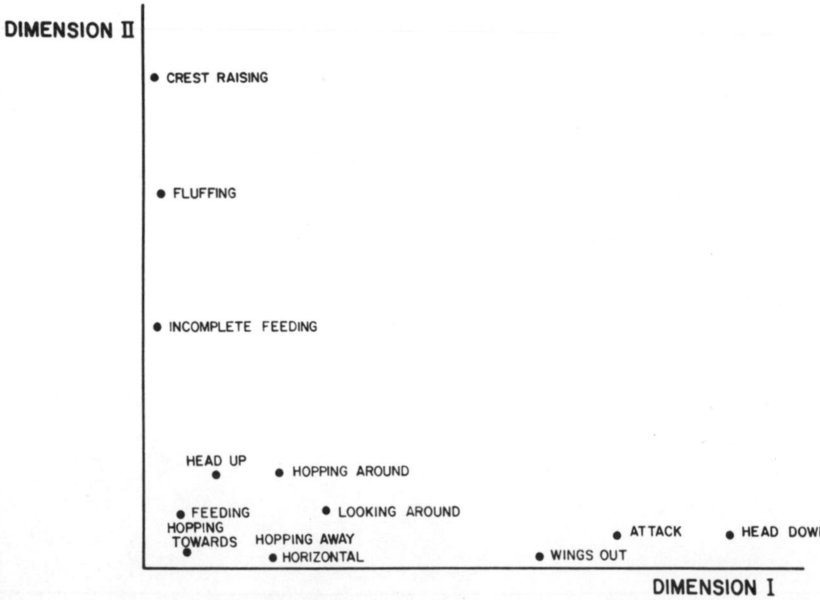

Figure 11. The stimulus weight space from the asymmetric ALSCAL analysis of the Blurton Jones data.

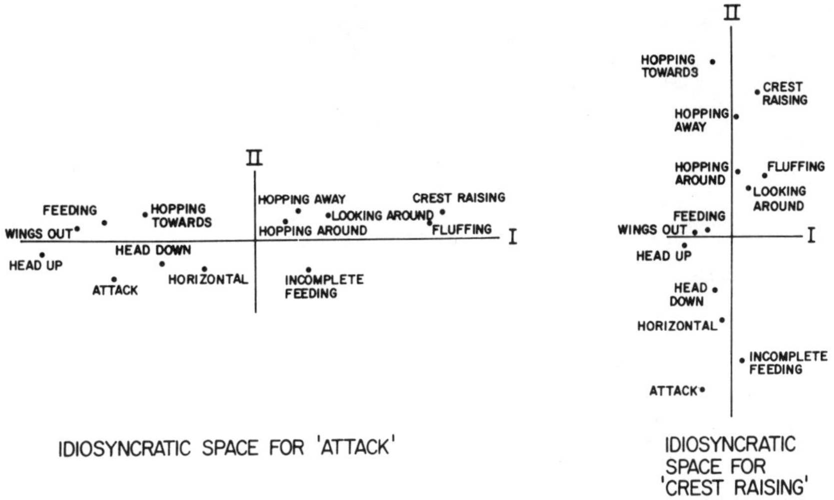

Figure 12. Two idiosyncratic spaces from the asymmetric ALSCAL analysis of the Blurton Jones data.

it is seen that feeding and head down are likely to follow attack, but looking around, crest raising and fluffing are not. Similarly fluffing, hopping toward, and hopping away are related to crest raising, but not head down, attack, or incomplete feeding.

A Bayesian Example

Ramsay's (1977) paper gives examples using the maximum likelihood procedure for both the weighted and unweighted models. The solutions are similar to those that might be obtained using metric or nonmetric scaling, or INDSCAL. Consequently, although I think that there are definite benefits to be obtained using the Ramsay approach, further examples of the same sort would be redundant. Therefore, the analysis presented here was obtained by using the Bayesian version of the program. The data were collected and analyzed by Ramsay and represent the judged similarity of outdoor sports by several subjects.

The common solution is shown in Figure 13. The most interesting feature of this plot is the inclusion of the Bayesian estimated marginal and conditional 95 percent credibility regions. These are analogous to confidence regions; thus the uncertainty in locating each point may be assessed. The conditional region represents the uncertainty in the location of a point given that the other points are fixed, whereas the

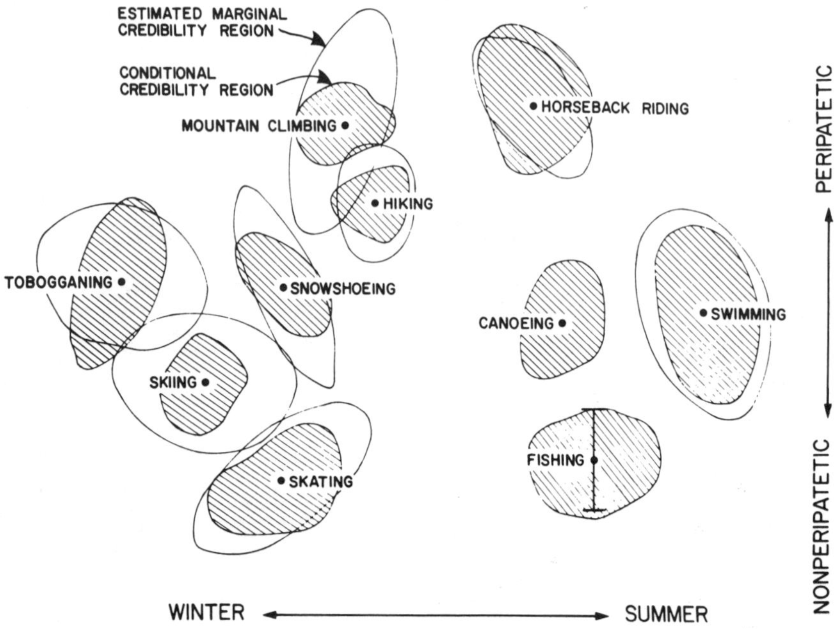

Figure 13. A two-dimensional representation of several outdoor sporting activities (After Ramsay 1977.)

marginal regions are simultaneously constructed. Actually due to the fact that insufficient degrees of freedom are available for estimating all regions, this has not been done for canoeing and only for one coordinate dimension with fishing. Of course, the analysis could be repeated with any two other stimuli replacing these two, if it is desired to obtain an idea of the size of these missing regions.

This method of assessing uncertainty in the location of points is not available in any other scaling procedure. The idea is similar to the confidence ellipsoids described in Gates, Powelson, and Berger (1974), and Seal (1964) for canonical analysis. It is my feeling that data analysts will make increasing use of such methods of determining the possible errors in locating the points.

4. INTERPRETATION AND LIMITATIONS

Interpretation

In the examples of this chapter, little effort has been made to interpret exhaustively the results of the analyses mainly because these illustrations were designed to display the capabilities of the algorithms. The serious user of any scaling procedure is likely to want to be much more detailed and comprehensive in his explanation of the results. Frequently, naming or interpretation of the discovered dimensions will be desired, and often it will be necessary to relate the obtained configuration to external data or other analyses.

Rotation of a configuration is frequently a prelude to interpretation. At least in simple scaling the orientation of the axes is indeterminate, as is the location of the origin. This is similar to the situation that prevails in principal components analysis and some varieties of factor analysis. Consequently, there is no reason why an experimenter should not use graphical or analytical rotation techniques in order to obtain a more pleasing orientation of axes. A popular criterion is that of "simple structure," and often the Varimax rotation procedure (see Chapter 8) is used as an approximation to that criterion. However, other analytic methods can be used. Another common device is to rotate to a hypothesized target configuration; there are many programs for doing this (e.g., Cliff 1966; Schönemann and Carroll 1970). This kind of Procrustean transformation of the solution is often useful in indicating to the user, in an informal way, whether his analysis is any good.

Weighted Euclidean models yield solutions that are not rotationally invariant. However, there is no reason why one cannot, or should not, rotate an INDSCAL group stimulus space, particularly if the rotation is small. It is true that the implied loss function will consequently increase, and the relationship of the weights to the configuration will change, but if it seems to make sense, in relation to the data or substantive considerations, by all means rotate.

Once the preferred orientation is established, the temptation is great to name the dimensions. This can always be done successfully! It is a happy feature of the human mind that structure or meaning can be divined in any spatial distribution of objects—no matter how random. Witness the constellations of the ancient astronomers, or the imaginative interpretations of cloud shapes by children, and at least two Shakespearean characters. Or, in a more sobering vein, consider how many flights of fancy have appeared in our scientific journals following the results of a factor analysis or multidimensional scaling. It is my opinion that the naming of dimensions, or interpretation of directions in the space, should not be done unless either (a) the interpretation is so obvious as to be uncontroversial, or (b) recourse has been to some kind of external analysis. Situations of the first kind sometimes turn up, especially where there are well known underlying physical or biological correlates. More often than not, however, an investigator will have to support his description of the results by relating the coordinates, dimension by dimension, to other variables. This can be done by use of multiple regression or canonical correlation techniques. A good description is given in Kruskal and Wish (1977) and examples of such analyses can be found in Carroll and Wish (1974) and Rosenberg and others (1968).

The question of noise in the data is not unimportant. Sometimes the error level can be so high that the data are essentially random. Scaling programs will produce results as easily with random data as with good data, although the configurations will be just as random as the data. In such cases *any* interpretation will constitute an addition to the genre of science fiction. Using the Monte Carlo results of Klahr (1969), Stenson and Knoll (1969), or Spence and Ogilvie (1973), may help experimenters to tell the difference.

Finally, it may be that the dimensions are not readily interpretable. It is possible that the best discussion of the configuration may be in terms of the distances between the points; for example, interpretable clusterings may be found in low dimensional representations, or the points may be arranged in some regular fashion, demanding interpretation of the arrangement rather than the dimensions (although see the discussion on degeneracy before you are tempted to do this). An excellent discussion of the kinds of structures that might appear, and a review of empirical exemplars, is to be found in Degerman (1972).

Dimensionality

Before any rotation or interpretation takes place, some consideration of the number of dimensions to be retained is required. Of course, the two issues are not independent, and decisions with regard to one will often affect the other. Clearly one should have as few dimensions as possible: both parsimony and visualizability are better served when the solution is of low dimensionality. Nevertheless this should not be an overriding concern—certainly, to dismiss solutions with dimensionalities greater than three, as some authors have advocated, is somewhat foolish.

The principal method in use for deciding the appropriate number of dimensions is based on the goodness of fit index. If the addition of an extra dimension does not appreciably improve the fit then the lower dimensional solution should be accepted. Substantial improvements in goodness of fit suggest the addition of dimensions. Thus, with most scaling programs, it is necessary to examine stress-like measures, or correlation statistics, as successive dimensions are added: what should be looked for is some kind of "elbow." This is a quite familiar exercise to old style factor analysts, some of whom are still living, and is called "root staring."

Of course, sometimes the elbow is not there, or is hard to see. Consequently, at least for nonmetric scaling, some procedures based on Monte Carlo experimentation may be of help. Isaac and Poor (1974) describe a simple method based on a comparison with random data results, and Spence and Graef (1974) have written a computer program, based on extensive simulations of experimental situations, that will automatically determine the appropriate dimensionality and also give an indication of the amount of noise, or error, in the original data.

Some authors have expressed misgivings about the use of Monte Carlo based procedures to aid in the determination of the appropriate dimensionality. Shepard (1974), for example, believes that such studies are invalid because of the possibility of a large number of local minimum solutions having been obtained. As explained in Spence (1974), this criticism is simply not valid since most investigators have gone to considerable trouble to ensure that this source of bias is *not* present. Shepard's other point is that Monte Carlo techniques disregard considerations of stability and substantive interpretability. The truth is that procedures based on simulation are neutral with respect to such issues; further, I doubt that any of those who have developed such aids would deny the importance of the collateral criteria that Shepard mentions. Consequently, this criticism is just not relevant to the usefulness of these procedures as another source of helpful information.

Shepard also claims that there is a widespread tendency to retain more dimensions than are really necessary, citing a study of Levelt, Van de Geer, and Plomp (1966) as an example. Shepard (1974) shows that their data probably need no more than one dimension, and he is almost certainly correct. I reanalyzed these data, using TORSCA, and then used the program M-SPACE developed by Spence and Graef (1974) and the criterion of Isaac and Poor (1974)—both clearly indicated only one dimension was necessary. As an interesting aside, Shepard's reproduction of the data (his Figure 3) perpetuates an error in the originally reported similarity matrix: the entry in the fifth row and third column must either be 13 or 23, and not 73. I detected this clerical error when I inspected the Shepard diagram output from TORSCA—the residual turns out to be implausibly large, and a rereading of Levelt and others (1966) confirms that 32 was the maximum possible score. Shepard diagrams can be useful!

I have performed several similar analyses on other sets of data that exhibit curved structures in spaces of two or three dimensions, like the ones Shepard describes. In all cases, the M-SPACE program indicated that only one dimension was necessary. Recently, in a biological context, Miller (1975) revised his estimate of dimensionality downwards as the result of using M-SPACE. Thus use of these techniques seems to counteract the tendency to extract more dimensions than necessary.

Robustness

Without going too deeply into algorithmic details, the most important consideration is probably robustness against suboptimal solutions. Most scaling algorithms are fairly robust in most situations. However suboptimal solutions do occasionally occur, especially in one dimension and also in non-Euclidean metrics, and the user should be constantly aware of this possibility. A suboptimal solution results when an algorithm converges to a local minimum or is in the area of a local minimum and cannot escape despite being allowed a large number of iterations to do so. The work of Spence (1972) and Lingoes and Roskam (1973) has shown that, of the simple nonmetric scaling programs, M-D-SCAL is most susceptible to this kind of problem. Consequently, its use is not recommended. Indeed some users, such as Shepard (1974) and Arabie (1973), seem to have developed an unreasonable prejudice against all nonmetric algorithms largely on the basis of their experience with M-D-SCAL alone. However, although there is still some disagreement on this point, the general consensus seems to be that programs that generate reasonably good starting configurations are fairly unlikely to produce seriously suboptimal solutions, and even if they do not converge exactly to the global optimum, they will not be too far off. From a practical point of view, users of TORSCA, SSA-I, MINISSA-I, and KYST will probably not be troubled much by poor solutions if the scaling is done in the Euclidean metric. This is largely a result of the good starting configurations generated by these programs.

Little systematic work has been done on the robustness of weighted Euclidean algorithms. The informal consensus seems to be that they are not as susceptible to such difficulties. The major problem here is the large number of iterations that may be required for convergence—INDSCAL, for example, can be quite slow. Ramsay (1975) and de Leeuw and Pruzansky (1975) have suggested methods of speeding up the process, and the former method has been incorporated in a maximum likelihood scaling program, whereas the latter procedure forms part of the ALSCAL program. Both seem to be more rapid than INDSCAL, but they are still capable of making quite a dent in your computer budget. In any event, as with the nonmetric programs, the use of a rational starting position (such as that of Schönemann 1972) is sensible: the algorithm will be less likely to converge to an undesirable local minimum, and will certainly be faster than if provided with a random start.

Degeneracy

One kind of suboptimal solution that the user must constantly watch for is the degenerate solution. This may occur when the ratio of the number of dissimilarities to the number of coordinates in the solution is quite small—less than about 3 : 1. The two danger situations are (a) when the order of the dissimilarity matrix is small, and (b) when an incomplete data matrix is being scaled. In the first case, the work of Young (1970) and others suggests that if we are scaling in one dimension we should have at least six objects, in two dimensions a minimum of 10 or 11, and in three dimensions 17 or 18. Attempts to perform nonmetric scaling with fewer objects can easily lead to degeneracy. When scaling incomplete matrices the advice of Spence and Domoney (1974) and Spence (1977) should be heeded.

Some signals of degeneracy are easy to spot. If solutions contain sets of points that occupy the same location, or are very close together, they are likely to be degenerate. If the points divide and line up like two opposing armies, or assume any other regular pattern with many of the interpoint distances equal, then the solution should be regarded with suspicion. Examination of the Shepard diagram will often show other signs of this pathology—usually a very large proportion of the fitted $\hat{d}_{ij}$ will be equal, and the "function" will have very few steps. Shepard (1974) has some other observations on the subject of degeneracy, and the reader should consult his paper for useful information on this topic.

Coda

The procedures described in this chapter are very powerful and useful data reduction techniques. Their use may facilitate interpretations that might not be obtainable in any other way. They do, however, have their drawbacks, the most salient of which is complexity—not conceptual but algorithmic—and this leads to potential difficulties for the user. Every scaling solution should be carefully examined and reexamined. If possible, alternative methods of analysis should be employed. If discrepant or unexpected results occur, the user must assume that something has gone wrong and attempt to discover the source of trouble. In this respect, a user who has a good knowledge of how the algorithms work will be very much better off. This will probably change in the future as more knowledge is accumulated regarding the plausibility of models and the efficiency and robustness of algorithms. But at present many of the above mentioned techniques are still experimental, with care and good common sense required for their successful use.

CHAPTER

8

Principal Component Analysis and Factor Analysis

DENNIS F. FREY and RICHARD A. PIMENTEL
California Polytechnic State University

Behavioral methodology, though flowing through divergent historical pathways, has proceeded from examining more general aspects of a phenomenon to more specific or detailed aspects. Early ethologists often devoted enormous amounts of time and energy to developing simple descriptive (sometimes causal) models of functional categories of behavior, such as the models central to "classic-ethology" (Crook 1970) that derived from the gray-lag goose egg retrieval experiments by Lorenz and Tinbergen (1939). Since then the complex division of behavioral activity into finer units emphasizing many facets of total behavior has become a major concern of ethology (see Barlow 1968; Altmann 1974; and Sustare in Chapter 10 of this book for more complete overviews of the area of behavioral units).

More recent examples of this multifactor approach are seen in the study of predation. Holling's (1959) work is significant in attempting to relate predation to several behavioral "components." At a somewhat finer level, in that it involves more subunits of overall behavior, is Beukema's (1968) study of predation in three-spined sticklebacks. There are numerous other examples of this approach in ethology.

The historical procession of a molar (e.g., natural history) focus followed ultimately by more micro approaches (e.g., neurobehavior) can also be equated with this tendency of dissecting behavioral phenomena. Part of the underlying rationale of this process lies in the conviction that it builds more realistic models of a phenomenon. In other words, there will be a more congruent fit of the model to the actual behavioral events that take place if the input is in the form of many small units.

Another general to specific trend can be traced in the evolution of factor analysis as a quantitative methodology. Spearman's (1904) work in proposing a "general factor" in the intelligence probe is the point of departure for tracing the development of factor analysis. Harman (1967) points out that ". . . the majority of mathematical foundations of factor analysis followed in the next twenty years." Intelligence theory eventually became multifactorial (i.e., more specific) in terms of models (Garnett 1919; Thurstone 1931).

It is almost paradoxical that in behavioral studies multivariate statistical analysis, specifically the factor analysis of Spearman, historically preceded univariate statistical analysis. Unfortunately the use of multivariate techniques has remained minimal in ethology. Perhaps the move to univariate statistics was another attempt to turn from the study of general factors to specific variables. Colgan aptly points out in the Prologue of this book that univariate statistics have played a paramount role in bringing at least some quantitative analysis to ethology. It seems, however, that an aura of misguided confidence grew out of the early "univariate era" since we automatically assumed that our variables represented reality when so treated. The appropriateness of variables chosen to represent a phenomenon were often evaluated on the basis of anticipated patterns of differences or similarities within a particular theoretical framework, ecological context, or treatment series. Thus elements of circularity were ever present.

The dissecting approach and the advent of univariate statistics in data analysis probably led researchers to use fewer variables as representative of the behavioral phenomenon under consideration. Hinde (1970) addressed this issue by criticizing the use of single variables in drive concepts. We are in full agreement with such a criticism. Furthermore, the majority of behavioral research during ethology's brief history has been in the analysis of phenomena that cannot be realistically studied with "single variable" approaches (which we contend also includes studies of many variables independently). With few exceptions, quantification of one or even a few variables never adequately represents the "shape" of a behavioral phenomenon. A simple example illustrates our point. Measuring the biting rate of one fish during intraspecific agonistic encounters does not measure or equate with the phenomenon of dominance establishment or even aggressiveness for that matter. In other words, we have known for some time that we should measure a behavioral phenomenon in a variety of ways. One of the salient features of multivariate analytical methods such as principal component and factor analysis is that, when used correctly, they tend to promote "phenomenon realism."

"Factor analysis" is a generic term that encompasses a variety of procedures. An initial appreciation of some of the differences between procedures can be achieved by examining the possibilities at each of three stages in the process of data analysis.

The first stage involves the choice of the form of the variation to be studied. In an R-type factor analysis a correlation matrix (or variance-covariance matrix) of pairs of variables is used. In a Q-type analysis generally a correlation matrix of pairs of sampling units (individuals, areas, etc.) is the input. Aspey and Blankenship (1976) have recently used this latter type of factoring to determine patterns of burrowing in the marine gastropod *Aplysia brasiliana* that are unique to specific classes of individuals. Used in this fashion Q-type procedures become classification tools or operate in much the same way as cluster analysis (see Chapter 5). Analysis based on association among variables (R-type) is the more common form and is illustrated in this chapter.

The second stage involves the determination (mathematical solution) of the minimum number of independent patterns of data variation from the data matrix. This is accomplished by transforming an original dispersion matrix (variance-covariance or correlation), which almost invariably displays intercorrelations, to uncorrelated principal components or factors. If the new variables are exact mathematical transformations (as in principal components analysis), then the components are said to be *defined*.

In the case of other factor analysis methods, the data (i.e., the correlation matrix) are assumed to consist of two or three parts: (1) a theory matrix that directly applies to the phenomenon under study, and (2) a residual matrix containing error or noise. Sometimes so-called "unique factors" specific to each variable are extracted from the theory matrix. This latter source of variation theoretically is independent of the "common factors" remaining in the theory matrix. Since the residual variance supposedly does not contribute to the meaningful intercorrelations of the variables, it is extracted prior to determining the factor solution of the theory or common factor matrix. Attempts to partition the variance of the data require estimates of the redundancy measured by the variables (referred to as *communality*). Kim (1975) points out that the ". . . determination of communalities remains one of the most difficult and ambiguous tasks in factor analysis." Harman (1967) devotes an entire chapter (Chapter 5) to various methodologies and problems of communality estimation. Since partitioning of the theory matrix is defined by some *a priori* judgment of the investigator in classical factor analysis, the factors resulting from the theory matrix are *inferred* factors. A number of different factoring procedures have been developed, each with different underlying objectives and each with its particular group of supporters. (See Harman 1967, Chapter 6, and Kim 1975 for a good overview of this area.)

The third stage that sometimes follows in this process is the transformation of the initial component or factor solution into some terminal reference solution. Kim (1975) notes that ". . . the exact configuration of the factor structure is not unique; one factor solution can be transformed into another without violating the basic assumptions or mathematical properties of a given solution." (Also see Harris 1975 for an excellent discussion and examples on this point.) This procedure is referred to as factor rotation and has historical roots in Thurstone's (1935) search for "simple structure" solutions. "Simple structure" is a somewhat nebulous phrase, and various authors have summarily enumerated Thurstone's intent in operational terms (e.g., Harman 1967, p. 98; Overall and Klett 1972, p. 118). The basic choices at this point are whether to choose a method that forces the factors to remain uncorrelated (orthogonal) or allows them to assume varying degrees of correlation (oblique).

Orthogonal rotations are more widely used probably because they facilitate an easier interpretation of the output. Intuitively, however, one might argue that oblique solutions provide a more realistic description of the data. Those interested in an ethological research example of oblique rotation procedures will find Lefebvre and Joly's (1976) work worth reading.

Suppose one is convinced that a particular set of p-variables should be measured on each sampling unit to account adequately for or outline the "shape" of the behavioral phenomenon. This results in our dealing with the following population parameters: p means, p variances, and $(p^2 - p)/2$ covariances or correlations. As p increases the number of parameters become so numerous as to be very difficult to interpret. If $p = 14$, for example, 119 parameters result. In such cases it virtually becomes mandatory that the number of parameters be reduced or their interpretation simplified. One often-cited use of factor analysis is to reduce the "analytical space" and parameters to be interpreted. Kim (1975) notes that the ". . . single most distinctive characteristic of factor analysis (and principal component analysis) is its data-reduction capability." These techniques eliminate $(p^2 - p)/2$ covariances or intercorrelations. For example, principal component analysis is a method whereby a set of variables is transformed into a new set of composite variables (components) that are uncorrelated (orthogonal) to one another. The ability of the methods to remove correlations between variables is often the sole basis for their use. Too frequently the mere fact that several variables are involved in a single phenomenon will bring about meaningless apparent relationships. For this reason, the worker might well be suspicious of any correlations in poorly understood behavioral patterns and thus prefer to deal with uncorrelated variables.

Three major functions or specific uses apply to principal component analysis or factor analysis. The first function deals with the use of components or factors as descriptive statistics. In principal component analysis, the only method to be developed in some detail here, as many independent patterns of variation are extracted as there are variables. It is possible that all patterns of variation will be biologically meaningful and the analytical space is not reduced; however, this should prove rare in ethology. In the method of component analysis, the first pattern of variation extracts the maximum variance along a single dimension among individuals in the sample. Each successive pattern of variation also extracts the maximum variance remaining but is subject to the constraint of being independent of all previous patterns of variation. Since these patterns of variation are inherent properties of the phenomenon being studied, they generally amount to what the ethologist is seeking from his sample. The number of components or factors (i.e., dimensions of the data) considered interpretable are usually fewer than the number of original variables and this results in what Harman (1968) refers to as "scientific parsimony."

The generation of new hypotheses concerning the behavioral phenomenon is a second major function of principal component analysis or factor analysis (Eysenck 1953). This is sometimes referred to as a heuristic property of multivariate methods.

A third use of these procedures is in the generation of indices to be used as variables in further analysis. Education psychologists and workers dealing with psychiatric disorders have been in the vanguard in using derived or inferential factor scores in subsequent data manipulations (see Cooley and Lohnes 1971 and Overall and Klett 1972, for example). We illustrate each of these uses in Section 4 of this chapter.

This chapter deals primarily with principal component analysis. Major attention is given to defined components of initial solutions. Little is said of rotation of component axes or classic factor analysis. This is done for three reasons. First, etholgoists rarely have sufficient *a priori* or even *a posteriori* knowledge to create the necessary theoretical and residual matrices required for factor model analysis. Second, other factor analysis procedures are not likely to produce factors leading to interpretations appreciably different from components as presented here. Finally, factors, rotated factors, and so on are generally in a form less meaningful than components (Harris 1975).

The following sources should prove helpful to those seeking additional information in the area of classic factor model analysis: Cooley and Lohnes (1971), Harman (1967, 1968), Overall and Klett (1972), Harris (1975), and Kim (1975). Further information relative to rotation of axes and seeking "simple structure" can be found in Cooley and Lohnes (1971), Harris (1975), and Kim (1975).

This chapter is organized into five sections. The first lists the assumptions necessary for principal component analysis as presented here. The second provides the methodology of component analysis. The third provides interpretive aids (certain statistics generally helpful in unraveling the relationships represented by components). The fourth shows how components and their interpretive aids can be applied in an ethological study, specifically fighting behavior in a fish. The final section presents limitations of the method of principal component analysis.

We wish to call the reader's attention to the excellent introduction to matrix algebra in Cooley and Lohnes (1971). Matrix algebra is the language of multivariate analysis so the reader must have some familiarity with the subject. Those completely new to the subject should be aware that it is applied as a shorthand to indicate mathematical models and operations that otherwise would require most complex symbolism. Many of the operations are familiar arithmetic procedures. The new procedures that must be learned are the nature of an inverse, determinant, and characteristic equation. The brief section in Cooley and Lohnes is sufficient.

1. ASSUMPTIONS

Multivariate techniques in general appear to be very robust to departures from assumptions (see Harris 1975, Chapter 8 for a detailed discussion). Specifically, the assumptions for principal component analysis are as follows:

1. Multinormal (i.e., multivariate normal) distribution of the population being sampled. This is necessary if any tests of significance (e.g., significance of eigenvalues) are made. All such tests involve obtaining critical values from statistical tables, and the latter are based upon normal distributions.
2. Components can be described in terms of linear relationships. Formulas basic to all component analyses derive component scores y_i from raw data vectors x_i or from standardized normal data vectors z_i by linear transformations (see model below).
3. Random sampling is required so that samples are representative of the populations from which they are drawn.

2. METHOD

Recall that our primary purpose is to disclose independent patterns of variation in terms of variables (or individuals, areas, etc. in the case of Q-type analysis) that also are independent of one another. In more mathematical terms, we wish to eliminate correlations by transforming original measurements to new uncorrelated variables. We can appreciate the method that will be used by considering a bivariate case.

Consider a possible distribution of two variables, X_1 and X_2, in terms of regression (Figure 1*A*). The line (l_1) in Figure 1*A* represents the line of regression

$$X_2 = \alpha + \beta X_1 \tag{1}$$

where α is the y intercept or value of X_2 when X_1 is zero, and β is the slope coefficient of the line of regression. One equation for the correlation coefficient ρ is

$$\rho = \beta \frac{\sigma_{X_1}}{\sigma_{X_2}} \tag{2}$$

where σ_{X_1} and σ_{X_2} are the standard deviations of X_1 and X_2 respectively.

A first step in transforming original data vectors X to component score vectors y is to transform raw scores to deviation scores

$$x = X - \bar{X} \tag{3}$$

By subtracting the mean vector, $\bar{X}$, from each data vector, X, the new mean vector $\bar{x} = 0$ has zero for all variables (i.e., is a null vector). This results in coordinate axes x_1 and x_2 in which the mean for both axes is their origin (Figure 1*B*). This first step is for the purpose of placing the origin of the later derived components axes y_1 and y_2 at the means of y_1 and y_2. Transformation to deviation scores changes the regression model to

$$x_2 = \beta x_1 \tag{4}$$

since $x_2 = 0$ when $x_1 = 0$. The values for β and ρ will remain unchanged since subtracting a constant from each observation does not change the values of variance or covariance (i.e., $\sigma_{x_1} = \sigma_{X_1}$ and $\sigma_{x_2} = \sigma_{X_2}$).

The purpose of the second step of the transformation from deviation axes x_1 and x_2 to component axes y_1 and y_2 is to eliminate correlation between the variables. The way this is done is analogous to making the slope coefficient β equal to zero. It is possible to do this by creating new axes, y_1 and y_2, one axis parallel to the line of regression and the other perpendicular to it. In addition, for later convenience, let us retain μ_{y_1} and μ_{y_2}, the transformed means, at the origin of the new reference axes (see Figure 1*B*). In essence we have done the following: (1) The relationship between original data points is maintained. (2) The location of the final mean vector is the same as that for the deviation scores (i.e., at the origin). (3) Original raw data variable axes were rotated to new positions where there is no correlation between variables.

Mathematically we have proposed

$$y_{1i} = \cos\theta_1(X_{1i} - \mu_{x_1}) + \cos\theta_2(X_{2i} - \mu_{x_2}) \tag{5}$$

and

$$y_{2i} = -\cos\theta_2(X_{1i} - \mu_{x_1}) + \cos\theta_1(X_{2i} - \mu_{x_2}) \tag{6}$$

to transform each set of original data coordinates $(\mathbf{X}_{1i}, \mathbf{X}_{2i})$ of a sample of

N individuals to a new coordinate reference pair of ($\mathbf{y}_{1i}$, $\mathbf{y}_{2i}$) transformed points (termed component scores) by the simple procedure of first subtracting the mean from each original **X** observation and then rotating the resulting coordinate axes through the angle θ_1 from the $\mathbf{X}_1$ axis and θ_2 from the original $\mathbf{X}_2$ axis. However, $\theta_2 = 90° - \theta_1$ so rotation is from one set of axes 90 degrees apart to another pair 90 degrees apart. Therefore, a fixed set of axes are moved to an origin via ($\mathbf{X}_{1i} - \mu_{x1}$) and ($\mathbf{X}_{2i} - \mu_{x2}$) and then rotated θ_1 degrees from the original $\mathbf{X}_1$ axis which amounts to θ_2 degrees from the original $\mathbf{X}_2$ axis (Figure 1*C*).

We have developed a model for accomplishing our purposes of independent variables **y** from correlated variables **X**. The statistical form of the model in matrix algebra form is

$$\mathbf{y}_i = \mathbf{A}'(\mathbf{X}_i - \bar{\mathbf{X}}) = \mathbf{A}'\mathbf{x}_i \tag{7}$$

where $\mathbf{x}_i$ are data vectors of deviations of multiple measurements from their means and **A** is a matrix of vectors. Each vector **a** of **A** consists of a set of coefficients (one coefficient applying to each original variable and each coefficient being the cosine of the angle between a new y axis and an original X or x axis).

The properties of the coefficients a_{lj} (where l is one of the coefficients of the p-variables and j is one of the p-components) of these vectors (principal components) are very important. From this point i refers to any of the N individuals in a sample, j and k to components (eigenvectors) and eigenvalues, l to component coefficients or their corresponding raw variables, and p to the number of variables. For any one vector the sum of the product of corresponding elements will be one, and for any two vectors this product will be zero. In terms of the jth and k^{th} component vectors

$$\mathbf{a}_j'\mathbf{a}_j = 1 \text{ or } \mathbf{a}_k'\mathbf{a}_k = 1, \qquad (j = 1, 2, \ldots, p;\ k = 1, 2, \ldots, p) \tag{8}$$

and

$$\mathbf{a}_j'\mathbf{a}_k = 0, \qquad j \neq k \tag{9}$$

The reason for this stems from the fact that if $\mathbf{a}_j'\mathbf{a}_j = 1$ we have the angle between a vector and itself, or zero degrees, in terms of a cosine function. Then, if $\mathbf{a}_j'\mathbf{a}_k' = 0$ (the value of the cosine of an angle of 90 degrees), all vectors (principal components) are 90 degrees apart (orthogonal) and component scores are mathematically independent of one another.

The importance of component scores from different component axes being independent of one another cannot be overemphasized. The method allows consideration of j independent patterns of variation (components), and there are corresponding independent variables and component scores. Note that in the model a component score is a transformed variable, a transformation of an original data vector. Since this is the case, component scores each pertain to facets of the original variables, facets that are often termed size or shape variables.

The model as considered thus far

$$\mathbf{y}_i = \mathbf{A}'\mathbf{x}$$

subject to

$$\mathbf{a}_j'\mathbf{a}_j = 1 \quad \text{and} \quad \mathbf{a}_j'\mathbf{a}_k = 0, \qquad j \neq k$$

allows an infinite possibility of solutions. One further restriction provides for a single solution: each successive solution, although limited to the above restrictions, will account for the maximum variance possible (see Figure 1). The mathematical solution to this model is the *characteristic equation*

$$| \mathbf{S}^2 - \lambda\mathbf{I} | = 0 \tag{10}$$

where $\mathbf{S}^2$ is the variance-covariance matrix of the original raw data and λ is a vector of eigenvalues. There are p eigenvalues, p principal components, p coefficients in a principal component, and p variables in the original data vectors. The quantity I is the identity matrix (i.e., a $p \times p$ matrix with ones along the diagonal from the upper left hand to lower right hand corner and zeros elsewhere). The effect of multiplying an identity matrix by a scalar λ is to place the values of the scalar along the diagonal of the identity matrix. Therefore, $L = \lambda\mathbf{I}$ is a matrix having eigenvalues along the diagonal and zeros off the diagonal and is formed to allow the subtraction in the characteristic equation. The bivariate case is represented by the following relationships

$$\mathbf{L} = \lambda\mathbf{I} = \lambda \begin{bmatrix} 1 & 0 \\ 0 & 1 \end{bmatrix} = \begin{bmatrix} \lambda & 0 \\ 0 & \lambda \end{bmatrix}$$

More important than the mathematics, the characteristic equation transforms the variation of correlated variables, $\mathbf{S}^2$, to the variation of uncorrelated variables, $\mathbf{L}$. It will be found that there are two solutions to λ, λ_1 and λ_2, that is, λ is a scalar variable. The properties of $\mathbf{L}$ are emphasized by the fact that there are variances, λ_1 and λ_2, but zero covariances and hence zero correlations.

Once the p eigenvalues are known, the eigenvector function is used as follows to obtain each of the p component vectors (i.e., the eigenvectors), $\mathbf{a}_j$, one associated with each eigenvalue λ_j from

$$(\mathbf{S}^2 - \lambda_j)\mathbf{a}_j = 0 \tag{11}$$

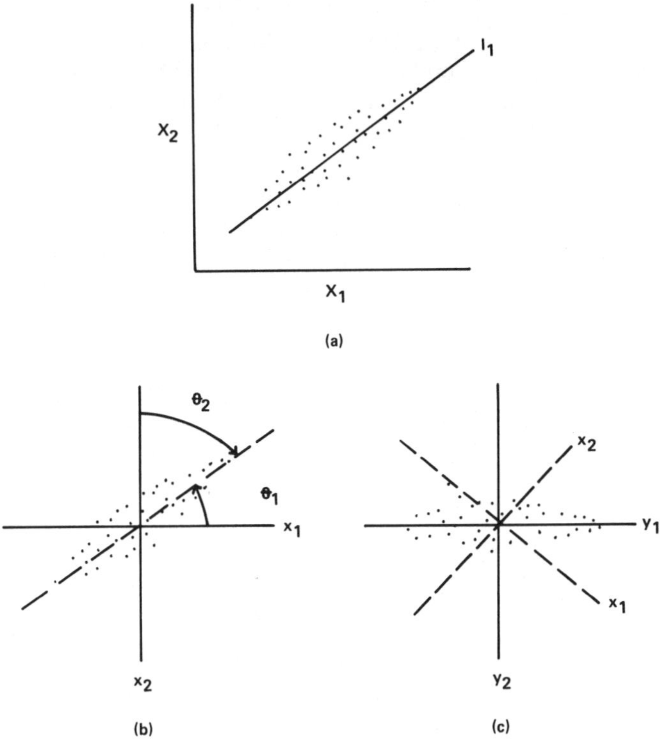

Figure 1. Bivariate geometric representation of the steps involved in transforming raw data axes to component axes. (*A*) Hypothetical raw data points in terms of bivariate axes X_1 and X_2. (*B*) Deviation scores $(X_1 - \bar{X}_1)$ and $(X_2 - \bar{X}_2)$. Data points retain their original orientation but x_1 and x_2 axes are centered on $\bar{X}_1$ and $\bar{X}_2$. (*C*) The data following transformation to component axes. The transformation is accomplished by a rotation counterclockwise of x_1 and x_2 so they conform to y_1 and y_2 respectively.

The characteristic equation and eigenvector function provide the following conditions

$$\mathbf{y}_i = \mathbf{A}'\mathbf{x}$$

$$\bar{\mathbf{y}} = 0$$

$$\mathbf{S}_y{}^2 = \mathbf{A}'\mathbf{S}\mathbf{A} = \mathbf{L} = \lambda\mathbf{I}$$

$$\mathbf{a}_j{}'\mathbf{a}_j = 1$$

$$\mathbf{a}_j{}'\mathbf{a}_k = 0, \qquad j \neq k$$

where $\bar{\mathbf{y}}$ is the mean vector of $\mathbf{y}$, a null vector of zero elements, and $\mathbf{S}_y{}^2$ (variance-covariance matrix of $\mathbf{y}$) is typified by $\lambda_1 > \lambda_2 > \cdots > \lambda_p$.

The above model considers raw data and hence *absolute* variation of the data. This model is appropriate when the various variables are measured in equivalent units, a condition seldom found in behavioral studies. For this reason, it usually is preferable to consider *standardized* variation where the variation of each variable is equal to that of all others. This is accomplished by transforming each variable to a z score (standardized normal scores with mean equal to zero and standard deviation equal to one). In this model, since the variance-covariance matrix of z scores is the correlation matrix of the raw data X the solution is via

$$| \mathbf{R} - \lambda\mathbf{I} | = 0 \tag{12}$$

and again for each eigenvalue, λ_j, there is a unique eigenvector, $\mathbf{v}_j$ —"v" being used rather than "a" to distinguish between Equation 11 and this model

$$(\mathbf{R} - \lambda_j\mathbf{I})\mathbf{v}_j = 0 \tag{13}$$

and the formulae indicating conditions are

$$\begin{aligned} \mathbf{y}_i &= \mathbf{V}'\mathbf{z} \\ \bar{\mathbf{y}} &= 0 \\ \mathbf{S}_y{}^2 &= \mathbf{V}'\mathbf{R}\mathbf{V} = \mathbf{L} = \lambda\mathbf{I} \\ \mathbf{v}_j{}'\mathbf{v}_j &= 1 \\ \mathbf{v}_j{}'\mathbf{v}_k &= 0, \qquad j \neq k \end{aligned}$$

where $\mathbf{V}$ corresponds to $\mathbf{A}$ in the previous model.

3. INTERPRETATION

Interpretation of results of either model are in terms of each component, or pattern of variation, resulting from a stimulus that involves a certain percent variation (absolute or standardized) in the phenomenon. Although the first few patterns of variation might relate to aspects of behavior familiar to the ethologist, frequently many or perhaps even all components will relate to unfamiliar aspects. Also, there is no way of testing the significance of individual coefficients of any eigenvector or of being sure that a small coefficient indicates that the corresponding variable is unimportant. For these reasons, a number of interpretative aids are helpful. These aids apply to both models but for consistency with the example, aids will be discussed in reference to the standardized variation model.

The first aid is component correlations, the product moment correlations between each variable and each component. In the sense that a correlation coefficient is a measure of goodness of fit, component correlations are measures of the goodness of fit of original variables to principal components. The calculation of the correlation of the lth variable with the jth component is

$$r_{lj} = v_{lj} \sqrt{\lambda_j} \tag{14}$$

The second information source in interpretation is the percent variance of each variable involved in each component. This measure indicates the relative influence of a variable from component to component. The calculation for the lth variable of the jth component is

$$s_{lj}^2 = 100\left(\frac{v_{lj}^2 \lambda_j}{p}\right) \tag{15}$$

where p is the number of original variables.

A third useful step in data analysis is Bartlett's test of sphericity or equality of the last $p - k$ eigenvalues by the following relationship

$$\chi^2 = \left[N - k - \frac{2(p-k) + 7 + 2/(p-k)}{6} + \sum_{j=1}^{k}\left(\frac{\bar{\lambda}}{\lambda_j - \bar{\lambda}}\right)^2\right] \times$$

$$[-\ln(\lambda_{k+1}\lambda_{k+2} \cdots \lambda_p) + (p-k)\ln\bar{\lambda}] \tag{16}$$

which approximates the chi-square distribution with $(p - k - 1)(p - k + 2)/2$ degrees of freedom, so $p - k$ must be ≥ 2 (Cooley and Lohnes 1971). In the above N is the sample size, p is the number of measurements on each individual, $(p - k)$ are the number of eigenvectors tested for equality, and $\bar{\lambda}$ is the mean of the last $(p - k)$ eigenvalues. This so-called test of sphericity is applied in sequence from $k = 0$ to $k = p - 1$. The first set of $(p - k)$ variables not deemed significantly different indicates equality of those $(p - k)$ eigenvectors, a condition indicating equality and arbitrary placement of the last $(p - k)$ component axes. Such equal axes define a spheroid and cannot legitimately be interpreted in terms of behavioral meaning. On the other hand, the k significant axes can be interpreted but some may be related to noise in the system. However, in practice, biological meaning frequently ceases prior to significance. Another "rule of thumb" commonly followed by many psychometricians is to interpret components whose eigenvalues are 1.0 or more (Guttman, 1954).

The fourth aid is the use of component scores of the N individuals

$$\mathbf{y}_i = \mathbf{v}'\mathbf{z}_i$$

on each of the p components in an effort to unravel the biological implications of each component. It is also possible to plot component scores of individuals for two or three axes at a time for ordination of individuals in a cluster analysis (see Chapter 5).

Finally, the percent variance of an individual accounted for by each component is sometimes helpful. The total variance of the ith individual in algebraic form is

$$s_i^2 = \sum_{l=1}^{p} \sum_{j=1}^{p} z_{jl}^2 \tag{17}$$

The percent variance of an individual accounted for by each component is

$$F_{ij} = 100 \frac{\sum_{l=1}^{p} v_{lj}^2 z_{jl}^2}{s_{ik}^2} \tag{18}$$

and discloses the relative contribution of each individual to each component. This statistic can be used to learn how many components account for most of the variance of all individuals. If only a few individuals contribute to a seemingly unimportant component, the component might be important or the individuals outliers. Note that both Equations 18 and 15 are percent variances accounted for by each component. However, Equation 18 is the percent variance of an individual data vector and Equation 15 of a single variable. The two will be shown to have different uses in the example.

The foregoing summarizes most of the interpretative aids included in the computer program used as the basis for our example.

4. EXAMPLE

The following data are taken from a study on the fighting behavior of the anabantoid fish, *Trichogaster trichopterus* Pallas. Further details can be found in Frey and Miller (1972). The research attempted to describe some of the dynamics involved in the phenomenon of dominance establishment. The experimental protocol (Experiment II in Frey and Miller 1972) involved taking males from a common stock tank, placing them in adjoining compartments of aquaria, and isolating them physically and visually for 10 days. The fish were then allowed to fight until an operationally defined dominance relationship was established (Frey and Miller 1972, pp. 16–19). Many variables were recorded on each fish during the period of dominance establishment. In the original paper 44 variables were recorded. Twelve of those variables plus four previously unreported variables are used here as representative of the "shape" of dominance establishment for this species. The variables range from average rate of occurrence of six different operationally defined states to the interopponent conditional probabilities associated with biting and lateral displaying. Table 1 is a brief synopsis of the 16 variables of this example.

Forty-one pairs of fish were studied so the data consist of 82 records (individuals): a sample of 41 winners and 41 losers. For each pair, the loser and winner were determined by standard criteria. Although this interjects subjective judgment of the observer, the judgment is *a priori* in terms of the analysis, and the analysis serves as a check upon this judgment. The intra-individual correlation matrix is shown in Table 2.

All principal component calculations presented here constitute part of that option of a multivariate statistical program (DISANAL) developed by Pimentel. We know of no "canned" computer program that either presents as many interpretive aids as DISANAL or as many as discussed here. The Biomed series—BMD01M (Dixon 1970a), BMDX72 and BMD08M (Dixon 1970b and 1973, respectively)—falls far short of what is shown here. Much better in terms of output is the SPSS program FACTOR (Kim 1975).

Table 3 displays the output one would normally examine first. It shows the eigenvalues, λ_i of Equation 12, of each of the first four principal components, the percent contribution of each eigenvalue to the total variance, and the eigenvectors, v_i of Equation 13(i.e., columns of direction cosines relating variables to components). The eigenvector matrix is often referred to as the eigenvector or factor pattern, the factor matrix, or the factor loading matrix. The cumulative percent of variance accounted for by successive components is also shown.

As pointed out previously, p-components (16 in this case) result from this sort of analysis and one is faced with the decision of how many to attempt to interpret. We have shown four components here only for purposes of illustration, not because this is the maximum interpretable number. Bartlett's sphericity test, Equation 16, can be used in this program (DISANAL) to test for significant differences between successive eigenvalues. In this case the first 12 eigenvalues reached significance levels ($P < .05$). Table 3 shows that the first four components accounted for 60 percent of the variation. If we included the next two components (i.e., all components with eigenvalues of 1.0 or more) we account for 74.5 percent of the variance. It should be remembered that each succeeding component accounts for a lesser and lesser proportion of the variance.

Table 1. Variables used in Principal Component Analysis of Fish Fighting Data

Variable Name	Symbol	Brief Description
1. Lateral display frequency	L_f	The average frequency per fight min of the state, lateral display.
2. Lateral display duration	L_d	The average duration per fight min of the state, lateral display.
3. Opercle spread frequency	O_f	The average frequency per fight min of the state, opercle spread.
4. Opercle spread duration	O_d	The average duration per fight min of the state, opercle spread.
5. Fin tug frequency	F_f	The average frequency per fight min of the state, fin tug.
6. Fin tug duration	F_d	The average duration per fight min of the state, fin tug.
7. Bites per bite session	B/BS	The average number of bites per cluster in a bite session.
8. Bite frequency	B_f	The average frequency per fight min of the event, biting.
9. Bite session frequency	BS_f	The average frequency per fight min of the state, bite session.
10. Surfacing frequency	S_f	The average frequency per fight min of the state, surfacing.
11. Pause frequency	P_f	The average frequency per fight min of breaks in the flow of behavior exceeding 4 sec in duration.
12. Conditional probability of biting following biting	P(B/B)	The interopponent conditional probability of biting following biting.
13. Conditional probability of biting following displaying	P(B/L)	The interopponent conditional probability of biting following lateral displaying.
14. Conditional probability of displaying following biting	P(L/B)	The interopponent conditional probability of lateral display following biting.
15. Sequence entropy	H	The average intra-individual sequence entropy estimate.
16. Interindividual behavioral independence	C	The ratio of interfish conditional sequence entropy (Dingle 1969) to the average intra-individual sequence entropy (H). This variable can have values 0 to 1, where zero indicates maximum behavioral "cuing" and unity indicates complete interfish independence.

Table 4 presents the correlation between components and variables using Equation 14. Since each component correlation represents the product of an original eigenvector coefficient and the square root of the corresponding eigenvalue, this table resembles Table 3. However, here one can consider interpretations in terms of associations between original variables and components. Component correlations are also termed factor structure.

Table 5 shows the percent variance of each variable involved in each component as computed from Equation 15. For example, only 2.15 percent of the variance of variable O_f is contributed to PC_1.

Approximately 25 percent of the total pattern of variation is accounted for by the first principal component (Table 3). Eight variables have their largest association with PC_1 (see Table 5, variables C, F_f, F_d, B_f, BS_f, S_f, H, and P_f). Tables 3 and 4 reveal that seven of these variables have positive directional associations with the first component while one (P_f) has a negative loading.

Components where a number of variables load strongly and predominantly in the same direction are referred to as *general components*. Often general components are left more or less at this "descriptive" level. A powerful aid in unraveling such components lies in analysis of individual component scores y_i for that component.

Table 6 displays the individual component scores for the first three components for all 82 fish. The scores of PC_1 between winners and losers for each of the 41 fights are strongly correlated ($r = 0.83$; $P < 0.01$) yet 31 of the 41 winners have relatively higher scores than did their opponents (i.e. individuals 1 versus 42, 2 versus 43, 3 versus 44, . . . , 41 versus 82). This significant difference ($P < 0.001$) leads us to speculate that PC_1 represents a pattern of agonistic behavior that might be termed "successful dominance strategy." This data pattern is clearly related to specific contact behavior such as fin tugging and biting, and sequential ordering of activity. In the univariate analysis of 12 of these variables, Frey and Miller (1972) found only fin tug measures to differ significantly between winners and losers.

To illustrate the second and third functions of principal component analysis, as outlined in the introduction of this chapter, we will continue our analysis of PC_1. Let us look at another aspect of the dynamics of fighting in these fish, namely length of fight. The fighting durations required to establish dominance relationships naturally varied among fights. The total number of acts (i.e. total lateral displays, opercle spreads, fin tugs, bite sessions, surfacings, or pauses) that made up a data record also varied among fights. The total number of acts (for both fish) ranged from 24 (for a very brief encounter) to 956 (a prolonged

Table 2. Intra-Individual Correlation Matrix

	P(B/B)	P(B/L)	P(L/B)	C	L_f	L_d	O_f	O_d
P(B/B)	0.999	−0.162	−0.331	−0.233	−0.003	0.115	−0.278	−0.173
P(B/L)	−0.162	0.999	−0.038	0.586	−0.035	−0.200	0.049	0.139
P(L/B)	−0.331	−0.038	1.000	0.279	−0.061	0.388	0.002	−0.205
C	−0.233	0.586	0.279	0.999	0.091	0.067	−0.028	−0.096
L_f	−0.003	−0.035	−0.061	0.091	0.999	0.334	0.283	0.062
L_d	0.155	−0.200	0.388	0.067	0.334	0.999	0.139	−0.016
O_f	−0.278	0.049	0.002	−0.028	0.283	0.139	0.999	0.813
O_d	−0.173	0.139	−0.205	−0.096	0.062	−0.016	0.813	0.999
F_f	0.163	0.168	−0.226	0.295	−0.003	−0.264	−0.213	−0.141
F_d	0.138	0.212	−0.206	0.399	0.101	−0.168	−0.264	−0.204
B/BS	0.039	−0.088	0.205	0.093	−0.024	0.162	0.108	−0.079
B_f	0.044	0.180	0.029	0.302	−0.008	−0.051	0.093	−0.074
BS_f	0.005	0.337	0.038	0.401	0.089	−0.143	0.149	0.016
S_f	−0.165	−0.020	0.263	0.406	0.066	−0.164	−0.067	−0.132
H	−0.168	0.189	0.301	0.366	−0.052	0.047	0.392	0.334
P_f	−0.141	−0.109	−0.406	−0.434	0.103	−0.193	0.031	0.231

	F_f	F_d	B/BS	B_f	BS_f	S_f	H	P_f
P(B/B)	0.163	0.138	0.039	0.044	0.005	−0.165	−0.168	−0.141
P(B/L)	0.168	0.212	−0.088	0.180	0.337	−0.020	0.189	−0.109
P(L/B)	−0.226	−0.206	0.205	0.029	0.038	0.263	0.301	−0.406
C	0.295	0.399	0.093	0.302	0.401	0.406	0.366	−0.434
L_f	−0.003	0.101	−0.024	−0.008	0.089	0.066	−0.052	0.103
L_d	−0.264	−0.168	0.162	−0.051	−0.143	−0.164	0.047	−0.193
O_f	−0.213	−0.264	0.108	0.093	0.149	−0.067	0.392	0.031
O_d	−0.141	−0.204	−0.079	−0.074	0.016	−0.132	0.334	0.231
F_f	0.999	0.895	0.102	0.381	0.600	0.232	0.439	−0.187
F_d	0.895	0.999	0.015	0.255	0.528	0.250	0.362	−0.166
B/BS	0.102	0.015	1.000	0.798	0.287	0.268	0.371	−0.574
B_f	0.381	0.255	0.798	1.000	0.718	0.341	0.376	−0.589
BS_f	0.599	0.527	0.287	0.718	1.000	0.320	0.527	−0.367
S_f	0.232	0.250	0.268	0.341	0.320	1.000	0.316	−0.477
H	0.439	0.362	0.371	0.376	0.527	0.316	1.000	−0.339
P_f	−0.187	−0.166	−0.574	−0.589	−0.367	−0.477	−0.339	1.000

Table 3. Eigenvectors (Direction Cosines) of Principal Components, Eigenvalues, Percent Variance, and Cumulative Percent Variance

	Component			
Variable	PC_1	PC_2	PC_3	PC_4
P(B/B)	−0.080	0.121	−0.029	0.534
P(B/L)	0.163	−0.015	0.158	−0.131
P(L/B)	0.123	0.199	−0.210	−0.399
C	0.326	0.113	0.076	0.029
L_f	0.072	−0.144	−0.300	0.495
L_d	0.098	0.113	−0.434	0.331
O_f	−0.073	−0.515	−0.374	−0.016
O_d	−0.133	−0.535	−0.164	−0.097
F_f	0.337	−0.171	0.348	0.214
F_d	0.334	−0.138	0.375	0.240
B/BS	0.202	0.254	−0.297	−0.053
B_f	0.372	0.090	−0.144	−0.045
BS_f	0.357	−0.225	0.027	−0.003
S_f	0.287	0.034	−0.166	−0.173
H	0.302	−0.390	−0.046	−0.187
P_f	−0.320	−0.151	0.279	−0.013
Eigenvalue	4.03	2.10	1.98	1.49
Variance (%)	25.2	13.1	12.4	9.3
Cumulative Variance (%)	25.2	38.3	50.7	60.0

fight). Figure 2 is a plot of individual component scores (PC_1) and $\log_{10}$ of total acts (fight length). PC_1 is clearly related to fight length. In this way, we have used the immediate output of the component analysis in a further analytical sense (that is, the third major function of component or factor analysis discussed in the beginning of this chapter).

We cannot conclude at this point that these elements are the only aspects of this primary source of data variation (PC_1), but because this methodology lends itself to hypothesis generation (the second major function mentioned in the chapter introduction) we can propose the following:

1. PC_1 reveals that prolonged dominance encounters become more "intense" (as measured by increasing frequency of contact behavior such as biting and fin tugging as opposed to no such increase in displaying) and have greater degrees of sequence unpredictability (as measured by the information variables C and H).

Table 4. Correlations between Components and Variables.

Variable	Component 1	2	3	4
P(B/B)	−0.162	0.176	−0.040	0.651
P(B/L)	0.328	−0.022	0.222	−0.159
P(L/B)	0.248	0.288	−0.296	−0.487
C	0.654	0.164	0.107	0.035
L_f	0.144	−0.208	−0.423	0.604
L_d	0.198	0.164	−0.612	0.404
O_f	−0.146	−0.746	−0.528	−0.020
O_d	−0.267	−0.775	−0.231	−0.119
F_f	0.678	−0.248	0.490	0.261
F_d	0.672	−0.200	0.529	0.292
B/BS	0.406	0.368	−0.418	−0.065
B_f	0.746	0.131	−0.202	−0.055
BS_f	0.717	−0.326	0.038	−0.004
S_f	0.576	0.049	−0.233	−0.211
H	0.606	−0.564	−0.065	−0.229
P_f	−0.643	−0.219	0.394	−0.016

Table 5. Percent Variance of Each Variable Involved in Each Component

Variable	Component 1	2	3	4
P(B/B)	2.609	3.087	0.163	42.441
P(B/L)	10.778	0.050	4.929	2.538
P(L/B)	6.137	8.307	8.790	23.741
C	42.751	2.687	1.149	0.121
L_f	2.080	4.328	17.922	36.505
L_d	3.901	2.677	37.412	16.340
O_f	2.146	55.625	27.829	0.040
O_d	7.142	60.096	5.329	1.408
F_f	45.957	6.135	24.048	6.822
F_d	45.124	3.988	27.971	8.555
B/BS	16.490	13.546	17.474	0.420
B_f	55.694	1.704	4.092	0.302
BS_f	51.360	10.617	0.143	0.001
S_f	33.171	0.242	5.449	4.443
H	36.755	31.850	0.426	5.230
P_f	41.393	4.797	15.491	0.026

Table 6. Individual Component Scores for All 82 Cases

Winner's Component Score				Loser's Component Score			
Individual	PC_1	PC_2	PC_3	Individual	PC_1	PC_2	PC_3
1	−2.190	−3.796	0.870	42	−2.345	−0.280	−1.009
2	−0.846	−1.384	0.909	43	−1.341	−0.391	0.287
3	1.977	−1.328	0.844	44	0.262	−0.471	−1.328
4	1.924	0.625	1.367	45	0.640	0.953	−0.371
5	2.466	−0.505	1.749	46	1.542	−1.036	1.558
6	1.553	0.166	0.654	47	2.676	0.438	1.010
7	3.498	−0.058	1.997	48	2.269	−0.522	1.062
8	1.499	0.687	1.605	49	1.413	0.110	1.333
9	1.863	−0.526	2.483	50	0.338	−0.839	0.175
10	−3.542	2.375	2.997	51	−2.289	1.547	2.539
11	−3.476	0.482	2.696	52	−4.981	2.690	1.682
12	0.921	−2.600	0.860	53	0.165	1.579	−0.506
13	−1.493	−0.600	1.157	54	−2.197	0.634	0.098
14	1.289	−0.201	−0.951	55	0.080	−0.938	−1.742
15	−0.624	0.957	−0.220	56	−0.743	−1.770	0.216
16	1.969	1.281	0.180	57	1.912	1.458	0.298
17	1.187	−0.862	−0.178	58	−0.825	−1.759	−1.159
18	1.775	−1.558	0.227	59	0.753	−0.843	0.728
19	0.423	0.973	1.249	60	0.367	2.127	−0.405
20	0.783	−0.228	0.934	61	−0.513	0.520	−0.124
21	1.921	0.504	1.306	62	3.581	−1.074	3.409
22	1.417	−0.854	−0.438	63	−0.234	−0.221	−0.759
23	0.522	−1.731	−1.786	64	1.242	−0.431	−1.580
24	−0.337	−1.426	0.156	65	−0.987	1.604	0.019
25	1.586	1.390	−0.451	66	1.490	1.451	−2.736
26	2.160	3.384	−1.727	67	0.219	1.392	−1.898
27	−0.921	−0.931	−1.696	68	−0.779	−1.888	−2.305
28	2.266	0.455	0.464	69	2.771	0.466	−0.261
29	1.778	−0.986	−1.002	70	1.308	1.152	−0.801
30	0.035	2.554	−1.712	71	0.546	−0.176	−2.062
31	−0.817	−1.779	−1.585	72	−1.616	−0.883	−2.036
32	2.832	−0.817	2.913	73	−1.186	−0.943	0.537
33	2.368	−0.549	−0.415	74	0.677	0.343	−1.862
34	2.290	0.822	0.047	75	1.095	1.289	−2.194
35	0.544	−1.358	−1.030	76	−0.696	1.280	−3.545
36	−0.483	−1.000	0.210	77	−1.745	2.415	−1.722
37	−2.553	−0.689	2.012	78	−1.226	1.135	1.347
38	−2.300	−1.902	−0.167	79	−5.006	2.566	0.258
39	−4.070	−0.771	0.402	80	−4.056	0.553	1.234
40	−3.307	0.947	−1.630	81	−2.812	−0.955	−1.283
41	−1.563	2.839	0.019	82	−4.093	−4.286	−0.352

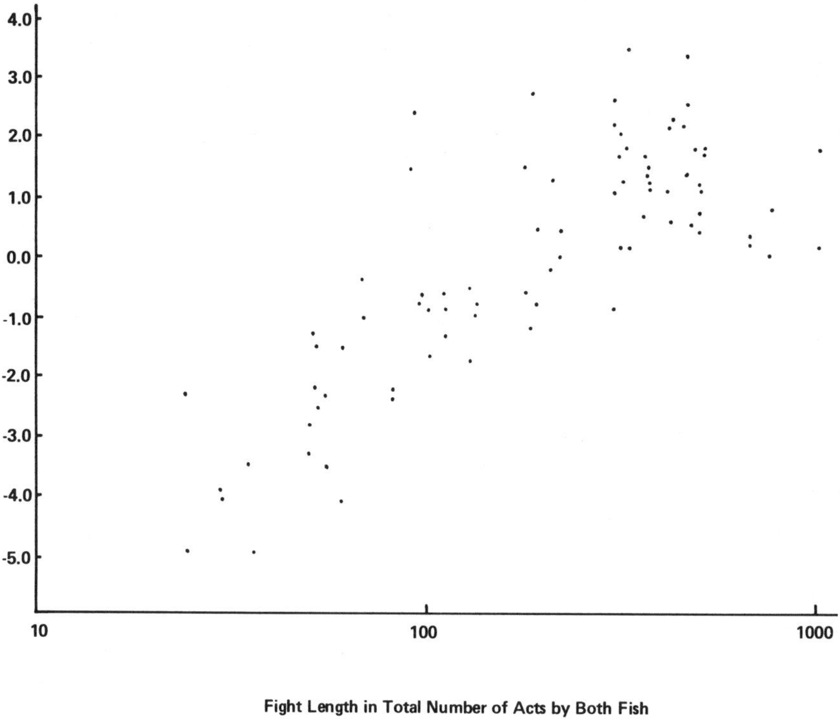

Figure 2. The relationship between fight duration and PC_1 scores for all 82 individuals.

2. Winners show this increase in "intensity" more strongly than do losers.

The foregoing conjectural points concerning PC_1 could easily be tested in terms of examining the minute to minute dynamics of the fighting trajectory in each case [e.g., does activity such as biting increase as bouts proceed, and do the variables loading heavily onto PC_1 (Table 3) show time courses different for eventual winners and eventual losers?].

In the example, the component scores clearly indicate the behavioral relationships. This is not always true. When more of the role of individuals is required to interpret components, the percent variance of each individual for each component (18) is consulted (not tabulated here). In essence these variances disclose how each individual contributes to each component: the larger the percent variance, the greater the contribution. Usually of more use is which individuals contribute to each component. The latter application generally is a major clue to how many components should be interpreted. For example, an attempt should be made to interpret all components having appreciable variances of a number of individuals. Unfortunately there is no statistical rule for determining what constitutes appreciable variance. It is up to the ethologist to do this.

The second independent pattern of variation (PC_2) accounts for about one-half the variation of PC_1 (i.e. 13.1 percent vs. 25.2 percent, Table 3). Two variables have high negative loadings on PC_2 (O_f and O_d, see Tables 3, 4, 5). The variable H also loads moderately heavily onto PC_2. The individual component scores of Table 6 for this component do not reveal the winner-loser dichotomy, nor the association with bout duration that was revealed in PC_1. We are thus forced into a more speculative role concerning this component.

In this instance it seems especially fruitful to refer to the sequencing of behaviors relative to opercle spreading. This behavior is strongly "linked" sequentially to the lateral display activity of an opponent. Yet the two behaviors appear causally distinct (Frey and Miller 1972, p. 50). This hypothesis (proposed by the univariate approach of Frey and Miller (1972)) is born out here since lateral displaying variables (L_f and L_d) load more strongly on PC_3 and PC_4 (see Tables 3, 4, 5). The assignment of specific behaviors to unique causal systems is illustrated by one of the initial ethological studies to use a factoring technique (Wiepkema 1961). Similar approaches are found among the many papers on open field rodent behavior (e.g., Poley and Royce 1973; Furchtgott and Cureton 1964; Whimbey and Denenberg 1967).

The third principal component represents 12.4 percent of the total pattern of variation. Reference to Table 5 reveals four variables whose loadings exceed 20 percent for this component (L_d, O_f, F_d, and F_f). Examination of their direction cosines from the eigenvector pattern (Table 3) or from their correlation with PC_3 (Table 4) indicates that two of them show negative associations with PC_3 while the other two are positively associated with this component. Components of this type (i.e. those with several variables loading positively on a component or factor and several loading negatively on a component or factor) are said to be *bipolar* (Harman 1967). Such components are often more readily interpreted than general factors.

The same relationship between the component scores for winners and losers was found for PC_3 as for PC_1: winners have relatively higher scores than losers (see Table 6). As one might expect the regression of PC_3 factor scores on bout length revealed no significant relationship. Since the coefficient signs for L_d and O_f are negative while F_f and F_d are positive, we are inclined to suggest that PC_3 is associated with a "switch" or "shift" to a more vigorous type of fighting in some fights (i.e. more fin tugging and less displaying). This component is also associated with "dominance success." Unlike the "increase in tempo" factor suggested in PC_1, this component seems relatively independent of bout duration. The analysis as we have carried it out thus far suggests that the process of dominance determination in these fish is quite complex and understood only with multidimensional models. Using principal components analysis we achieve both the rigor initiated by the univariate approach and the comprehensiveness of the Gestalt concept as introduced by early ethologists.

Rather than continue interpretation of components, we close this section by briefly discussing rotation of axes. Tables 7 and 8 display part of the Varimax rotated factor matrix as obtained using BMD and SPSS programs. The BMDX72 program FACTOR (Dixon 1970b; or 08M in the later edition, Dixon 1973) was run to correspond to the conditions associated with the initial DISANAL run. In other words, the input was the correlation matrix, axes were orthogonal, and the commonly used Varimax criterion for "simple structure" was employed. The SPSS program (PA1) was subject to the same constraints. While eigenvector coefficients for factors 1 and 2 vary slightly between BMD and SPSS they give rise to essentially the same interpretations as did the unrotated eigenvector pattern (Table 3).

Table 7. SPSS Varimax Rotated Factor Matrix

Variable	Factor 1	Factor 2	Factor 3	Factor 4	Factor 5
P(B/B)	−0.111	−0.156	−0.666	0.144	0.201
P(B/L)	0.212	−0.043	0.126	0.133	0.020
P(L/B)	−0.087	−0.149	0.729	0.134	0.021
C	0.397	−0.327	0.274	0.193	0.320
L_f	0.100	0.173	−0.130	−0.057	0.831
L_d	−0.105	−0.041	0.073	0.196	0.822
O_f	−0.031	0.892	0.017	−0.036	0.189
O_d	−0.069	0.884	−0.113	−0.097	−0.049
F_f	0.889	−0.116	−0.076	0.029	−0.033
F_d	0.883	−0.181	−0.065	−0.016	−0.008
B/BS	−0.130	−0.059	−0.004	0.884	0.049
B_f	0.353	−0.010	0.056	0.862	−0.003
BS_f	0.716	0.135	0.215	0.238	0.069
S_f	0.282	−0.094	0.574	0.244	0.220
H	0.620	0.440	0.374	0.190	−0.032
P_f	−0.229	0.126	−0.251	−0.622	−0.292
Eigenvalue	3.96	2.11	2.01	1.46	1.26

Table 8. BMD Varimax Rotated Factor Matrix

Variable	Factor 1	Factor 2	Factor 3	Factor 4	Factor 5
P(B/B)	−0.084	−0.146	0.138	0.219	0.647
P(B/L)	0.230	−0.041	0.156	0.012	−0.074
P(L/B)	−0.125	−0.164	0.128	0.005	−0.716
C	0.387	−0.320	0.202	0.319	−0.287
L_f	0.095	0.188	−0.056	0.832	0.100
L_d	−0.120	−0.037	0.191	0.820	−0.085
O_f	−0.063	0.893	−0.036	0.179	−0.047
O_d	−0.077	0.884	−0.089	−0.058	0.118
F_f	0.899	−0.093	0.054	−0.023	0.038
F_d	0.896	−0.157	0.008	0.001	0.031
B/BS	−0.143	−0.068	0.885	0.048	0.023
B_f	0.330	−0.007	0.871	0.000	−0.079
BS_f	0.687	0.148	0.251	0.070	−0.272
S_f	0.233	−0.095	0.238	0.213	−0.618
H	0.581	0.448	0.205	−0.040	−0.417
P_f	−0.191	0.125	−0.619	−0.293	0.292
Eigenvalue	3.95	2.11	2.01	1.46	1.26

Factor scores of the 82 individuals support the interpretations revealed from the unrotated solution. Subsequent eigenvectors, however, reveal some departure from the unrotated DISANAL output, and furthermore, the two programs (BMD and SPSS) differ between themselves slightly. For example, Factor 3 of the BMD program (Table 8) is the fourth factor of the SPSS program (Table 7). Likewise, Factor 3 of the SPSS is extracted as the fifth dimension in the BMD program. It seems as if the Varimax rotation in this case results in loss of behavioral interpretive value in that the factors become more descriptive and less heuristic than the unrotated components.

5. LIMITATIONS

Principal component analysis gives every indication of being robust. The method is likely to be valid for most if not all data analyses. However, it is based upon the following assumptions and/or decisions:

1. Variation arises from independent stimuli producing independent actions. It is assumed that the model leading to uncorrelated component scores (variables that might be identified as biological features, e.g.,

successful dominance strategy, etc. in the example) are more meaningful than original variables whose responses are influenced by one another. In essence, the method is based on the fact that correlation coefficients are unreliable estimates of cause and effect relationships (Snedecor and Cochran 1967).

2. The variables describe the phenomenon; included variables are meaningful and excluded variables are not critical to describe the phenomenon. Associated with this point is the problem of experimental dependency. This phenomenon is probably best illustrated by studies such as open field activity in rodents. Subjects are generally tested in several different experimental situations. Specific variation patterns often emerge that are clearly a function of apparatus constraints rather than an element of the actual phenomenon. (See Archer 1973 and Royce et al. 1970 for clarification in this specific area.)

3. The use of the variance-covariance matrix to study absolute variation or correlation matrix to study standardized variation generally is determined by the scales of measurement of the variables. If all variables are measured in the same or comparable scales, the former probably should be used. If the scales of measurement are not comparable, generally the correlation matrix should be used.

4. The choice of which components and which coefficients within a component to interpret is always a problem. None of the interpretative aids is absolutely reliable.

5. The entire approach can be called "fishing for results." There is no experimental design leading to specific null hypotheses to be tested. Satisfying such criticism is not easy. The fact that the results of such studies make good biological sense avoids, but does not answer, the criticism.

6. The methodology requires large sample sizes. Unfortunately, the minimum permissible sample size is unknown (Cooley and Lohnes 1971).

7. Analyses become useless if covariances or intervariable correlations approach equality or zero. In the former case all eigenvalues will have very similar values. In the latter case the S^2 or R matrix approaches the L matrix and the analysis is superfluous.

CHAPTER
9

Multivariate Analysis of Variance and Discriminant Analysis

RICHARD A. PIMENTEL and DENNIS F. FREY
California Polytechnic State University

In principal component analysis and factor analysis the purpose is to examine variation *within* a single population (Chapter 8). However, animal behavior research often investigates differences or similarities in patterns of variation *between* populations. The concept of "species specific" behavior patterns has developed from observed uniqueness of certain activities in different species. Also, the verification of behavioral criteria to erect or modify phylogenetic or taxonomic schemes required recognition of both differences and similarities between the groups (Heinroth 1911; Whitman 1919). The term *groups* in this chapter refers to operationally defined or biologically distinct units such as animals of a particular age class, geographic locale, given genetic background, or size class. More recent between-population studies have involved multivariate methods. Bekoff and others (1975) used a behaviorally oriented multivariate approach (i.e. Distance Analysis) to examine the affinity of the so-called New England Canid to wolves and coyotes. Another major work of this nature is Michener's (1974) analysis of the social biology of bees to establish grouping of species for comparison to morphological taxonomies. Finally, ethologists often attempt to unravel the nature of differences occurring between similar groups that occupy different habitats or manipulative regimes. (For variations on this approach see, among others, MacArthur 1958, Wecker 1963, Grant 1970, and Altmann 1974.) Riechert (1975) has illustrated the usefulness of discriminant analysis to evaluate microhabitat features that distinguish "web-sites" and "nonweb sites" in the spider *Agelenopsis aperta*.

As we pointed out in Chapter 8, single variable approaches to examine patterns of behavioral variation are unrealistic. Multivariate analysis of variance and, more important, discriminant analysis techniques provide powerful quantitative tools to examine the "shape" of intergroup behavioral phenomena. In another sense, the methodology also provides objective bases for judgments regarding the merits or usefulness of variables for future reference (see subsequent sections of this chapter for clarification of objectivity). For example, Aspey (1974) used discriminant analysis to identify which of 20 behavioral acts distinguished dominant from subordinate wolf spiders in male-male interactions.

Until the last few years, there was virtually no history of the use of discriminant analysis in ethology. This is understandable since discriminant analysis is the most complex of the methods of multivariate analysis, and the required computations for a study of meaningful scope were prohibitive for most workers prior to the advent of high speed computers and computer programs.

The first section of this chapter states the assumptions of discriminant analysis, the second section presents major elements of methodology, and the third section deals with interpretation. Section 4 provides an example, Section 5 discusses the limitations of the method, and Section 6 briefly discusses computer programs for discriminant analysis.

1. ASSUMPTIONS

The assumptions for discriminant analysis are those for principal component analysis (Chapter 8) plus the assumption of equality of all expected covariance matrices of the populations sampled. Empirical studies and Monte Carlo simulations (see Chapter 11) suggest that the methods of discriminant analysis appear to be robust from departures from assumptions even when no real precautions beyond care in sampling techniques are taken by the investigator (Cooley and Lohnes 1971). Ito and Schull (1964) have demonstrated in the two-group case that, when covariance matrices are quite unequal, actual significant levels tend to match true significance levels *if group sample sizes are large and equal*. Unfortunately no study has determined how large a sample size must be, but many statisticians would agree that 25 individuals per group might be adequate.

2. METHOD

Discriminant analysis is confusing to ethologists unfamiliar with multivariate techniques, for many reasons. Two prime sources of confusion are that discriminant analysis includes many different methods, and the terminology applied to some of the methods is often interchanged. Space prevents a detailed consideration of the methods. Our procedures in this chapter follow those identified by Hope (1969) and Cooley and Lohnes (1971), and interpretation approximates that outlined by Blackith (1965).

Discriminant analysis consists of three major procedures, or areas of interpretation, that involve five possible methods of multivariate analysis. The first procedure is to test the null hypothesis that the mean vectors (centroids) of the populations sampled are equal. This is accomplished by a single method called the *multivariate analysis of variance* or *manova*, a multivariate extension of the analysis of variance which includes a test of the null hypothesis of equality of population covariance matrices (Cooley and Lohnes 1971). Rejecting the hypothesis of equality of centroids is necessary to establish differences between populations prior to studying the differences.

The second procedure is *classification* of individuals to groups. Again a single method is involved. Classification assigns each individual to the group it most closely approximates in form. Each individual is assigned either to its own group, a *hit*, or to another group, a *miss*. Although a single method of doing this is applied in a given analysis, two different methods are in wide use and we shall apply yet another method. One method involves the *square of the generalized distance* (square of Mahalanobis' distance D^2) of each individual from each group centroid. The square of the generalized distance of the ith individual of the total sample of all individuals of all groups from the jth group is

$$D^2_{ij} = (\mathbf{X}_i - \bar{\mathbf{X}}_j)'\mathbf{S}^{-2}(\mathbf{X}_i - \bar{\mathbf{X}}_j)$$

where $(\mathbf{X}_i - \bar{\mathbf{X}}_j)$ is the deviation of the ith individual, $\mathbf{X}_i$, from the centroid of the jth group, $\bar{\mathbf{X}}_j$, and $\mathbf{S}^{-2}$ is the inverse of the pooled within group dispersion matrix (equivalent to the univariate error mean square). In practice, the deviations often are simplified to $\mathbf{d}_{ij}$ and the above formula becomes

$$D^2_{ij} = \mathbf{d}'_{ij}\mathbf{S}^{-2}\mathbf{d}_{ij}$$

Of greater importance is the fact that D^2 approximates the chi-square distribution with p degrees of freedom (where p is the number of variables or measurements). For this reason it can be used as a classification chi-squared for the purpose of assigning a probability from zero to unity of membership of an individual to each group. This is the procedure used in the BIOMED Computer Program, BMD07M (Dixon 1973) and, in part, in the SPSS program, DISCRIMINANT (Klecka 1975). In addition, the generalized distance, D, can be used as a statistic to indicate the distance of each individual from each centroid. In effect, either D, D^2, or probabilities derived from D^2 can be used for actual assignment of individuals to groups.

The second classification procedure uses a so-called *classification function* to assign individuals. This method is incorporated in DISCRIMINANT of the SPSS package.

A more efficient method for assigning individuals to group membership is by *Geisser classification probabilities* (Geisser 1964), a method discussed at length by Cooley and Lohnes (1971). Geisser's method is much more robust to departures from assumptions than is D^2 or classification functions and performs much better for small samples. The computations for Geisser probabilities are complex but can be considered a correction of D^2 which is involved in the formula.

Fundamental to any classification is the fact that the actual group to which each individual belongs is known. The procedure, then, determines whether an individual is more like its actual group of membership or some other group. Also, if after an initial analysis new individuals are classified in terms of the prior analysis, it is assumed that the new individuals belong to one of the groups examined.

The final area of interpretation is to examine relationships between groups and individuals in *discriminant space*. Discriminant space is one derived by rotating the orthogonal variable axes of Euclidean space so that the angle between any possible pairs of variable axes has a cosine that is the correlation coefficient between the variable pair. The rotation of variable axes to discriminant space is much more complicated than rotation to component space (see Chapter 8, Figure 1). In essence, this provides a space in terms of the shapes of individuals and groups. (The generalized distance between individuals or groups is in discriminant space.)

One of two approaches can be implemented to analyze discriminant space. The choice of which one to use depends upon the complexity of the particular discriminant space, and hence the nature of the research problem under study. The simpler analysis occurs whenever discriminant space is two or three dimensional, or whenever there are no more than three variables or four groups. However, this analysis is also appropriate whenever three dimensions can summarize most of the relationships in total discriminant space. The only way to discover this is to attempt the analysis that might be termed *distance and discriminant vector analysis*. For the distance analysis part, as Blackith (1965) suggests, one starts by cutting rods, using any convenient scale, the lengths of which indicate the generalized distance D of each group centroid from all other group centroids. Small spheres of clay, styrofoam, and so forth can be used to represent group centroids. If it is possible to assemble the model of centroids in "discriminant space," this analysis is appropriate. Figure 1 is a photograph of a "distance model." The constructed model plus the discrimination vectors, one for each group, become the basis for interpreting relationships. The use of distance models and discrimination vectors will be shown in the interpretation section and example.

Prior to considering the alternative of discriminant space being too complex to interpret via distance and discriminant vector analysis, certain matters pertaining to terminology require clarification. To do this we must introduce the classification function

$$\mathbf{y}_{ij} = \mathbf{X}_j'\mathbf{1}_i - \tfrac{1}{2}\bar{\mathbf{X}}_i\mathbf{1}_i + P_i$$

where $i = 1, 2, \ldots, g$ groups of classification; $j = 1, 2, \ldots, g$ groups of actual membership; $\mathbf{X}$ is an individual's data vector; $\mathbf{1}$ is a vector that transforms relationships to discriminant space; $\bar{\mathbf{X}}$ is a group centroid; P is the probability of group membership; and y is a classification score for

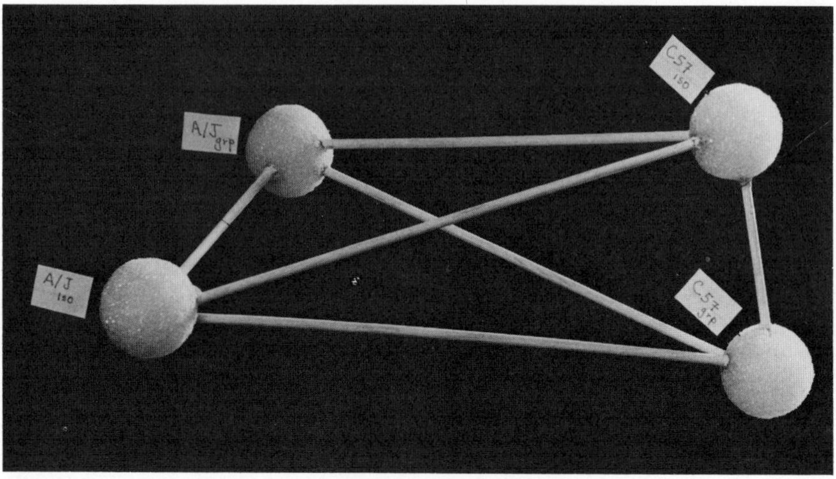

Figure 1. A photograph of a distance model.

an individual. In most American usage the above function is termed a classification function and the vector **1** is called a classification vector. However, in Fisher's (1936) original presentation of discriminant analysis, he designated a formula that is directly related to $\mathbf{X}_j'\mathbf{1}$ of the classification function as the *discriminant function*. Perhaps for this reason, some statisticians, outside the United States, such as Hope (1969), often call that entire term the discriminant function. Unfortunately, Americans typically call what are introduced later as canonical vectors, discriminant functions. This confusion is possible since in the simple two-group case introduced by Fisher, a single discrimination vector can be, and was, used, and such a vector is exactly equal to the only possible canonical vector. In view of the inconsistency, we shall avoid referring to discriminant functions; however, because of our particular use of the classification vector, **1**, to examine differences between groups we shall hereafter term it a *discrimination* or *discriminant vector*. The discriminant vector shall be applied to analyzing *qualitative* differences between groups and the generalized distance applied to *quantitative* differences between groups.

When discriminant space is too complex to be analyzed in total by D and **1**, we are forced to observe selected dimensions in that discriminant space. For example, three selected dimensions can be analyzed simultaneously and a generalized distance model pertaining only to those dimensions can be constructed. However, the dimensions to be examined are not selected at random. Selection is very specific and accomplished by a technique called *canonical analysis.* The exact nature of canonical analysis need not concern us since we are concerned only with the resultant properties of the data. The particular canonical analysis incorporated is one of discrimination. Many different but interrelated models for the latter are possible. The only one presented in this section, in descriptive terms, is a canonical analysis of discriminance of raw variables standardized for within or error covariance. It will be seen that raw rather than z-score variables are used and that the pooled within group covariance is the basis for standardizing the variance of the transformed variables y. The model is

$$\mathbf{y}_j = \mathbf{a}_j'\mathbf{x}$$

$$\bar{\mathbf{y}}_j = \mathbf{0}$$

$$s^2_{y_j} = \mathbf{a}_j'\mathbf{S}^2\mathbf{a}_j = 1$$

where $\mathbf{y}$ is a vector of scores, one for each dimension of canonical discriminant space, for an individual; $\mathbf{a}_j$ is a vector of coefficients, one for each variable, that locates individuals on the jth canonical axis of discriminant space; and $\mathbf{x}$ is an individual's vector of deviations of raw scores from the grand centroid $\bar{\bar{\mathbf{X}}}$ of all individuals of all samples ($\mathbf{x} = \mathbf{X} - \bar{\bar{\mathbf{X}}}$). The mean canonical score $\bar{\mathbf{y}}_j$ of all individuals on each canonical axis is zero, so the origin of the set of canonical axes is at the grand centroid which becomes a vector containing all zeros because $\bar{\mathbf{x}} = \mathbf{X} - \bar{\bar{\mathbf{X}}} = \mathbf{0}$. Since the variance of any $\mathbf{y}_j$ is unity, we see the basis for standardization based upon the within covariance. The variance for all $\mathbf{y}_j$ forms a matrix $\mathbf{S}_y{}^2 = \mathbf{A}'\mathbf{S}^2\mathbf{A} = \mathbf{I}$, the identity matrix. The identity matrix merely summarizes the fact that the variance of all canonical scores for any canonical axis is unity.

The canonical axes are selected so that the first will disclose the greatest possible difference between groups. Each successive axis is selected with the same criterion but with the added restriction that it is orthogonal to all other preceding axes. Solution of the matrix $\mathbf{A}$ becomes an eigenvalue-eigenvector problem. First

$$|\ \mathbf{BW}^{-1} - \lambda_j\mathbf{I}\ | = 0$$

provides $r = \min\ |\ g - 1, p\ |$ eigenvalues, one for each canonical axis that exists in the given discriminant space. Again, g is the number of groups; p is the number of variables. The eigenvalues λ are derived by transforming the product of $\mathbf{B}$, the among group cross-products matrix (the multivariate counterpart of the group sum of squares of deviations from the mean of the analysis of variance) and $\mathbf{W}^{-1}$, the inverse of the within group (error or residual) cross-products matrix (the multivariate equivalent of the error sum of squares). Since multiplying a matrix by the inverse of a second matrix is equivalent to dividing the first by the second (there is no direct method of division in matrix algebra), $\mathbf{AW}^{-1}$ is analogous to the F statistic of univariate analysis of variance. In other words, $\mathbf{AW}^{-1}$ is a measure of between group differences that is fractioned into successive maximum possible orthogonal portions.

Then,

$$(\mathbf{AW}^{-1} - \lambda_j \mathbf{I})\mathbf{a}_j = \mathbf{0}$$

provides r eigenvectors $\mathbf{a}_j$ each relating the rotation from original variable axes to canonical axes in discriminant space.

The canonical scores of individuals $\mathbf{y}$ can then be plotted, two or perhaps three axes at a time, on a canonical graph. Such a graph should also use the corresponding eigenvectors $\mathbf{a}_j$ to plot original variable axes in the canonical discriminant space to indicate the influence of variables in that space (Jolicoeur 1959).

The number of dimensions in discriminant space h is equal to the smaller of the number of groups g or the number of variables p. However, if the total of all group sample sizes is less than h, a situation that should never be allowed by the investigator, the number of dimensions is defined by the total number of individuals.

3. INTERPRETATION

Interpretation will be considered in terms of the three major procedures; however, all procedures pertain to how and why groups differ and are related to one another.

The test of equality of centroids, manova, does not clarify which groups differ from one another. In fact, even a multivariate multiple range test that might provide clusters of like populations would be of minimal use. However, it is often helpful to judge the merit of each measurement singly. For this reason, a univariate analysis of variance for each variable and application of the Student-Newman-Keuls multiple range test (Sokal and Rohlf 1969) to variables producing significant F-tests is an appropriate preliminary step to the analysis. In this manner, a crude concept of variable importance might be forthcoming. The importance of variables derived from a multiple range test is crude because variables are considered independently, while discriminant analysis compares "shapes," each resulting from a synergistic interplay of variables. The researcher must therefore realize that the multiple range tests can be misleading. More important, the multiple range tests can become a device for validating further techniques. Although evidence consistently indicates extreme robustness of multivariate analysis to departures from assumptions, the robustness of univariate procedures is also well documented. For this reason, demonstrated univariate differences between groups can further justify significant differences among centroids when population covariances are found to be significantly different from one another.

In DISANAL, a computer program developed by Pimentel, a procedure not previously discussed and not to be developed in the example is helpful here. The pertinent part of the program is based upon the fact that group covariances generally are found to be unequal. When such inequalities exist, there is justification for utilizing group covariances to examine the differences that exist between groups. Such comparison is accomplished by principal component analysis. The program performs a complete principal component analysis, including all details discussed in the previous chapter, for each group's covariance matrix and the pooled within group covariance matrix. The latter provides reference components that can be compared with each group's components. This allows examination of the patterns of variation present or absent in each group and the amount of variation involved in each pattern. The comparison of components is accomplished, primarily, by angular comparison.

Classification, in a sense, is a multivariate multiple range test. The hits and misses plus the probabilities of individuals being assigned to each group clearly indicate the degree of uniqueness and relationship between groups. In essence, classification is a measure of group affinities that uses individuals as the measurement.

Interpretation of relationships in discriminant space is direct. When consideration of total discriminant space is possible, the constructed generalized distances model becomes the source of examining quantitative differences between groups. On the model (Figure 1), in addition to observing quantitative distances between groups, one often can define the bases of discrimination along three arbitrary but generally orthogonal axes. For example, single axes of species, sex, or other differences often are recognizable. Therefore, the model typically leads to a definition of axes of discrimination.

To obtain qualitative differences between groups, angles between all possible pairs of group discrimination vectors are calculated. In general each discrimination vector indicates the orientation of its group in discriminant space, so the angular comparisons are of qualitative differences. The most convenient way to present angular comparisons is by a simple two-dimensional graph of the angles. If angular departures between groups tend to be oriented in a single plane in discriminant space, the angles between groups can be approximated on the two-dimensional graph, but if the group discriminant vectors depart sufficiently from a single dimension, the angles will not be additive. Lack of agreement implies distinct differences in the distributions of individuals in the various groups and marked qualitative differences between groups. An example of this point is shown in a subsequent section. In some cases such differences might be so great as almost to defy interpretation. Since this is not to be expected, when it occurs one should reexamine the variables for errors in measurement, in consistency from group to group, in being pertinent to individuals, in defining the phenomenon, and so forth.

When distance models fail, canonical analysis of discriminance becomes most useful. The approach of distance analysis can be applied to any two- or three-dimensional canonical graph. Again, arbitrary axes of biological meaning should be identified. However, canonical graphs can include two features not found in distance models. A canonical graph can include vectors of original variable axes as they occur after rotation into the particular canonical space. Such vectors, one for each variable, originate as a cluster from the origin of the graph (i.e. the position of the grand centroid). The vector for each original variable is plotted from the coefficients of the variable that correspond to the particular canonical axes in the graph. The scale of the vector plotting can be arbitrary. The vectors disclose the direction of positive "push" for each variable. In general, a group centroid that extends just beyond a long vector for a variable will represent a group having large measurements for that variable. Also, a group occurring in the opposite direction can have a low value for the measurement. This is not always true since all variables act synergistically in discriminant space and the location of a group can represent a complex interplay of variables. In spite of such possible confusing interplay of variables, the vectors when compared with original data centroids should reveal why a group is located as it is in canonical space. Also, the vectors indicate how original variables were rotated into canonical space and which variables tend to act together (i.e. have low angular departure) and which tend to interact in antagonistic relationships.

The second feature that can be added to a canonical graph is a confidence circle about each group centroid. Each confidence circle estimates the location of a population centroid with a certain confidence coefficient. The confidence radius of the ith group centroid is

$$\frac{t_{\alpha_i, N_i - 1}}{\sqrt{N_i}}$$

where N_i is the sample size of the ith group and where t_{α_i,N_i-1} is the tabulated value of Student's t distribution at the α percent for $N_i - 1$ degrees of freedom. Nonoverlapping confidence circles also can be interpreted as significant differences between groups but are limited to the space portrayed.

In a canonical analysis involving many axes the first two or three axes might summarize insufficient intergroup differences, a situation termed *hypermultivariate*. Such distributions indicate extreme complexity of motivational or ecological substrates to the behavioral phenomenon. When such distributions occur, much information still can be gained by analyzing canonical graphs.

4. EXAMPLE

Data for our example are from unpublished work by Frey on "open-field" behavior in male mice. For the study two strains were chosen because they were reported to be widely divergent in so-called "emotionality" levels. (The strains were C57BL/6J and A/J from Jackson Laboratories, Bar Harbor, Maine.) Archer (1973b) presents a comprehensive review of the concept of emotionality in rats and mice. One purpose of the investigation was to provide both quantitative and qualitative descriptions of phenomena heretofore differentiated in vague terms. Another purpose of the example here is to display the power of discriminant analysis.

The mice were about 3 weeks old when brought into the laboratory. They were housed individually in 25 × 15 × 15 cm cages for 5 to 7 weeks. After this phase they were allowed a two-day period of resocialization. Finally, they were assigned to one of two different housing regimes for an additional 8 weeks. The experimental treatments were as follows: In a random half of the mice of each strain, each individual was placed back into a physical-visual isolation condition (these mice hereafter will be referred to as C57iso and A/Jiso groups). In the other half of each group, pairs of mice were housed together (these mice hereafter will be referred to as C57grp and A/Jgrp groups). Therefore, the experimental treatments involved both a strain difference and a housing manipulation. At the end of the 8 weeks one mouse from each of the pairs was selected at random and tested, as were all isolated mice.

The testing apparatus was a modified version of that used by File and Day (1972) for rats (Figure 2). In each "exploratory chamber," the 30 cm × 45 cm wood floor was subdivided into 16 equal-sized rectangles and a 2 cm diameter hole was located in the center of each rectangle.

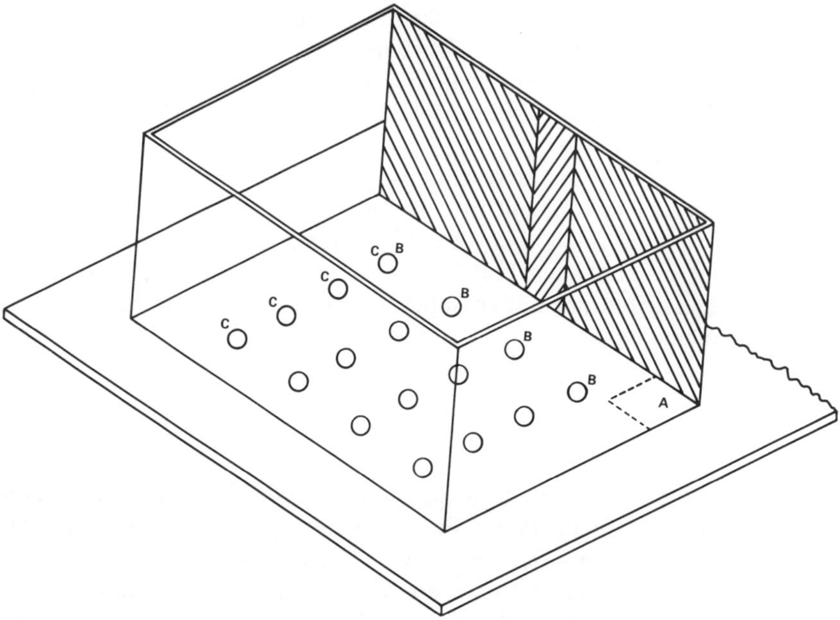

Figure 2. "Open-field" behavior apparatus showing the introductory corner (*A*), holes along the wooden wall (*B*), and holes opposite the introductory corner (*C*).

The floor was elevated 10 cm from ground level and the mice would typically investigate several holes by dipping their heads into the holes to varying depths. A clear view of activity in each chamber was possible through the three plexiglass walls.

For testing, each mouse was placed into the introductory corner of the board (A in Figure 2) and allowed to remain there for 5 minutes of activity data acquisition. The activities recorded as variables are briefly described in Table 1. The data consist of eight measurements ($p = 8$) on nine individuals for each of the four treatment groups ($g = 4$). Since the variables represent both different and hardly comparable scales of measurements, they were transformed to z-scores. The model remains the same as that shown previously for raw data, the only difference being that $\mathbf{z}$ is substituted for $\mathbf{X}$ and since the dispersion of z-scores is the correlation matrix of raw scores, $\mathbf{S}^2$ becomes $\mathbf{R}^2$. The analysis, therefore, becomes one of relative variation among variables between groups, the variation being relative since all variances of z-scores are equal to unity. The use of z-scores results in exactly the same distribution of individuals on a canonical graph as would result from raw data.

The only difference is in the vectors of variables—the interplay of standardized *z*-score variables in discrimination is different from that for raw variables.

The first step in analysis involves examining each variable for univariate among-group differences followed by manova. Table 2 presents the group centroids plus univariate *F*-ratios, probabilities of the *F*-ratios, and results of Student-Newman-Keuls multiple range tests for each variable. Notice that six of the eight variables exhibit statistically significant differences between means ($P < 0.05$). Only hole investigations to ear depth (EARS) and dips along the "top row" (TOP) (i.e. opposite the starting area, see Figure 2) were not statistically significant. A strain difference is verified for two variables, perimeter dips (EDGE) and dips along the wooden wall of the apparatus (LEFT). The remaining differences are more complex and the strengths of discriminant analysis will aid in understanding their nature.

The test of homogeneity of group covariance matrices is statistically significant ($F = 1.775$ with 108 and 2274 d.f., $P < 0.01$). Since the

Table 1. Open-Field Behavior Variables

Variable	Symbol	Brief Description
1. Holes examined	HOLES	Number of different holes in which head dipping occurred during 5 min observation (max = 16)
2. Nose level dips	NOSE	Number of superficial hole investigations per 5 min: head inserted only to nose depth.
3. Eye level dips	EYES	Number of intermediate hole investigations: head inserted between nose and eye level.
4. Ear level dips	EARS	Number of "deep" hole investigations: head inserted beyond ear level.
5. Perimeter dips	EDGE	Percent of total hole investigations into the 12 peripheral holes.
6. Dips along "top row"	TOP	Percent of total hole investigations along the top row opposite the start area (Figure 2, Area C)
7. Left Column dips	LEFT	Percent of total hole investigations along the wooden wall (Figure 2, Area B).
8. Grooming	GROOM	Seconds of auto grooming during the 5 min observation.

Table 2. Group Centroids and Some Univariate Statistics for Open-Field Mice Study

	Group Centroid (z-scores)						
Variable	Group 1 (A/J grp)	Group 2 (C57 grp)	Group 3 (A/J iso)	Group 4 (C57 iso)	F-Ratio	F P-level	SNK*
HOLES	−0.49	0.82	−0.09	0.76	23.9	0.00	(1) (3) (2, 4)
NOSE	−0.34	0.42	−0.70	0.61	4.5	0.01	(3) (1, 2, 4)
EYES	−0.42	0.80	−0.46	0.08	3.9	0.02	(1, 3, 4) (2, 4)
EARS	−0.40	0.54	−0.40	0.25	2.2	0.10	(1, 2, 3, 4)
EDGE	0.42	−0.82	0.83	−0.43	8.4	0.00	(1, 3) (2, 4)
TOP	0.29	−0.33	0.17	−0.14	0.7	0.55	(1, 2, 3, 4)
LEFT	0.55	−0.59	0.62	−0.58	5.9	0.00	(1, 3) (2, 4)
GROOM	0.88	−0.57	0.11	−0.20	4.5	0.01	(1) (2, 3, 4)

* SNK: Student-Newman-Keuls multiple range tests; nonsignificantly different groups are enclosed within the same parentheses.

methodology is robust we will ignore the inequality of covariance matrices. We will do this since sample sizes are equal (N_i = 9 for each group) and in spite of the fact that sample sizes are small. These data are part of a larger study whose analyses agree with these results. Also, as stated before, we will not include methods to analyze the differences between covariance matrices.

The test of equality of group centroids also reveals statistically significant differences among the groups ($F = 2.49$ with 24 and 73 d.f., $P = 0.002$).

The results of the next step, Geisser classification of individuals to groups, are presented in Table 3. Although the probabilities for each individual's membership are not shown, of the eleven misses (those off the diagonal of Table 3) only one individual displayed a low probability (0.04) for membership in its own group. All other misses displayed 0.15 to 0.48 probability of belonging to their actual group of membership. Only one of the 11 misses involved an interstrain misclassification. The other 10 misses, then, were intrastrain so misclassification was on the basis of housing manipulation. Therefore classification demonstrates that, in general, the primary difference in open-field behavior in this study was predominantly a function of genetic background rather than the housing manipulation.

The generalized distance D between the groups of mice are presented in Table 4. As might be expected, these distances verify the hypothesis generated from Geisser classification concerning the nature of intrastrain versus interstrain differences. For example, observe that intrastrain distances are less than interstrain distances. The "socially reared" A/J mice centroid is 1.721 standard deviation (SD) units in discriminant space away from their "isolate reared" A/J counterparts. Socially reared C57 mice are 1.118 SD units from the isolate reared C57 centroid. Both these distances are considerably less than any distances between strains.

Table 3. Summary of Hits and Misses From Geisser Classification

Actual Group Membership	Predicted Group Membership			
	A/J grp	C57 grp	A/J iso	C57 iso
A/J grp	5	0	3	1
C57 grp	0	6	0	3
A/J iso	2	0	7	0
C57 iso	0	2	0	7

Table 4. Generalized Distances between Group Centroids

Housing and Strain	Grouped A/J	Grouped C57	Isolated A/J	Isolated C57
Grouped A/J	0.0			
Grouped C57	3.070	0.0		
Isolated A/J	1.721	3.627	0.0	
Isolated C57	2.598	1.118	3.457	0.0

The valuable information resulting from the distance model reveals the three dimensionality of the discriminant space. It is obvious from Table 4 that both strain and rearing effects are present. The model (Figure 1), however, clearly shows that the rearing experimental manipulation influences open-field activity of the A/J strain in a manner different than in the C57 strain (i.e. the A/J intrastrain axis is orthogonal to the C57 intrastrain axis). Figure 3 is a diagrammatic representation of the distance model labeled to further aid in understanding this strain times rearing interaction.

The program DISANAL provides between-group distances based upon each variable in addition to the generalized distance. We can now use these data to determine which variables have contributed most to (1) the interstrain distances (see *a* and *b* of Figure 3), and (2) the intrastrain distances (see *c* and *d* of Figure 3). This might seem analogous to stepwise discriminant analysis but it is not at all the same. For those familiar with BMD07M of the BIOMED or DISCRIMINANT of SPSS, a few remarks might be of interest. Group distances based upon each variable do not constitute a stepwise procedure since variables are neither retained or eliminated by the individual variable's distance

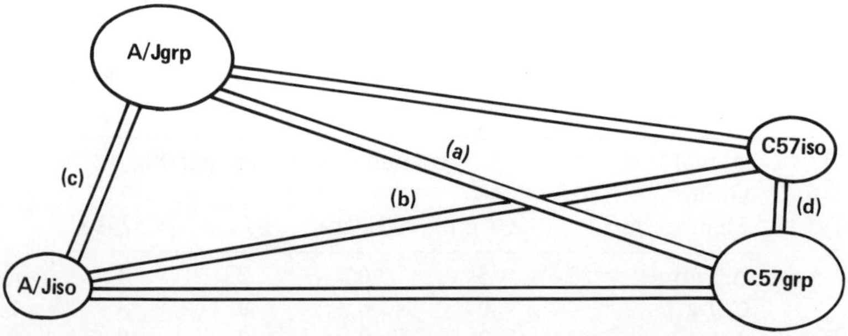

Figure 3. Diagramatic representation of the distance model, Figure 1. Interstrain distances *a* and *b* are shown to be greater than intrastrain distances *c* and *d*.

analysis. On the other hand, variable retention or elimination is the purpose of the stepwise procedure. We do not recommend the stepwise procedure since it evaluates variables independently rather than synergistically. For this reason it is not at all unusual for the method to retain "poor" variables and exclude "good" variables.

Our concern about stepwise procedures might surprise some readers. Snedecor and Cochran (1967) review many of the problems of stepwise procedures in reference to regression. In the case of discriminant analysis stepwise methods have the additional disadvantage of evaluating independent actions of variables in Euclidean space. In discriminant analysis, the transformation is from orthogonal original variable axes to variable axes having angular departures whose cosines are equal to their correlation coefficient; variable axes become oblique, associations are stressed, and independent contributions of variables are lost. In essence, a Euclidean space evaluation of variable merits might have little bearing on their discriminant space actions.

The influence of the eight variables in interstrain distances is summarized in Table 5. The data are presented as both the actual distance and the percent of the generalized distance that each variable contributes to the separation. (For example, D between A/Jgrp and C57grp, that is distance a of Figure 3, was shown to be 3.07 SD units in Table 4. The same distance based only on the variable HOLES was 2.269 SD units, or 74 percent of D.) Using these percent contribution data, it is readily seen that HOLES accounts for the major interstrain separation (i.e. 92 percent in isolates and 74 percent in socially reared mice). Additionally, EDGE and LEFT separate the groups regardless of rearing regime. The C57 mice visited on the average 13.9 holes while A/J mice visited only 5.3 holes per 5 minute observation period. The albino A/J mice made a greater percent of their investigations at holes on the periphery of the open field (EDGE) and also near the wooden panel (LEFT) than did the pigmented C57 mice. Other types of differences in open-field activity have previously been attributed to the "albino gene locus" in laboratory mice (DeFries 1969; DeFries, Hegmann, and Weir 1966). The other interstrain differences are specific to the rearing manipulation, for example, NOSE, EYES, and GROOM (Table 5).

Differences associated with the rearing manipulation within each strain are presented in Table 6. It should be kept in mind that generalized distances along these axes are considerably less than interstrain distances. Furthermore, the A/J intrastrain distance, c in Figure 3, is 35 percent greater than the C57 intrastrain distance, d in Figure 3. The variables HOLES, EDGE, and GROOM account for most of the A/J rearing separation, while EYES, EDGE, and GROOM predominantly

Table 5. Summary of Interstrain Distances between Groups Based on Single Variables

Housing Regime	Distances Based on Variable							
	HOLES	NOSE	EYES	EARS	EDGE	TOP	LEFT	GROOM
Isolation	3.184	1.489	0.605	0.676	1.612	0.312	1.435	0.099
	(92)*	(43)	(18)	(20)	(47)	(9)	(42)	(3)
Group	2.269	0.863	1.368	0.990	1.585	0.613	1.362	1.659
	(74)	(28)	(45)	(32)	(52)	(20)	(44)	(54)

* Values in parentheses are measures of distance as a percent of generalized distance.

Table 6. Summary of Intrastrain Distances between Groups Based on Single Variables

	Distance Based on Variable							
Strain	HOLES	NOSE	EYES	EARS	EDGE	TOP	LEFT	GROOM
A/J	1.025	0.410	0.045	0.000	0.519	0.111	0.080	1.134
	(60)*	(23)	(3)	(0)	(52)	(6)	(5)	(66)
C57	0.110	0.216	0.807	0.313	0.492	0.190	0.007	0.427
	(10)	(19)	(72)	(28)	(44)	(17)	(0)	(38)

* Values in parentheses are measures of distance as a percent of generalized distance.

Table 7. Discrimination Vectors for Each Group

	Discrimination Vector for Group			
Variable	A/J grp	C57 grp	A/J iso	C57 iso
HOLES	−1.543	2.902	−4.869	3.510
NOSE	0.391	−0.798	1.140	−0.732
EYES	−0.096	0.230	0.525	−0.660
EARS	−0.332	0.245	0.226	−0.139
EDGE	0.136	−0.453	0.037	0.280
TOP	0.058	−0.233	0.474	−0.299
LEFT	0.005	0.322	−0.231	−0.095
GROOM	1.203	−0.840	−0.184	−0.174

separate the C57 mice. Reference to Table 2 indicates that A/Jiso mice are more spatially restricted in the number of holes investigated compared to their socially reared counterparts. In both strains there is a tendency for the isolate reared mice to investigate relatively more peripheral holes (i.e. nearer the wall of the apparatus). This tendency would probably have been unnoticed on the basis of univariate differences. Finally, while the variable GROOM separates the rearing regimes within each strain, the direction of influence is opposite for each strain (i.e. the A/Jgrp mice spent much time grooming while the C57grp mice did very little grooming).

The discrimination vectors are presented in Table 7. The angles between these vectors are shown in Table 8 and diagrammed in Figure 4. The vector diagram presents angles in degrees between groups, angular departure from additivity in parentheses, and percent of angular departure across the two central vectors. Figure 4 relates to minor differences in the distance model since the main dichotomy based upon angles is in housing regime. The angles also substantiate that C57 mice are more alike than are A/J mice (25° vs. 42°). However, the latter agreement between angles and distance may be coincidental. No such agreement

Table 8. Angles between Group Discrimination Vectors

Group	A/J grp	C57 grp	A/J iso	C57 iso
A/J grp	0.0			
C57 grp	−22.99	0.0		
A/J iso	41.97	−23.13	0.0	
C57 iso	−39.23	25.28	−9.20	0.0

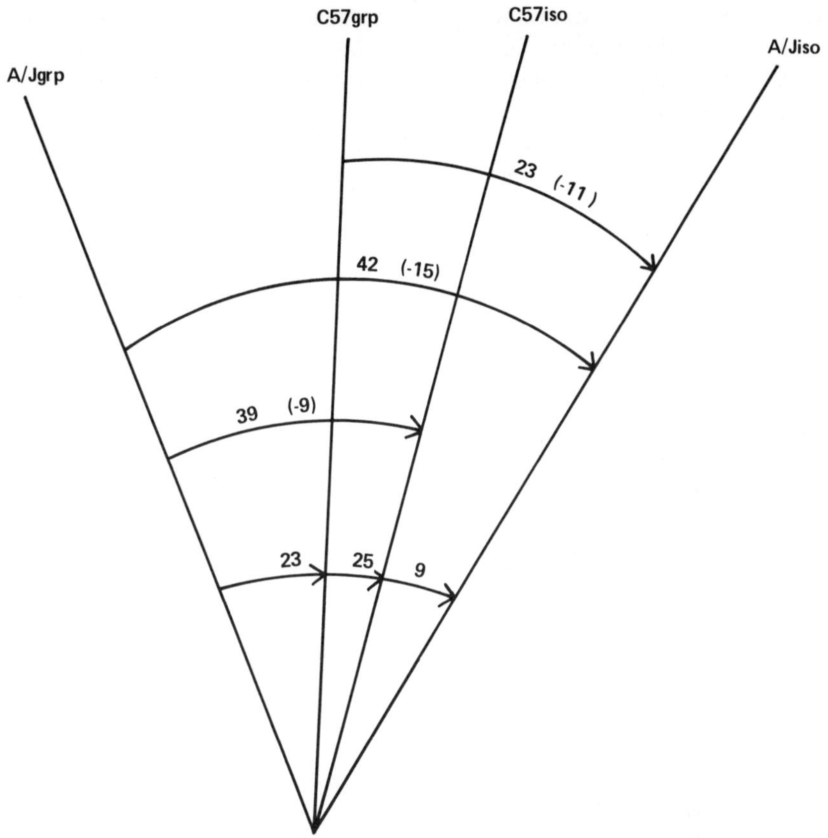

Figure 4. Angular comparison between group discrimination vectors. Departures of group vectors from a plane are given in parentheses as percentage discrepancy, for example, 39° between A/J grp and C57 iso departs from 48° (23° + 25°) by −9°. The 15° departure from a plane between A/J grp and A/J iso indicates that the vectors occur in three dimensions.

need occur. This is confusing only if one's concept of difference is fixed in quantitative terms as was the case in distance and all other analyses prior to angular comparison. The angles contrast the orientation of groups, independent of distance apart, in discriminant space. Since the two isolated groups are only 9 degrees apart, qualitatively they are much the same. The bases for the qualitative differences are beyond the scope of this chapter, but could be unraveled in conjunction with the aforementioned component analysis of all groups.

At this point we might conclude the analysis because examination of discriminant space was possible by distance and discriminant vectors.

However, we shall continue with canonical analysis of discriminance, both to provide an example of its use and to demonstrate how it relates to distance analysis. Table 9 presents the canonical vectors, eigenvalues, chi-squared for testing the significance of eigenvalues, chi-square degrees of freedom, chi-square probability, and percent trace (percent of the total difference between groups extracted by each canonical axis). Although only the first canonical axis displays statistically significant discrimination between groups [by a chi-square test provided by DISANAL and discussed by Cooley and Lohnes (1971) but not here], the second canonical axis is also used in the plot of individuals in canonical discriminant space (Figure 5).

The first canonical axis extracts 84.55 percent of the differences between groups in discriminant space. As one might expect, the first axis separates the strains. The second axis extracts 12.77 percent of group differences, but its lack of significance would lead to questioning its merits as a discriminator. For the purpose of developing the example, we shall temporarily ignore this fact. Figure 6 is a plot of group centroids, 95 percent confidence radii, and vectors of variables. A comparison of Figure 6 with Figure 3 (the distance model rotated to provide the same orientation as the canonical graph) should verify that canonical analysis is an analysis of distance in the same discriminant space.

Table 9. Canonical Vectors Standardized for Within and Related Statistics

	Canonical Vector		
Variable	1	2	3
HOLES	2.159	1.160	0.398
NOSE	−0.517	−0.193	0.113
EYES	−0.149	−0.524	−0.729
EARS	0.015	−0.404	−0.059
EDGE	−0.062	0.281	0.709
TOP	−0.185	−0.172	0.035
LEFT	0.088	−0.026	−0.545
GROOM	−0.273	1.004	−0.373
Eigenvalue	2.704	0.408	0.086
Chi Square	50.29	12.31	2.38
N.d.f.	24	14	6
Probability	0.001	0.581	0.892
Percent Trace	84.55	12.77	2.68

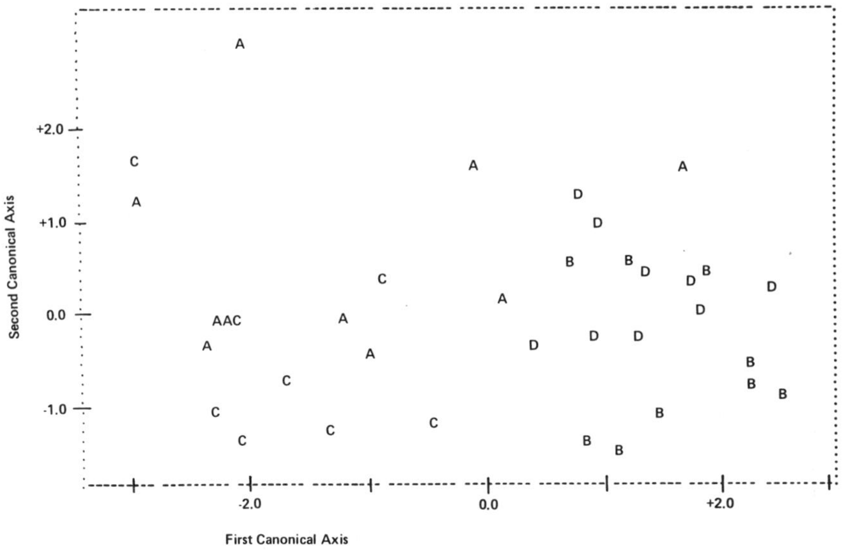

Figure 5. Plot of individuals along the first two axes of canonical discriminant space. The positions of A/J grp, C57 grp, A/J iso, and C57 iso mice are represented by the symbols A, B, C, and D, respectively.

Notice the direction of the various vectors; for example, HOLES is oriented toward the right of the graph and is positive for both canonical axes. A vector's direction is that of positive "push" of that variable; for example, NOSE increases into the negative quadrant of both canonical axes. The direction of some variables is consistent with their independent actions. For example, holes visited and ear dips "push" toward the C57 end of the first axis and are C57 traits (Table 2). GROOM pushes toward A/Jgrp, and Table 2 indicates the unique grooming frequency for this group of mice. One must remember, however, that variables act synergistically, a fact that explains seeming inconsistency on the graph. Thus, the pushes of NOSE, EYES, and LEFT are inconsistent with differences in each of these variables between groups. This apparent discrepancy is understandable in terms of the synergistic interplay of variables. For example, A/J mice occupy a certain position in discriminant space because they examine relatively few holes (HOLES) mainly along the wooden panel (LEFT). Their within-strain separation is mainly a function of positive grooming by A/Jgrp. One should not expect the analysis of vectors of variables in canonical space to summarize Euclidean space relationships between variables. Recall that the vectors are used to provide some clue as to the synergistic interplay of variables in discrimination.

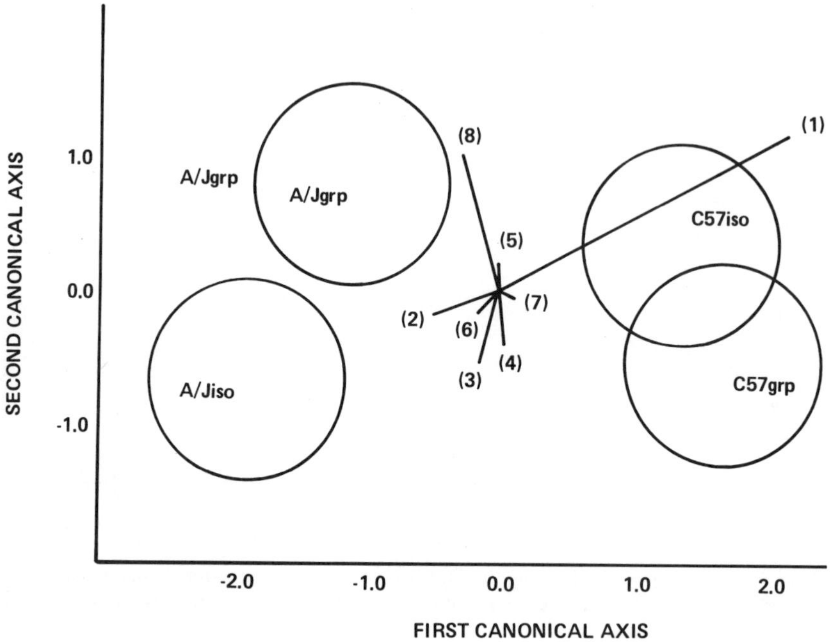

Figure 6. Two dimensional canonical discriminant space showing group centroids, 95 percent confidence radii, and vectors of variables. (1) HOLES, (2) NOSE, (3) EYES, (4) EARS, (5) EDGE, (6) TOP, (7) LEFT, and (8) GROOM.

When compared with distance analysis, the results of canonical analysis of discriminance are less comprehensive. For example, no difference between housing regimes is shown by canonical analysis. This is to be expected—canonical analysis provides orthogonal maximum distances between groups. Such distances need not be of fundamental biological importance. Therefore, the axes need not be the same as those an investigator would select from a distance model. On the other hand, the vectors of variables on the canonical graph are not a feature of distance analysis. The vectors clearly indicate, for instance, the major importance of hole seeking as a discriminator.

We believe that the explicit results of discriminant analysis in this restricted example are sufficiently detailed to demonstrate the power of multivariate analysis. We do not believe that observation or conventional analysis of the data would provide as much information. Moreover, the results are surprising only when first viewed. Although not readily apparent when viewing mice performing in their experimental treatment, pertinent behavior patterns are readily recognized once pinpointed by discriminant analysis.

5. LIMITATIONS

Like principal component analysis, discriminant analysis is very powerful statistically, based upon precise mathematical models, requires assumptions about the nature of the data, and is based upon decisions of the user.

Methodology

When using discriminant analysis one assumes that the best way to compare groups is by overall "form" of individuals. On the other hand, form might be a function of each variable independently. However, the premise that form represents a synergistic interplay of variables appears consistent with biological knowledge. Biologists generally accept as fact that living phenomena are greater than the sum of their parts. On the other hand, if implemented, the procedure of contrasting group components allows evaluation of independent facets of form. Furthermore, when compared, the differences between independent and synergistic form often become clear.

Variables

Of primary importance in discriminant analysis, variables must describe the form to be contrasted. All important variables must be included and extraneous variables must be excluded; variables must "cover" the phenomenon under study. For this reason, familiarity with the field of study is a necessity.

Although variables must cover the phenomenon in any definitive study, in many instances the investigator cannot predetermine the importance of variables. In such cases it is proper to select any variables that might seem appropriate and to use discriminant analysis to evaluate the performance of variables for a final analysis. Again, we do not recommend stepwise discriminant analysis for this procedure, since the stepwise method does not evaluate variables acting synergistically. Since any performance evaluation might well be influenced by happenstance performance of extraneous variables, results should be reexamined for repeatability before making any final conclusion about the importance of variables.

There are at least six criteria for selecting variables:

1. *Theoretical.* Knowledge of the phenomenon should suggest measurements that might be correlated with group differences. Most definitely, measurements of previous or similar studies should be considered.
2. *Accuracy, precision, and cost.* Measurements should be close to true values and repeatable; however, this criterion must be evaluated in terms of money, equipment, facilities, time, and effort required for the study.
3. *Nature of variation.* A good variable varies more between groups than within any single group. Also, no measurement can be a constant for all groups, even if the value of the constant changes from group to group. The methodology applies only to genuine variables.
4. *Direct.* A measurement should not be taken for an entire sample and then applied to each individual of the sample. For example, one should not record a percent response of a sample to every individual of that sample. More important in ethology is the inherent indirect nature of measures of behavior. A real challenge to the ethologist is to obtain good, direct measures of behavior.
5. *Inconsistency of variables.* No variable can change in meaning and specification from individual to individual. This pertains to the definition of populations. Each population must be so defined and sampled that each measurement has the same basis in all individuals of all samples.
6. *Completeness.* Each individual should consist of a complete set of measurements. Incomplete data virtually destroy the details of form.

Assumptions and Transformations

Conventional transformations of univariate data often do not satisfy multivariate assumptions so it might be best to use no transformations. How then can we use discriminant analysis? There are three bases for justification. First is the apparent robustness of manova. Since the most critical assumption is equality of group covariance matrices, recall that for two groups, if sample sizes are large and equal, inequality of dispersions has no real effect on the results of a manova. Therefore, there is justification for performing a discriminant analysis. However, the influence of inequality of group covariance matrices upon discrimination vectors, canonical axes, and generalized distances is poorly known. In spite of this, studies indicate that reliance can be placed upon these statistics as aids to interpreting relationships. Their shortcomings appear to exist mainly in the precise classification of individuals to groups. This leads to the second justification. Classification is by Geisser classification probabilities which Monte Carlo simulations indicate perform well. The third area of justification can come from angular comparisons of various vectors, especially discrimination vectors and canonical axes, with other vectors not so subject to assumptions. However, this is a subject beyond the scope of this chapter.

Evidence consistently indicates the robustness of methodology. However, there is a method for checking the influence of unequal covariance matrices: prior to analysis all original variables are transformed to logs. If the log analysis agrees with the raw data analysis, the raw data analysis generally is considered acceptable. The log transformation serves an additional purpose in that it often brings about linear relationships in otherwise nonlinear relationships between original variables (Blackith 1965). Discriminant analysis assumes a linear function among the original variables.

Objectivity of Methodology

Discriminant analysis, like all multivariate analyses, has been shown to be based upon rigid assumptions about the nature of variation. Also, an attempt has been made to verify the appropriateness of such approaches to examining variation. However, if one accepts that the assumptions are biologically realistic, individual studies in a sense become objective. The investigator cannot influence the results once data are collected. The data determine the outcome of the analysis. It is on this basis that we consider multivariate analysis to be an objective procedure for studying behavioral phenomena.

6. COMPUTER PROGRAMS FOR DISCRIMINANT ANALYSIS

The computations discussed in this and the previous chapter go beyond the scope of the computer programs with which we are familiar. In our opinion, the output from most programs often is difficult to interpret because important aids to interpretation are not included. For this reason, the computer program DISANAL was written.

DISANAL will perform a single-group component analysis or a multigroup discriminant analysis. In the single-group principal component analysis option, the program outputs all the interpretive aids described in the previous chapter plus certain other aids not generally so helpful. The discriminant analysis option provides output well beyond the scope of the present chapter. The goal behind writing DISANAL was to obtain maximum understanding of the form of phenomena in reference to variables, individuals, and groups. This goal is fundamental in a more detailed consideration of principal component analysis and discriminant analysis (Pimentel, in preparation). As DISANAL now exists the only version requires more than 500 K on an IBM 360/40 computer and runs in a partition of 100 K in a complex overlay system.

CHAPTER

10

Systems Diagrams

B. DENNIS SUSTARE
Clarkson College

Systems diagrams are figures that represent some features of the operation and relationships of a system. In ethology, "system" means many things, such as the drinking control system of a Barbary dove, the social system of a tribe of baboons, or the communications system of a dancing honeybee and her followers in the hive. Ashby (1956) defines a system as being a list of variables that are to be taken into account in an investigation. This definition is highly pragmatic, for although in theory one can "draw a circle" around anything in the universe and say that everything within the circle is part of the system, in reality there is an infinite number of variables that could be measured for any real object. Without a restriction on the variables of primary interest to the researcher, an impossible task would face the person who attempts to understand how a dynamic system operates.

Once a particular system is defined, there are three main approaches to the subsequent analysis of the system. These are the state, structural, and relational approaches. The state approach depends on the concept of the *state*, defined as the set of specific values of the variables in a system at a given instant in time (strictly speaking, the minimum such set required to define uniquely the state as distinct from all other states). Each time the system variables have those same values, the system is said to be in the same state. This method finds the relations between states by listing the states, analyzing sequences of states, and finally establishing a model with a good ability to predict future states of the system.

The structural approach breaks the system into smaller subsystems and components and determines the connections between these components with regard to the flow of energy, matter, and information. The relational approach determines first correlations then causal relations between variables in order to find what happens to the values of variables when the values of other variables are changed. A total study of a system might well involve all three methods, and in fact they overlap to a considerable degree, for to answer the questions raised by one approach often requires that another avenue be taken. Each of these three approaches to systems analysis may benefit from the use of systems diagrams, both at the stage in which the investigator is first formulating models and at the stage of publication where one is attempting to communicate models and results to other readers.

Most systems diagrams in ethology are drawn as figures containing boxes or circles interconnected by lines or arrows. No matter what the boxes and lines represent, one can relate this form of diagram to the mathematical theory of graphs. A graph (not to be confused with the familiar plotting of values of a dependent variable against an independent variable) is merely a geometric structure with a set of points (vertices) in space interconnected by a set of curves (edges) (Busacker and Saaty 1965). In the diagrams, a box can be thought of as a vertex, and a connecting line thought of as an edge. Graph theory can be used to help define types of system diagrams and to understand their uses.

In this chapter, Section 1 lists the assumptions usually made when drawing systems diagrams. Section 2 presents the methods used in the state, structural, and relational approaches. Section 3 gives a variety of examples of system diagrams, while Section 4, on interpretation and limitations, presents the relationships among the different types of diagrams.

1. ASSUMPTIONS

Because of the variety of diagram types available, there are few general assumptions for the appropriate use of these techniques. Possibly the most important assumption is that variables are operationally defined. There must be some way to assign values to variables through some measurement technique. Not all variables in ethological systems require interval or ratio scales, though there will be increasing use of such variables in the future as ethology becomes more like a physical science in its rigor of analysis. Nevertheless, it is not sufficient merely to name a variable without simultaneously specifying a method for determining its values.

The state of a system must be unambiguously defined by the values of the state variables of the system. This means that states must be mutually exclusive, and in general that the behavioral units used to determine the states—normally comprising part, if not all, of the system variables in an ethological system—should be discrete (discriminable from one another). This need not imply a deterministic system, however; in fact most behavioral systems turn out to be stochastic, at least with our present level of understanding.

For uniformity and clarity, it would seem best not to mix types of diagrams. Every box in a diagram should stand for the same sort of thing, as should every line or arrow. The same state or variable should not be represented by two or more separate boxes or arrows, although at times it may be difficult to accomplish this without losing information about the system.

There are further assumptions specific to certain types of diagrams. These are dealt with in the sections relating to those diagram types under Method (Section 2) and Examples (Section 3).

2. METHOD

State Approach

Kinematic Graphs. The kinematic graph is a figure used to represent a temporal or sequential ordering of states. In ethology a state will normally be some behavioral unit, the type of unit depending on the level of analysis for the investigation. An operational definition for each behavioral unit will include the values of the state variables. A specified set of these values defines a single state. Since the system can be in only one state at a time, each state as defined must exclude all other states. Furthermore, the states must be discrete (i.e. discriminable from one another). If, for example, the behavioral units being used are the action patterns of duck courtship, then "grunt-whistle" and "down-up" could be two different states since they are mutually exclusive and readily discriminable.

Once the states are adequately defined and listed, one may begin an analysis of the sequences of these states (see Chapter 4). Each time one state ends and another begins, a *transition* occurs. It is most convenient to consider transitions as instantaneous, or more precisely that a transition takes less time than the smallest unit of time being measured. Therefore at one instant the system must be in one and only one state; the system is never between states. New states can always be defined so as to attain this condition, if necessary splitting "slow" transitions into intermediate states separated by "zero-time" transitions. Such transitions, as well as intermediate states of shorter duration than the smallest unit of time measured, are also known as *events*.

In the kinematic graph a box or circle is drawn for each state. An arrow is drawn connecting two boxes when a transition may occur between those two states. The direction of the arrow shows the temporal sequences of states, always pointing toward the state following the transition (the transform, in Ashby's 1956 terms). Recurrent arrows (from one state back to itself) are drawn when a series of events are recognizable as distinguishing separate occurrences of the same state in succession. The size of the box may be used to show the proportion of time actually spent in each state.

When all the arrows are drawn, the diagram (Figure 1) gives a picture of the behavior of the system. One can start in any state and imagine a point traveling through the graph in the direction of the arrows, one step at a time. The history of the travel of this *representative point* will describe the behavior of the system. For example, in Figure 1, one

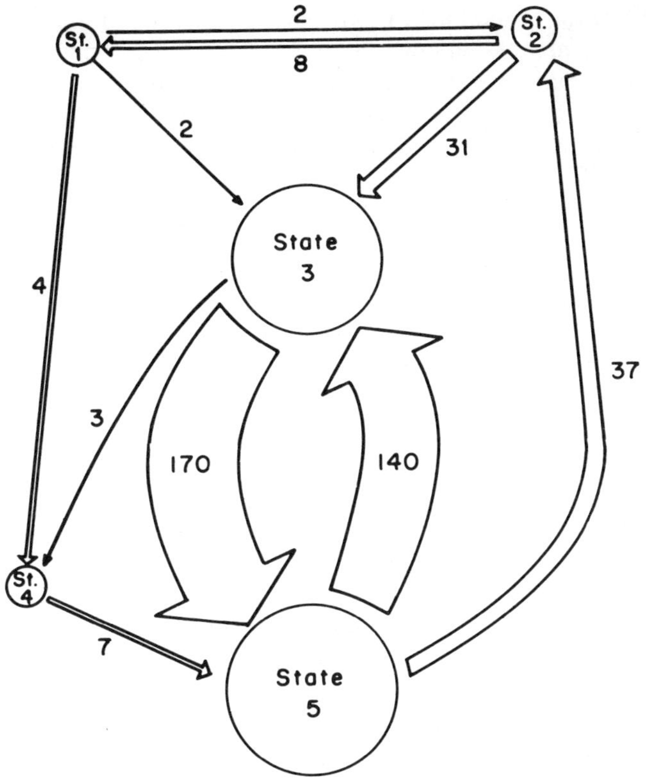

Figure 1. Kinematic graph. Circles: states; arrows: transitions. The size of circle represents amount of total time spent in that state or number of occurrences of that state. Width of arrows and their labeled numbers represent the frequencies of transitions.

sequence might be

$$1 — 4 — 5 — 3 — 5 — 3 — 5 — 2 — 3$$

where each number represents a state and each dash a transition, the whole being a *protocol*, or specific example of a sequence. In a deterministic system, at most one arrow will leave from each box. From any starting state in such a system, the representative point will eventually come to rest in a persistent state (one with no arrows leaving) or will enter a continual cycle from which it can never leave. The kinematic graph makes analysis of deterministic systems very simple. Note that the familiar computer flowchart is nothing more than a kinematic graph; each block represents an activity, or current state, for the computer processing unit.

More generally, some states may have two or more arrows leaving them as with most of the states in Figure 1 (i.e. a knowledge of the existing state is not in itself sufficient to say what the next state in the sequence will be). When such transitions occur, one is dealing with a stochastic process. This is another way of saying that some transitions in the system are probabilistically determined. Diagrams of stochastic processes may have the arrows labeled with fractions to show the probability of each path being traced, or the actual observed frequencies of transitions may be used as labels (as in Figure 1). The graph may also be drawn with the width of the arrows used to show frequencies (again, as in Figure 1) to give the viewer a better feel for the main pathways through the graph.

Since the animal can be in only one state at a time, in theory the total frequency of all transitions leaving a state should equal the total frequency of all transitions entering that same state. Similarly, if probabilities of transitions are used to label the arrows, the sum of probabilities from a state should equal 1.00. In any real case there may be several causes of deviation from these ideals. First, some states may not be included if they are rare and of little importance to depict. Second, some uncommon transitions may be omitted, to clarify the diagram and accentuate the main transitions in the system. Finally, sampling from a real population is likely to result in numerous short sequences rather than one very long sequence, so that the frequencies would no longer be expected to balance. In particular, if the starting point of sequences is chosen at random, sequences will be most likely to begin with those states in which most time is spent.

Further Applications of Kinematic Graphs. One often wishes to compare two systems; for example, to compare the sequential behavior of two species of birds. Two deterministic systems are *isomorphic* if it is possible to make a one-to-one transformation of each of the states of the first system into a corresponding state of the other system. Even if one system contains more states than the other, it may still be possible to do some ''lumping'' and show a resemblance between the two. Two deterministic systems are *homomorphic* if it is possible to make a many-to-one transformation from the states of the more complex system into a less complex form that is isomorphic with the other system. These comparisons can often be done by inspection of the kinematic graphs of the two systems.

With two stochastic systems, simply having the same form of kinematic graph does not mean the systems are isomorphic. One must examine the transition probabilities to judge the degree of resemblance between the two systems. For such a comparison the transition frequencies may be thought of as the values of variables so that the two distributions might be compared by using, for example, a Kolmogorov-Smirnov two-sample test. This is a test as to whether two independent samples have been drawn from the same population or from populations with the same distribution. To perform the test, one would rank the transitions of the first system in any (arbitrary) order, and rank the transitions of the second system in exactly the same order (according to the presumed isomorphic correspondences already determined). Now the frequencies for each transition are summed to yield the cumulative distribution needed for the Kolmogorov-Smirnov test. Refer to Siegel (1956) for the details of the test.

Similarly, having the transition frequencies of, say, a large number of individuals of male Prairie Chickens on a booming ground, one might use the multivariate techniques previously discussed (Chapters 5, 8, and 9) to discern differences between males or to group them based on their sequential behavior. There is no reason why other variables could not be similarly combined with transition frequencies so as to relate behavioral differences to environmental or morphological differences. In this case, these might include size or age of each animal, ambient air temperature, mean distance to adjacent males, and other such variables.

Higher-order kinematic graphs may be used to display longer-term relations between states as revealed by sequential analysis. For example, a second-order kinematic graph can be drawn from the frequencies of triads in behavioral sequences. A triad is a string of three states in succession, thus a combination of two successive transitions. In such a graph, the arrow passes from the box for the initial state to the box for the terminal state, being labeled with the identification of the intermediate state in the triad. The width of the arrow may be used as before to represent the frequency of transitions. A simplified counterpart of a second-order kinematic graph sums the frequencies of all triads that share the same initial and terminal states and uses only one arrow to represent this combined second-order transition. The method may be extended for even higher-order graphs.

State-Space Diagrams. Another way to represent the behavior of a deterministic system is to plot the transitions between states as vectors in n-dimensional space. The n axes of the space are each assigned to one of the state variables (those variables whose values are necessary and sufficient to define the state of the system). Each state is plotted as a single point in the space, and each transition is plotted as a vector from one point to another point. When all such vectors are plotted, the result is a state-space diagram.

The most easily visualized case is obviously when only two variables (i.e. two dimensions) are required to specify the states, so that a plane may be used as the state-space, as in Figure 2. The shaded areas represent states that may never be attained, such as those outside the physiological limits of the animal. The point x_1, y_1 in Figure 2 represents a persistent state in this system. Not only is this point stationary (i.e. a vector of length zero is associated with it) but it is locally stable (i.e. any small deviation from this point results in a return to the same point).

If there is a large number of states, as is the case when the state variables can take many values, it is usually sufficient to plot only some of the vectors, with the assumption that nearby vectors are similar in direction and magnitude (Figure 2). Chains of vectors may also be connected to form trajectories, or paths of behavior. Such trajectories may be plotted by starting at some state and letting the system run on

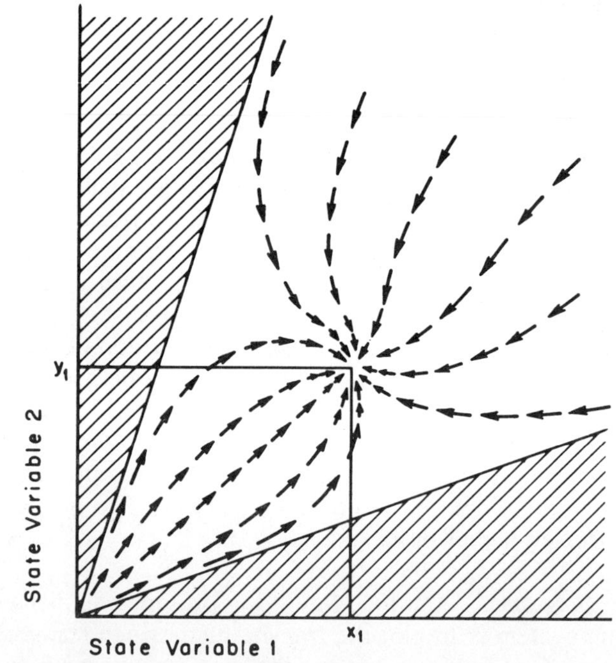

Figure 2. State-space diagram. Each point in the plane represents a state (as determined by the two state variables). Each arrow is a transition vector from one state to the succeeding state. x_1, y_1 is a persistent state, both stationary and locally stable.

from there, recording intermediate states through which the system passes.

With any real example of plotting the state-space for a complex system, the plot itself may aid the researcher in determining what additional data are needed. The procedure is to choose some starting state and plot the trajectory of behavior leaving that state until one of four conditions is met: (a) a limit cycle is entered, (b) a persistent state is reached, (c) the trajectory approaches one of the first two conditions as an asymptote or converging oscillation, or (d) the trajectory is unstable to the extent of leaving the region of interest of the state-space (as with a positive feedback situation). Then another starting state is chosen in a way to give maximum information about the remaining portions of the state-space (areas through which the preceding trajectories have not passed).

Structural Approach

Information Networks. In the structural approach, the first step is to subdivide the system into subsystems and components. Each of these components should be a clearly defined unit. Examples might include the individual animals in a society, when one studies their communication system, or the physical components presumed to make up a behavioral control system in a single animal. The initial stage of knowledge about the system might be only that there is an exchange of information between certain components. The relations between components in such systems may be shown by an information network.

In an information network each box in the diagram represents a single component or subsystem. The arrows show information channels that may be physical connections, such as axons in a diagram of nerve cells, or may be communication channels in the environment, such as the air for the passage of auditory signals. The direction of the arrow shows the direction of information flow through the channel; there may be arrows going in both directions between two components. In diagrams of communication networks, arrows may be labeled with channel capacities (i.e. the maximum amount of information that may be transferred over the channel per unit time). Figure 3 shows an information network representation of neural connections. Here the circles stand for cell bodies and dendritic fields, and arrows represent axons.

In contrast to the kinematic graph, there is no representative point that follows transitions from one box to the next in a clear sequence. Rather, information will often be flowing in many parts of the network at any given instant. As a result, the boxes can also signify points of

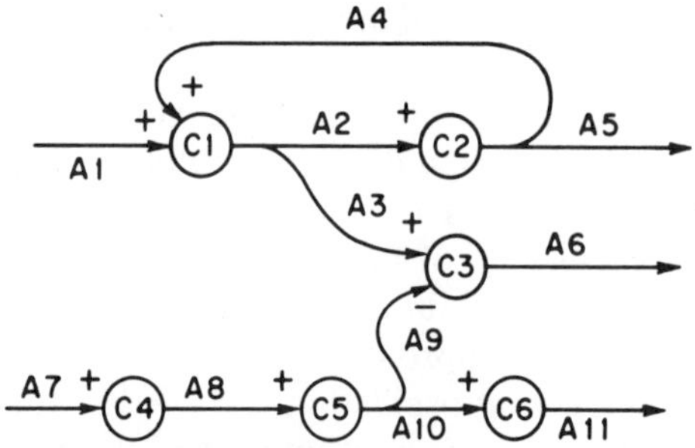

Figure 3. Information network representation of neural connections. Circles: cell bodies and dendritic fields (labeled C1 through C6); arrows: axons (A1 through A11); + and − are used to indicate excitatory and inhibitory synapses.

integration, where signals from several inputs are combined to determine what the output from that point will be. It is important to remember that the output in such a case will not (in general) be equal to the sum of the inputs to the same vertex in the network. For example, consider a logical network in which one component, acting as a filter, is continually comparing two inputs and will only respond (with a signal at the output) when there is a simultaneous arrival of signals at each input. The actual information content (in bits) of the two input signals impinging on this component might be considerable, and yet much time might pass before the condition of simultaneity occurs and the component responds.

Further Application of Information Networks. A particular form of information network useful in ethology is the *sociogram*. This diagram identifies each individual in a social group as a separate component (thus a separate box in the diagram) and uses arrows to show interactions between pairs of individuals. If the arrows are labeled with, for example, the number of aggressive encounters between individuals (with the "winner" of each operationally defined) the result is a sociogram showing a dominance hierarchy of the group. The arrows might similarly be labeled with the number of incidents of allogrooming, frequency of reproductive displays between individuals, or whatever pairwise interaction is of interest.

When studying a social group as a communication network, one may use the information network diagram for two types of analysis. In the first case, a cost is associated with each arrow, or information channel. An example of such a cost might be the energy needed to transmit a message through that channel; it might also be the time it takes to send the message. (Note that in the latter case the delay imposed at each relay point, or individual animal, might be of much greater importance than the actual transmittal time in the channel; on the other hand, a long message might require quite a long transmission on each leg.) Once the costs have been assigned, one problem that might be addressed is to find the lowest-cost route for the transmission of information between any two pairs of individuals in the group. For a simple network, this problem may be easily solved by inspection. For large, complex networks the solution can be much more difficult and require successive approximations using a computer. Consider also the complications likely to arise in any real network: each link may have a probability of error associated with it, redundancy (e.g. sending the message over several routes at once) may be built in to assure reliable communication, and the network structure itself may change through time. One might also wish to compare two networks to discover which is the more efficient design (considering average message cost, reliability, etc.). Fortunately, these wide-ranging problems have an extensive literature, especially in the areas of operations research and dynamic programming. Some good introductions to these methods include Singh (1968), Bellman and others (1970), and Siemens and others (1973).

The second type of analysis, closely related to that just discussed, puts some upper limit on the ability of each link to transfer information. This would be the channel capacity for a communication link. If large numbers of messages are to be transmitted back and forth through the network, these limits may become important. One might then wish to find where the bottlenecks are in the system, and how the network might be redesigned to operate more efficiently. Again, the methods of operations research would be appropriate.

Flow Networks. Very similar to an information network is a flow network. Here the flow of interest is some substance, either energy or matter, rather than a flow of information. The critical difference arises from the laws of conservation of matter and energy—stating that matter and energy cannot be created nor destroyed (though transformation of form, even between matter and energy, is possible). No such laws of conservation of information exist. As a result, certain relations must hold true in flow networks, relations that allow more powerful analytical methods than those available for information networks.

First consider two types of variables, the A(across)-variable and the T(through)-variable. An A-variable is measured between two points in a system, usually as some difference between those points, and includes such variables as velocity (for a mechanical system), voltage (electrical systems), and pressure (hydraulic systems). A T-variable must be measured by having some force of flow expressed through the point of measurement, and includes variables such as force (mechanical systems), current (electrical systems), and fluid flow (hydraulic systems).

In a flow network, two useful and interesting conditions will hold true: (1) the algebraic sum of all the A-variable values around any cycle (a closed path along which one always progresses in the direction of the arrows) must equal zero; and (2) the algebraic sum of all the T-variable values of paths entering and leaving a single vertex must equal zero. One modification to these rules is that there may be storage of a substance at a vertex. If such storage is possible, the two types of algebraic sum must include the amounts of matter or energy entering or leaving storage. In flow networks, the boxes or vertices in the diagram are often called compartments; strictly speaking, to be a compartment there must be no time involved for flow within a compartment (i.e. mixing is instantaneous), time being required only for flow between compartments (along

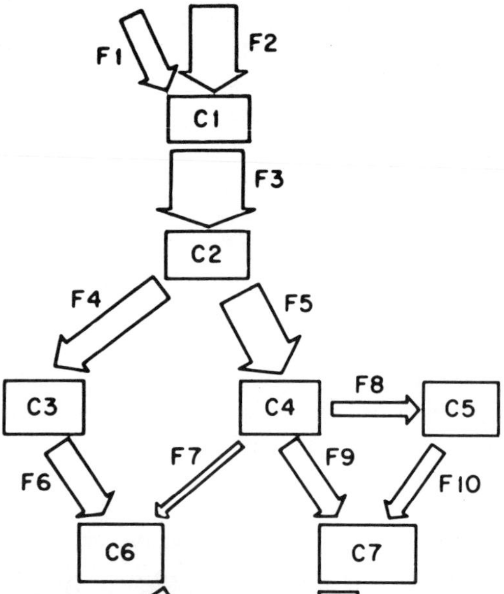

Figure 4. Flow network. Boxes: compartments (C1–C8); arrows: flow pathways for matter or energy (F1–F12). The width of arrows is proportional to the amount of flow in that pathway.

the branches in the network). In electrical systems the two conservation rules are known as Kirchoff's Laws; in physiology the second rule is Fick's Principle.

Figure 4 shows the form of a flow network, with the width of arrows proportional to the amount of substance flowing through those branches. A flow diagram should be labeled to show the amounts, or at least the percents of material or energy flow through each branch. Care should be taken that the conservation rules are met in the diagram. Flow should be in only a single direction within any one branch (i.e. only net flow is shown).

These diagrams are probably of less use in ethology than are information networks. When considering behavior whose function is the movement of material, such as some of the grooming behavior of an animal, the flow network can be used to represent the flow. If it is not possible to measure directly the amount of material transferred, the diagram might be labeled with the frequencies of each behavioral pattern used to transfer the substance (note that these frequencies need not meet the conservation rules). Such a diagram, showing the presumed material flow due to grooming behavior of houseflies (Sustare and Burtt 1976), is shown in Figure 5. Since frequencies of patterns and not amounts of material transferred are plotted in the diagram, the conservation laws need not apply.

Consideration of the movements of individuals between groups in a society also could be depicted in this manner. In addition, behavioral ecologists have used flow networks to show the movement of energy and nutrients through an ecosystem, as in Odum and Ruiz-Reyes (1970).

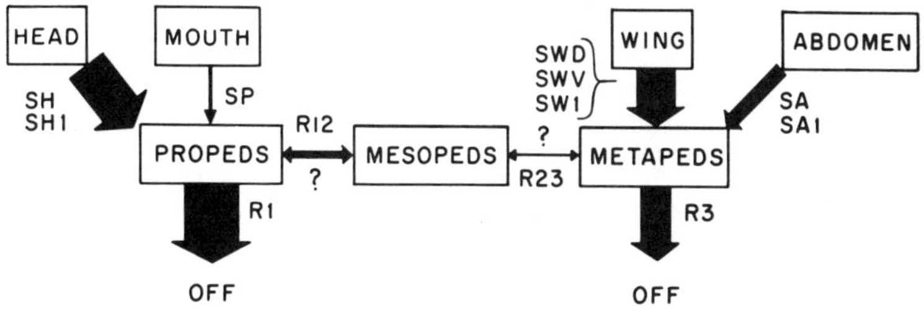

Figure 5. Flow network of the presumed material flow as a result of grooming behavior by houseflies (*Musca domestica*). Boxes indicate the parts of the body being cleaned; arrows represent behavioral patterns that transfer material from one body part to another. Arrows are labeled with code designations of various grooming patterns; the width of arrows are proportional to the summed frequencies of all the patterns indicated. "Off" indicates the points at which material leaves the fly because of mutual leg rubbing. (Sustare and Burtt 1976.)

Relational Approach

Association Diagrams. The association diagram is used to show statistical correlations between variables. In the diagram, the variables are indicated by boxes, and lines are used to connect each pair of boxes that show a significant correlation. Lines may be labeled or drawn in different styles to indicate positive or negative correlations, or to show the degree of correlation. There is no direction given for the lines (no arrowheads) since a simple correlation gives no ordering to the pair of entities so correlated. Figure 6 is an association diagram showing correlations between 10 variables in a system. Note that the graph is not connected, there being no strong correlations between any of variables 1 to 7 and the variables 8 to 10. It is often convenient to treat the components of such disconnected graphs as separate subsystems.

An association diagram also may be used to indicate the degree of similarity (measured by some sort of similarity index, see Chapters 5 and 8) between the entities represented by the boxes. For example, a behavioral state may be thought of as one value of an n-dimensional

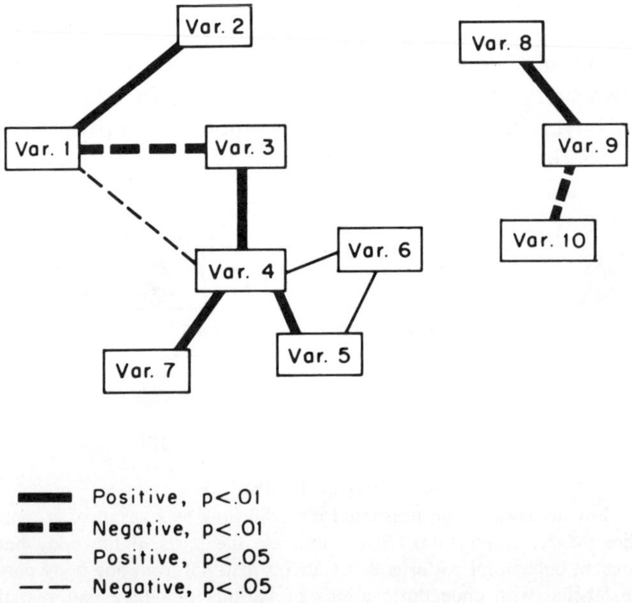

Figure 6. Association diagram. Boxes: system variables. Lines indicate significant correlations between pairs of variables. Positive or negative correlations and degree of significance is represented by the style of line, as indicated.

vector from the origin, and the boxes in the diagram would represent these vector variables, giving a way to depict correlations between entire behavioral systems or distinct motivational systems (in the sense of McFarland 1971). Often the distances between boxes in the plane (or space of greater dimensions) are used to indicate similarity, with the smallest distances representing the greatest similarity. In such a case it is often convenient to reduce the boxes to labeled points, as in an ordination diagram (Stephenson 1974).

An association diagram may be used to help see patterns of relationships between variables or behavioral units, and may suggest the direction of additional investigation. Further study may allow one to modify an association diagram into the following diagram type, the diagram of immediate effects.

Diagrams of Immediate Effects. The causal relationships between variables in a system can be shown with a diagram of immediate effects (sometimes called a symbol-and-arrow diagram, Riggs 1963). In this diagram each variable is represented by a box. Arrows show the qualitative effects that each variable has on the others. In theory, causal relationships are determined by making a primary change in a single variable and examining all other variables to see if changes occurred in them together with or just after the primary change in the first variable. If so, those variables showing change are dependent on the first variable, and the diagram will show an arrow passing from the independent variable to each dependent one. In the case of a monotonic relationship between two variables a solid arrow may be used to represent a direct relationship, whereas a broken arrow indicates an inverse relationship. These designations may also be accomplished by using + and −, respectively. Figure 7 is a diagram of immediate effects showing the relations among seven variables. The loop of 4 — 5 — 6 — 7 — 4 indicates a negative feedback situation (i.e. a small increase in

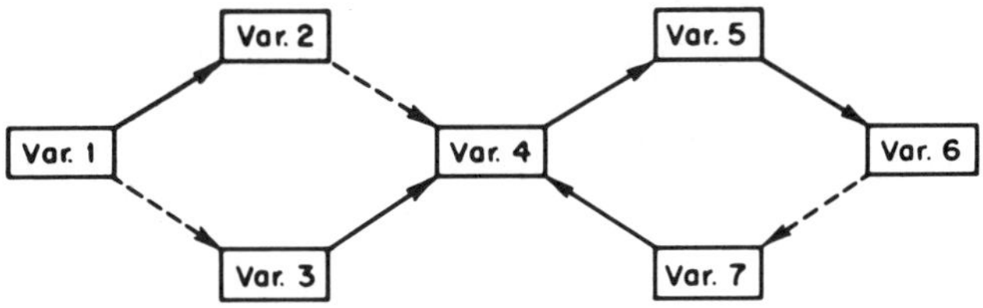

Figure 7. Diagram of immediate effects. Boxes: system variables. Arrows show primary causal relationships between pairs of variables. Solid arrows: direct relationship; broken arrows: inverse relationship.

variable 4, operating through the influence of intervening variables, acts to diminish the magnitude of that same increase).

This diagram could be redrawn as a diagram of ultimate effects (Ashby 1956), to show causal relationships even if they operate through long chains of causality. To do this, Figure 7 would be modified by the addition of arrows between all variables connected by chains of arrows in the same direction. For example, a solid arrow would be drawn from variable 4 to variable 6. The type of arrow to be drawn is determined by counting the number of broken arrows (inverse relationships) in the chain; if the total is even (including zero) draw a solid arrow, otherwise a broken arrow. The result of redrawing Figure 7 is seen as Figure 8. Unfortunately, there are two problems with this type of diagram; first, the large number of arrows (in this case 26 instead of the 8 in Figure 7) becomes confusing, and second, when there are multiple paths possible (as between variables 1 and 4) there can be ambiguous results depending on the path followed (though not in this instance).

Riggs (1970) points out that the transients caused by changing a variable are often so rapid as to be regarded as instantaneous, so that the diagram of immediate effects really shows the resultant effects after steady state is reached. This would seem to imply that the diagram of ultimate effects should be used in these cases, but in fact throughout his book Riggs makes use of diagrams of immediate, not ultimate, effects.

Signal Flow Graphs. One of the limitations of the diagram of immediate effects is the restriction that only monotonic relationships (direct or inverse) can be depicted. If there is a more complex functional relationship between two variables, there is a need for a diagram that can depict this system. The signal flow graph meets this need. In this diagram, each circle (called a node of the graph) represents a variable. The relationships between variables are shown by the arrows (called branches of the graph), as in the diagram of immediate effects. Here, however, the

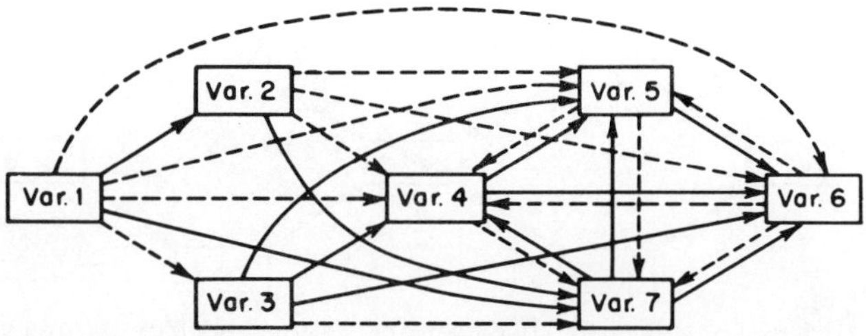

Figure 8. Diagram of ultimate effects. This is the same as Figure 7, except that causal chains are traced as far as possible through intermediate variables. See text.

arrows are labeled with transmission operators. An operator denotes the cause-effect relationship between two variables (Huggins and Entwistle 1968); it indicates the operation that must be performed on the input variable (variable at the preceding node) to yield an output signal (at the following node). Now it is possible to display any sort of functional relationship, rather than being limited to saying the relation is direct or inverse. An example is shown in Figure 9.

The value of signal flow graphs resides in the mathematical methods that have been developed for subsequent analysis of the system through the diagrams. I will briefly mention some of the rules of signal flow graph algebra and manipulation. The *addition* rule states that the value of a variable represented by a node is equal to the sum of all signals entering the node. In other words, each immediately preceding node has its variable serve as the input variable for the transmission operator (the connecting arrow), producing an output signal. Each of these signals is then added, with the result being the value of the variable of the node in question. All signals being added must, of course, be measured in the same units.

The *transmission* rule states that the value of the variable at a node is transmitted on every branch leaving that node. This input value will be operated upon by the various transmission operators of each branch, so that typically the output signals at the end of the departing branches will not all be the same. The *multiplication* rule allows simplification of a graph, since a pathway of two or more branches in succession can be replaced by a single branch with a transmission operator equal to the product of the old individual operators.

Numerous other rules (Mason's rules) exist for the reduction or manipulation of signal flow graphs. For a masterful explanation of these techniques, see Huggins and Entwistle (1968). Numerous examples of the use of signal flow graphs in ethology can be found in McFarland (1971).

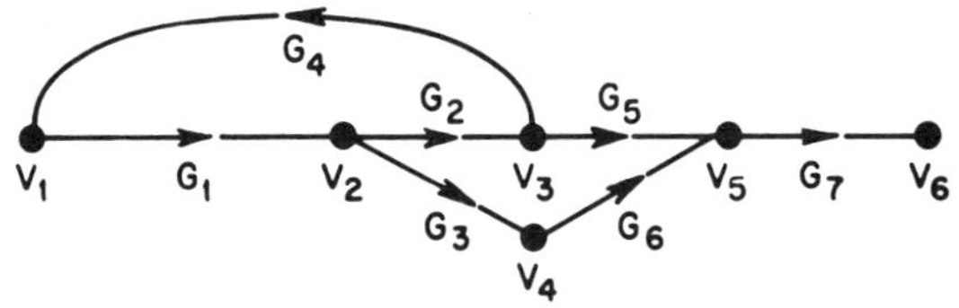

Figure 9. Signal flow graph. Solid circles (nodes): system variables (V_1–V_6); arrows (branches) show the relationship between variables, and are labeled with transmission operators (G_1–G_7). A transmission operator shows the operation that must be performed on the input variable (preceding node) to obtain the output signal (at the following node).

Block Diagrams. A block diagram shows the functional organization of a system, by having blocks (boxes) representing the components (or subsystems) of the system, and arrows showing the interconnections between components and the direction of signal flow along those connections. The diagram is similar to a signal flow graph, except that in the block diagram the arrows are labeled as the variables, and the blocks are labeled with parameters or transfer functions (see below).

Block diagrams may contain other features besides blocks and arrows; namely, *splitting points* and *summing points*. Splitting points allow one variable to be considered as input to more than one block, and are shown as one arrow dividing into two or more. Nothing happens to the value of a variable at a splitting point; in this sense the diagram is *not* like a division of current flow in an electrical circuit diagram. The entire signal flows along each of the branches from the splitting point.

A circle is used to represent a summing point where two or more signals are combined to form a new variable whose value is equal to the algebraic sum of the values of all variables entering the summing point. This summing of course means that all of the variables entering the summing point must be measured in the same units (though *many* examples are seen that fail to meet this restriction). See Figure 10 for an example of a block diagram. Here, variable 9 might also be considered a parameter, providing input to the system, if it is an independent variable external to the system.

The most beneficial use to the ethologist of the block diagram will often be as a heuristic tool, aiding in the visualization of the system and the devising of critical experiments to further understand the operation of the system. For this use, one need not be too strict about the form of the block diagram, even though engineering usage demands more rigid

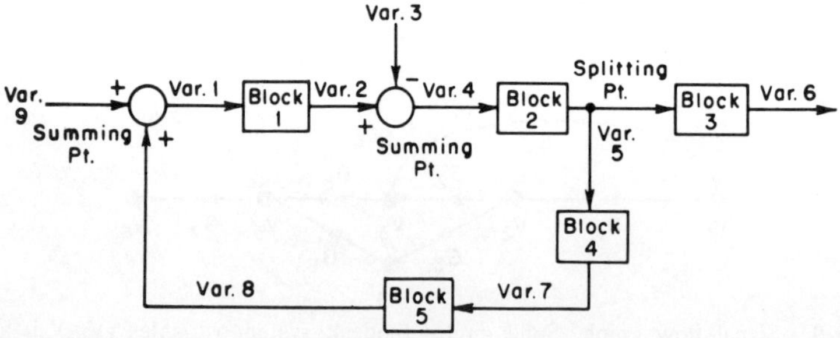

Figure 10. Block diagram. Boxes (blocks): system components or subsystems; arrows: system variables; open circles: summing points (+ and − represent addition or subtraction of variables); solid circle: splitting point.

adherence to certain rules. For example, one might draw a block representing the unknown activities of the central nervous system, with numerous inputs and outputs. Blocks may also indicate nonlinear processes, or even hypothetical processes for which there is only speculative evidence. It can be valuable to use block diagrams in this way for their communicative function, but one should be firmly reminded that fuzzy thinking can be hidden in such diagrams, and misinterpretation may result. Accordingly, I wish to discuss briefly some of the restrictions imposed on block diagrams in engineering usage, restrictions that allow the diagram to be an analytical tool as well as a communicative or heuristic device.

The most important restriction is that the system be linear; this means that the output of a sum of inputs is equal to the sum of outputs when each input is applied separately. The system should be stationary (characteristics not changing through time). Each block should have a single input and a single output (though these may be vector variables). When these conditions are all true, the transfer function of one block is equal to the ratio between the Laplace transforms of the output and input of that block.

The Laplace transform is a difficult concept. The Laplace transform of some function of time is a transformation into a function of s, the complex frequency. The variable s is related to the frequency at which exponential changes occur and the frequency at which sinusoidal changes occur. Particular examples of the complex frequency s are the roots (or eigenvalues) of the characteristic equation found as the solution of a differential equation. Systems whose dynamic behavior is best described by differential equations are often better dealt with in the s-domain (by using Laplace transforms) than in the t-domain (with functions of time) (Riggs 1970).

It is a simple matter to find the output for any input once the transfer function for that block is determined for one particular input and output. Multiply the Laplace transform of the input by the transfer function, and find the inverse Laplace transform of the result to obtain the output. This procedure may be used for a single block or for any portion of the system that may be reduced to a single block with its own transfer function. The easiest way to determine Laplace transforms and their inverse is by looking them up in a table, such as in a recent edition of the Handbook of Chemistry and Physics (Weast et al. 1964). Various Laplace operations are also listed in the Handbook to facilitate use of the tables.

Blocks may represent subsystems or components that are conveniently labeled with nonlinear or with nonalgebraic (e.g. differential or integral) operators; often one sees such a block containing a graphical plot showing the relationship between the input and output. Since these representations prevent algebraic combination of blocks (by multiplying together the transfer functions, as can be done in a linear system), Riggs (1970) advises against such usage and recommends reliance on Laplace transform representation. Only rarely will ethological models be so fortunate to avoid non-linear components, and it seems overly harsh to deny use of block diagrams to ethologists simply because many of the analytical advantages of the diagrams are absent. The linear portions of these diagrams may be dealt with normally. An extensive exposition on block diagrams and control systems analysis can be found in McFarland (1971, 1974a). An excellent introduction to both linear and nonlinear systems analysis, though concentrating on physiological systems, is to be found in Riggs (1970).

3. EXAMPLES

In addition to showing specific examples from the ethological literature, I will use a group of imaginary animals, "Digras," to illustrate several of the diagram types. My reasons for this choice are twofold. First, it is difficult to find ethological examples of each diagram type that are simultaneously able to show the features of interest and are simple and error-free. Second, I am able to show how different types of diagrams are used to bring out different aspects of the same system. Within each section I also refer to some of the better examples appearing in the literature.

Digras form small colonies during the reproductive season. Each female builds a nest, whose thickness is varied so as to maintain a constant light intensity in the nest interior at midday. The females judge the internal light intensity by the degree to which their bodies heat up while they are in the nest. Males bring nest material to the nest site, and the females induce the males to pass nest material to them by either allogrooming the male or by passing food to them (food that the female previously brought back for this purpose from foraging excursions). Several features of this hypothetical system are illustrated by use of systems diagrams.

State Approach

A kinematic graph of the main activities of a female Digra is shown in Figure 11. Each box represents a single category of behavior (e.g. Build

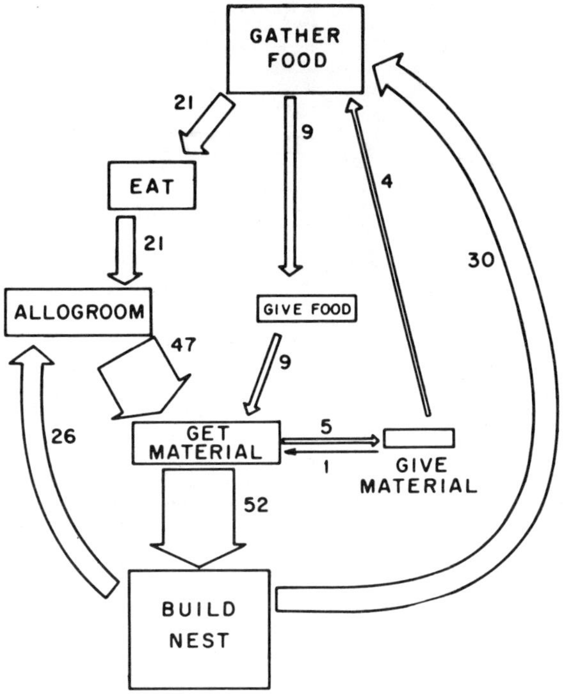

Figure 11. Kinematic graph of activities of a female Digra. Size of boxes is proportional to amount of time spent in each state. Arrow widths and labels represent transition frequencies.

Nest). The size of each box is proportional to the amount of time spent in that activity. The arrows are labeled with the frequencies of transitions between categories, with the arrow width also representing the transition frequency. Note that arrows may pass in both directions between a pair of categories (e.g. between Get Material and Give Material).

As noted above transition frequencies may not balance between input and output at all states, because of sampling conditions. Here, the larger number of transitions *from* Build Nest than *to* Build Nest might be due to having many recorded sequences begin during nest building, the most common state (as far as time spent).

Examples of kinematic graphs in the ethological literature include diagrams by Koopowitz (1970), Figure 12; Parker (1970), Figure 13; Yadava and Smith (1971), Figure 14; Roper (1973); Lerwill and Makings (1971); and Coulon (1971). State-space diagrams have been used less

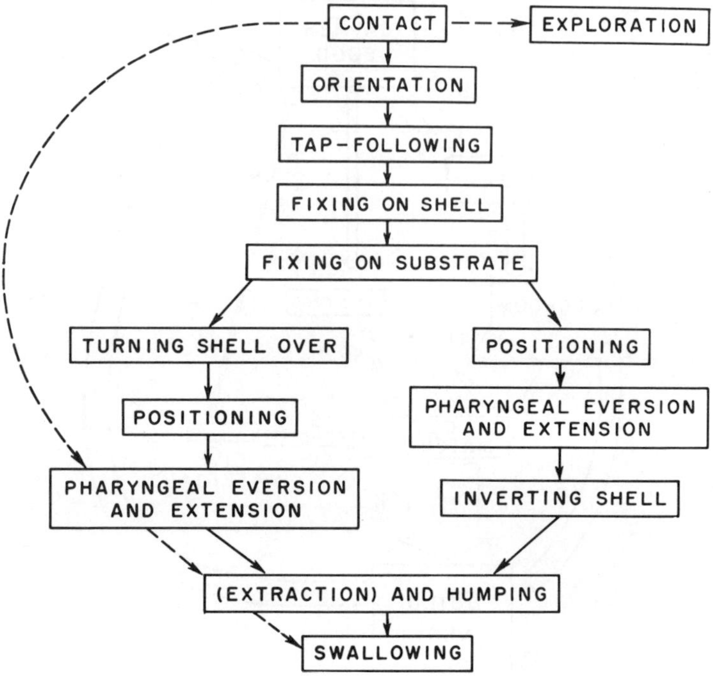

Figure 12. Kinematic graph of prey-catching behavior in flatworms (*Planocera gilchristi*). Dotted arrows indicate the feeding sequence of decerebrate animals. (After Koopowitz 1970.)

frequently, but may be seen in Sibly and McFarland (1974) and McFarland (1974b).

Structural Approach

An information network is shown in Figure 15, depicting the frequency of directed interactions (nest material exchange, allogrooming, food exchange) between individuals of a colony of Digras. This is an example of the use of the information network as a sociogram, showing how the members of a social group interact. Each arrow is labeled with the total number of interactions directed from one individual to another (with the arrow pointing in that direction). The diagram indicates that in this system interactions tend to be reciprocal, in that the magnitudes of frequencies are similar in both directions between any given pair of individuals (though the specific acts may vary). The network also points

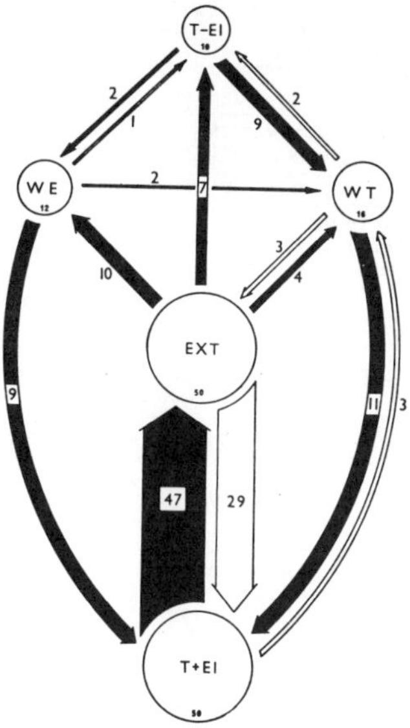

KINEMATIC GRAPH

Figure 13. Kinematic graph of a female fly's (*Scatophaga stercoria*) behavior at the oviposition site. Each circle represents one action pattern of the female; the small number within each circle shows the number of times each state was observed. Arrows represent transitions between states, with the width proportionate to the frequency; arrows are also labeled with the frequency. When transitions may go in either direction between two states, the most common transition has its arrow colored black. (Parker 1970.)

out the preponderance of male-female interactions compared to female-female interactions; no male-male interactions are shown. Examining a diagram of this sort can give a better mental picture of the relationships within a social group than a mere listing of frequencies in a table.

Figure 16 is a flow network, somewhat similar to the last diagram, but here concentrating on the actual flow of nest materials within the group. The widths of the arrows are proportional to the amount (in grams) of nest material transferred between individuals as well as that built into the nests. Here, the sum of material entering a compartment (individual animal, in this case) should equal the sum leaving the same compartment. This would not be true if there were instances of storage or if some of the material were being lost from the system. In the latter case, arrows should be included to show the additional flows, say by a female Digra dropping some of the nest material she has received from a male.

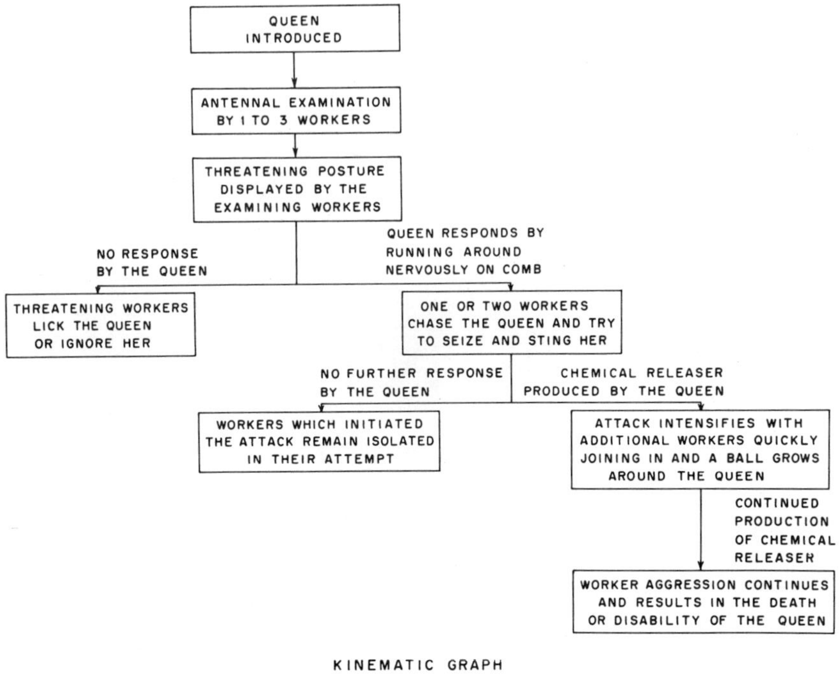
QUEEN INTRODUCED
ANTENNAL EXAMINATION BY 1 TO 3 WORKERS
THREATENING POSTURE DISPLAYED BY THE EXAMINING WORKERS
NO RESPONSE BY THE QUEEN
QUEEN RESPONDS BY RUNNING AROUND NERVOUSLY ON COMB
THREATENING WORKERS LICK THE QUEEN OR IGNORE HER
ONE OR TWO WORKERS CHASE THE QUEEN AND TRY TO SEIZE AND STING HER
NO FURTHER RESPONSE BY THE QUEEN
CHEMICAL RELEASER PRODUCED BY THE QUEEN
WORKERS WHICH INITIATED THE ATTACK REMAIN ISOLATED IN THEIR ATTEMPT
ATTACK INTENSIFIES WITH ADDITIONAL WORKERS QUICKLY JOINING IN AND A BALL GROWS AROUND THE QUEEN
CONTINUED PRODUCTION OF CHEMICAL RELEASER
WORKER AGGRESSION CONTINUES AND RESULTS IN THE DEATH OR DISABILITY OF THE QUEEN
KINEMATIC GRAPH

Figure 14. Kinematic graph showing the sequence of aggressive behavior patterns performed by honeybee (*Apis mellifera*) workers when a queen is introduced into the hive. The boxes are labeled with worker behavioral patterns. Varying responses may be given when queen behavior varies; arrows (transitions) are labeled with the behavior of the queen. (After Yadava and Smith 1971.)

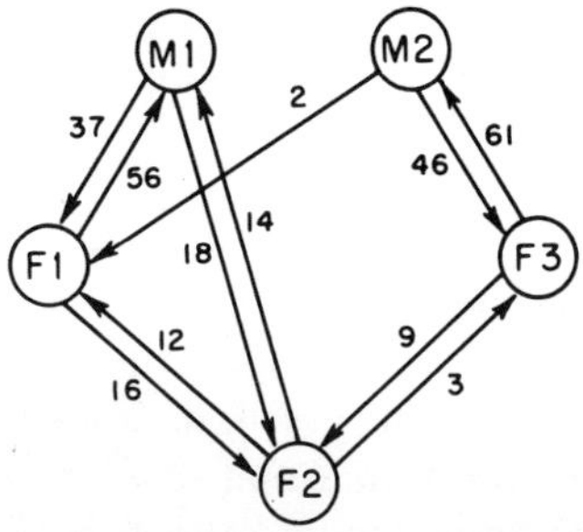
M1
M2
F1
F2
F3
37
56
2
61
46
14
18
12
16
9
3

Figure 15. Information network (sociogram) showing frequencies of directed interactions between members of a colony of Digras. M1 and M2 are males; F1, F2, and F3 are females. Arrows are labeled with interaction frequencies and point in the direction of the directed interactions.

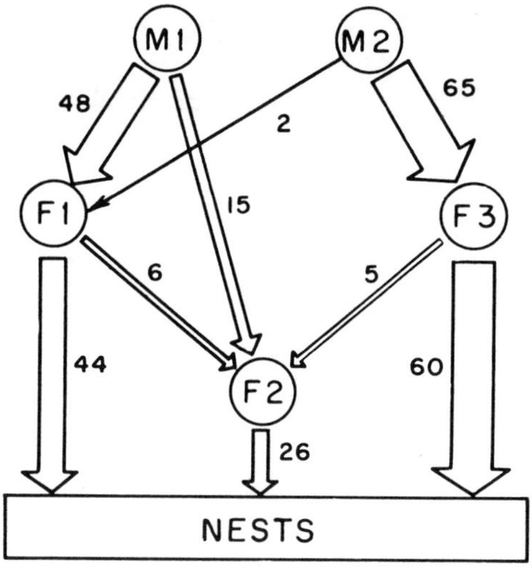

Figure 16. Flow network showing the paths of flow of nest materials within a Digra colony. M1 and M2: males; F1, F2, and F3: females. Labels and width of arrows represent the amount of nest material (in grams) transferred between individuals or built into nests.

By comparing Figures 15 and 16, one can attempt to relate, for example, the frequency of female-male interactions to the amount of material passed from the male to that female. It may also be possible to form operational definitions of pair bonds within the colony based on the information in these diagrams. The diagrams could also be used in conjunction with data on reproductive success to determine optimal strategies for directed interactions by Digras, both male and female.

Examples of sociograms include Wirtz (1967), Figure 17; Kummer (1971); and Wilson (1975), diagram based on Murchison (1935). Other information networks have been used by Dowling (1970), Figure 18; Schmidt (1971); Arbib (1972); and Griffith (1971). An ethological example of a flow network is found in Sudd (1967), Figure 19.

Relational Approach

Assume that you are interested in the system that controls nest building by Digras. Figure 20 is a diagram of immediate effects showing the relationship of some of the variables that seem to be important in this system. The diagram shows (solid arrows) that an increase in either the frequency of allogrooming or the frequency of food transfer from the

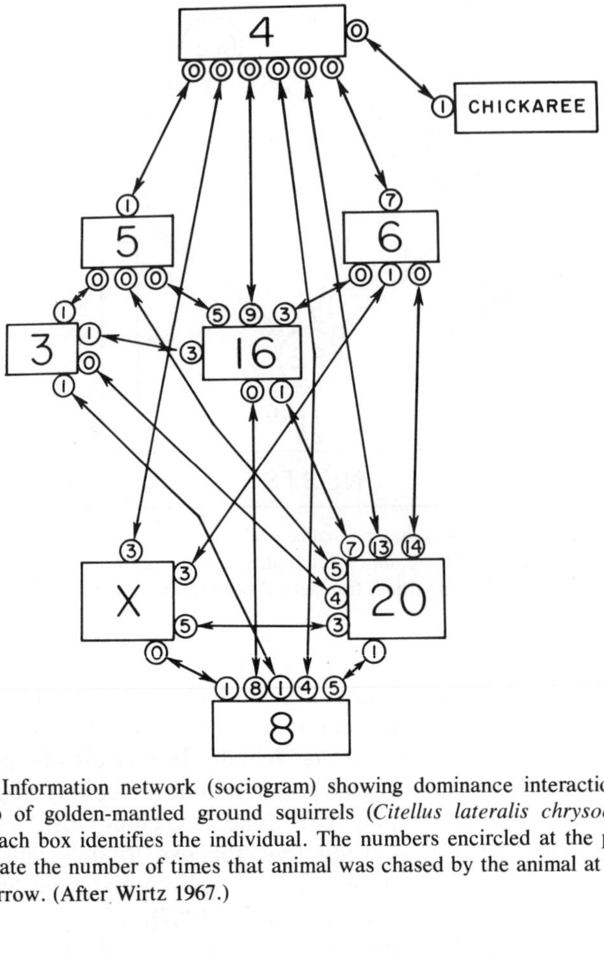

Figure 17. Information network (sociogram) showing dominance interactions within a social group of golden-mantled ground squirrels (*Citellus lateralis chrysodeirus*). The number in each box identifies the individual. The numbers encircled at the points of the arrows indicate the number of times that animal was chased by the animal at the opposite end of the arrow. (After Wirtz 1967.)

female to the male results in more nest material being received, and consequently the nest being built to a greater width. A wider nest results in a decrease (broken arrow) in the light intensity within the nest, and a decrease in internal light intensity means a corresponding decrease (solid arrows) in both allogrooming and food transfer. Thus there is negative feedback in this system, allowing regulated control over nest building. Examination of the diagram will point out areas in which the relationships need more study. For example, does a visual response directly mediate the effect of light intensity on allogrooming rate, or is some other mechanism involved?

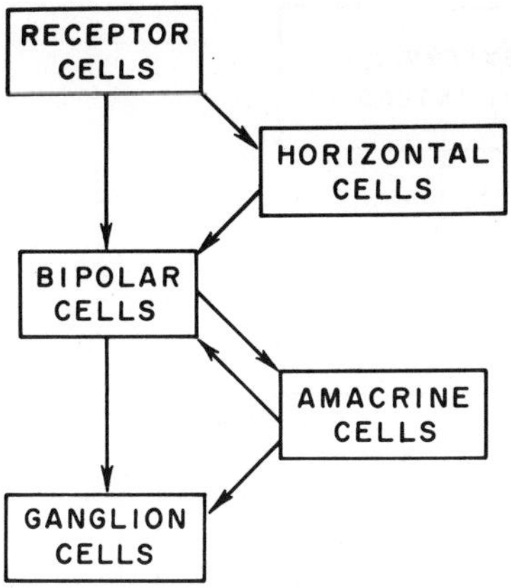

Figure 18. Information network showing one of several possible arrangements of synaptic contacts in a vertebrate retina. Arrows point in the direction of information flow. (After Dowling 1970.)

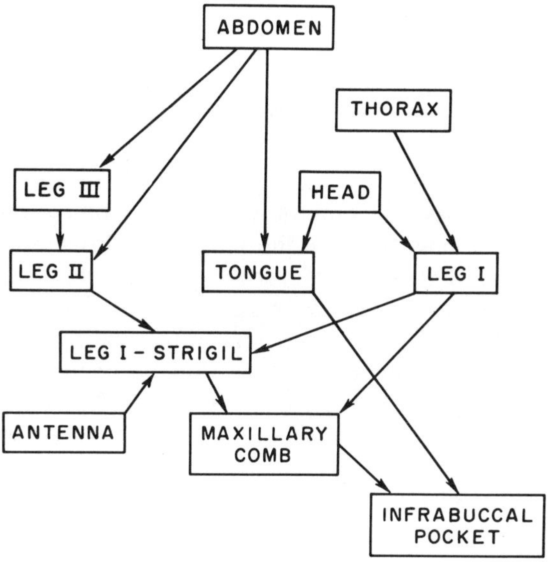

Figure 19. Flow network showing the routes by which material is transferred when an ant cleans itself. Each arrow shows the transfer of material from the part cleaned to the part that cleans it. No representation is made of the amount of material actually transferred. (after Sudd 1967.)

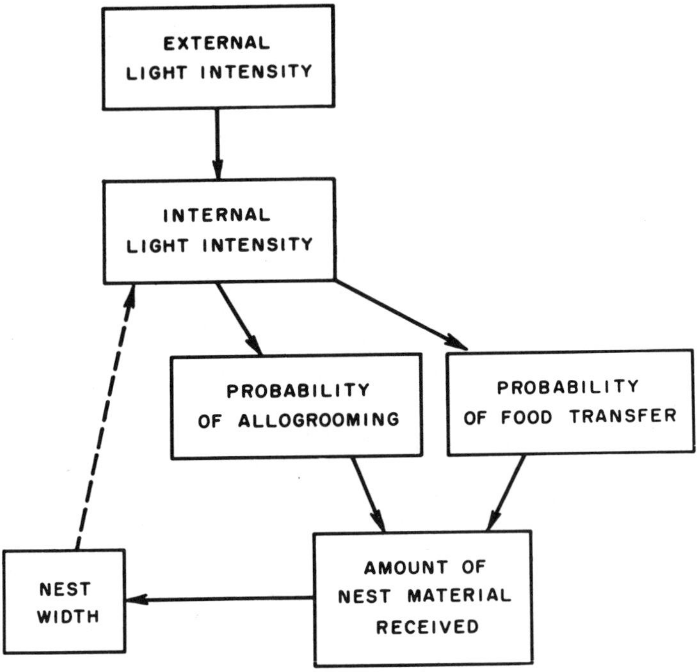

Figure 20. Diagram of immediate effects showing the relationships between pairs of variables of the nest building system of Digras. Boxes: system variables; solid arrows: direct relationshps; broken arrow: inverse relationship.

Figure 21 is a block diagram showing more completely the system controlling nest building. Some of the same variables shown in Figure 20 are now used as the labels of arrows in the block diagram, with additional intermediate variables being added. Each block is a subsystem, here being labeled with parameters of the male, the female, or the nest, precise transfer functions not yet having been determined. One splitting point and two summing points are shown. The external intensity of light at the nest is given as an input, or external parameter, to the entire system.

Notice the value of this diagram in directing future research with the system. Predictions can be made from this model and critical tests of the model can be devised based on this diagram. For example, if the negative feedback system works as indicated here, an artificial addition to the width of the nest by the experimenter should cause a reduction in the amount of material the female adds to the nest compared with what she normally would have done. If, on the other hand, she continues to

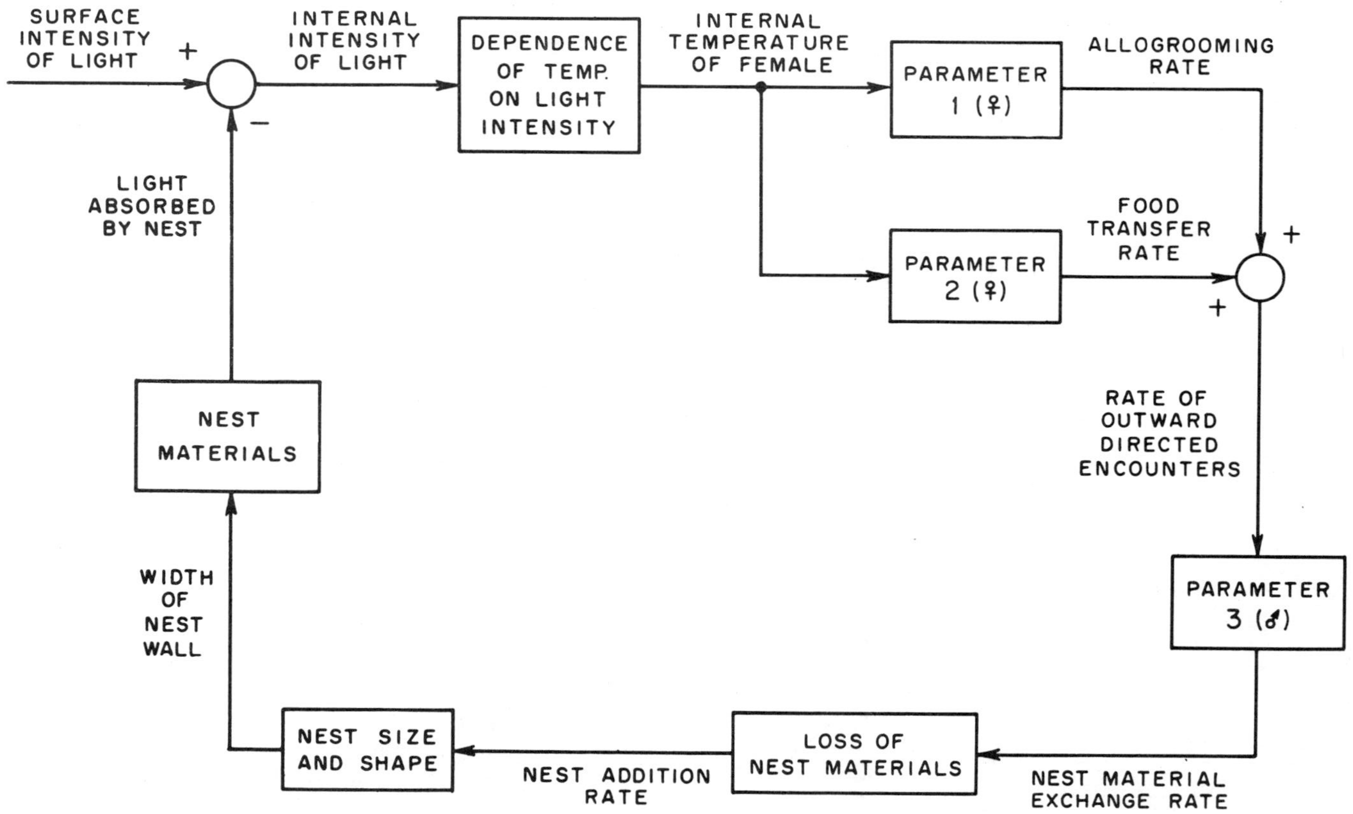

Figure 21. Block diagram of the system controlling nest building by Digras. Boxes (blocks): system parameters; arrows: system variables; open circles: summing points; solid circle: splitting point.

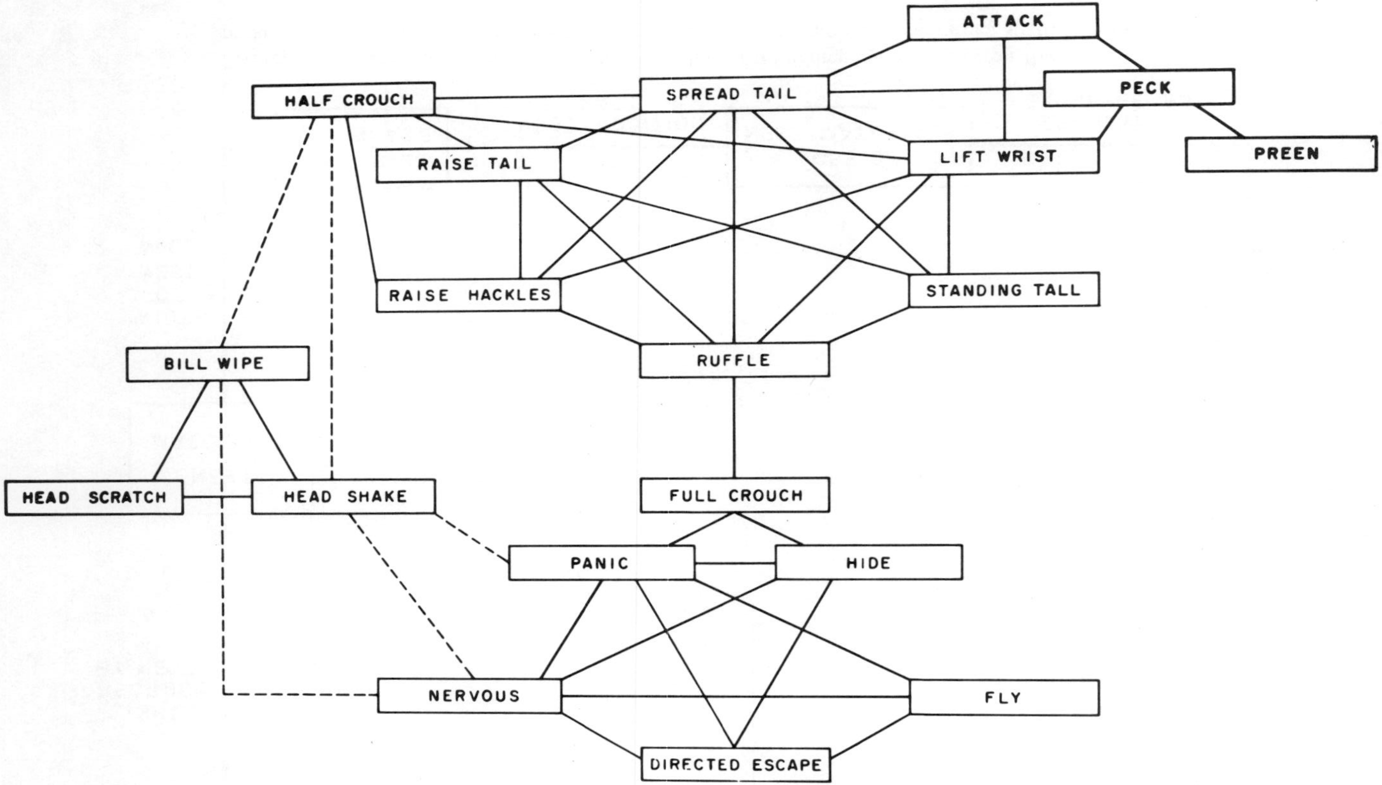

Figure 22. Association diagram of the frequency of coöccurrence of behavioral activities in brain-stimulated chickens (*Gallus gallus*). Solid lines indicate observed coöccurrence frequencies significantly higher than chance level; broken lines, those occurring together at less than chance level. Note that there is no direction given for the lines. (After Phillips and Youngren 1971.)

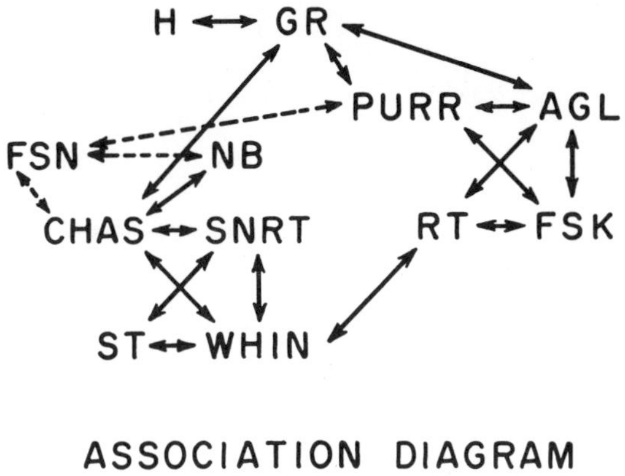

Figure 23. Association diagram showing the relationship between maternal, agonistic, social, and maintenance behaviors in socially housed green acouchi (*Myoprocta pratti*) mothers. Broken lines represent significant negative correlations and solid lines show significant positive correlations. Letters are codes for various behavioral patterns by the mother. Note that the arrowheads are not necessary, and could be omitted. (After Kleiman 1972.)

add material so that the nest finally becomes extra large (and the internal light intensity lower than expected), the model can be rejected (at least in this form). This type of model testing makes the block diagram useful even though the diagram is not complete, in the sense that all the transfer functions have not yet been determined. Even though Riggs (1970) terms neatly labeled but quantitatively uncharacterized black boxes an "affliction," the heuristic value of block diagrams whose parameters are not yet rigidly defined is great enough to justify their continued use by ethologists.

Examples of association diagrams include those in Phillips and Youngren (1971), Figure 22; and Kleiman (1972), Figure 23. An association diagram refined to the form of an ordination relating the age/sex classes of macaque troops is shown in Stephenson (1974). Diagrams of immediate effects can be found in Colgan (1973), Figure 24; McFarland (1965), Figure 25; Ashby (1956); and Riggs (1963). Signal flow diagrams are utilized by McFarland (1965), Figure 26; Hubbell (1973), an ecological example; and Huggins and Entwistle (1968). Many examples of block diagrams are now appearing in ethology. These include figures by Brockmann (1975), Figure 27; McFarland (1971); Metz (1974); Heiligenberg (1974); Geertsema and Reddingius (1974); Sibly and McFarland (1974); Van Sommers (1974); Brown and others (1974); Oatley (1974); and Toates and Oatley (1970).

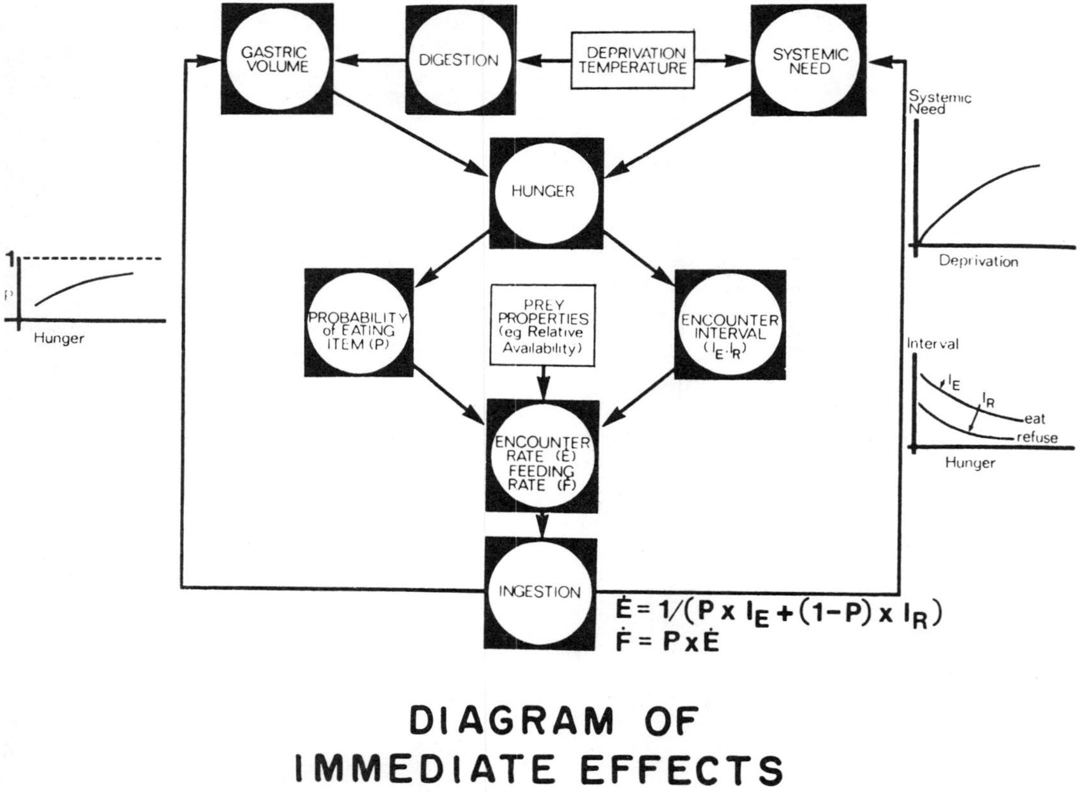

Figure 24. Diagram of immediate effects representing a model for monophagous fish feeding. Independent variables are in rectangles; dependent variables are in circles. The relations between certain variables are shown by small graphic plots to the sides of the diagram. (After Colgan 1973.)

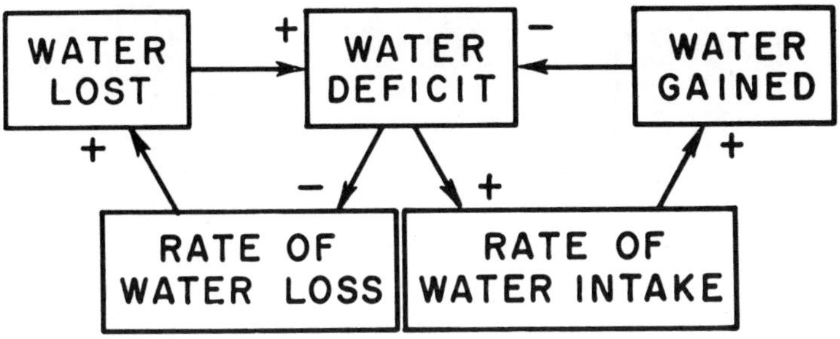

Figure 25. Diagram of immediate effects representing the functional relationships of variables in a hypothetical drinking control system in doves (*Streptopelia risoria*). Plus indicates a direct relationship; minus, an inverse relationship. (After McFarland 1965.)

4. INTERPRETATION AND LIMITATIONS

Table 1 gives an outline of the different types of diagrams and their uses. Kinematic graphs and state-space diagrams concentrate on changes of state through time. The kinematic graph has the advantages of being able to use any scale of measurement for the state variables, and being usable with stochastic as well as deterministic systems. The state-space diagram requires a deterministic system whose state variables are measured on interval or ratio scales. In addition, the problems of depicting a state-space of more than three dimensions become large. Nevertheless, in complex systems the state-space diagram may allow easier consideration of system stability and determinations of equilibrium points than do incomplete kinematic graphs that must of necessity show only a small fraction of the states and transitions of a system with many states. For a further discussion of systems that are only partly specified see Levins (1974).

Information networks (including sociograms) and flow networks emphasize the movement of information or materials, respectively, along channels in the system. The conservation laws applying to matter and energy require that in flow networks the amount of material flowing into

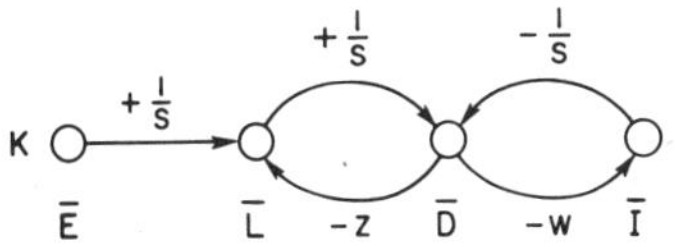

Figure 26. Signal flow graph of the same system shown in Figure 25. L: rate of water loss; D: water deficit; I: rate of water intake; E: environmental input; and K: the value of E. (After McFarland 1965.)

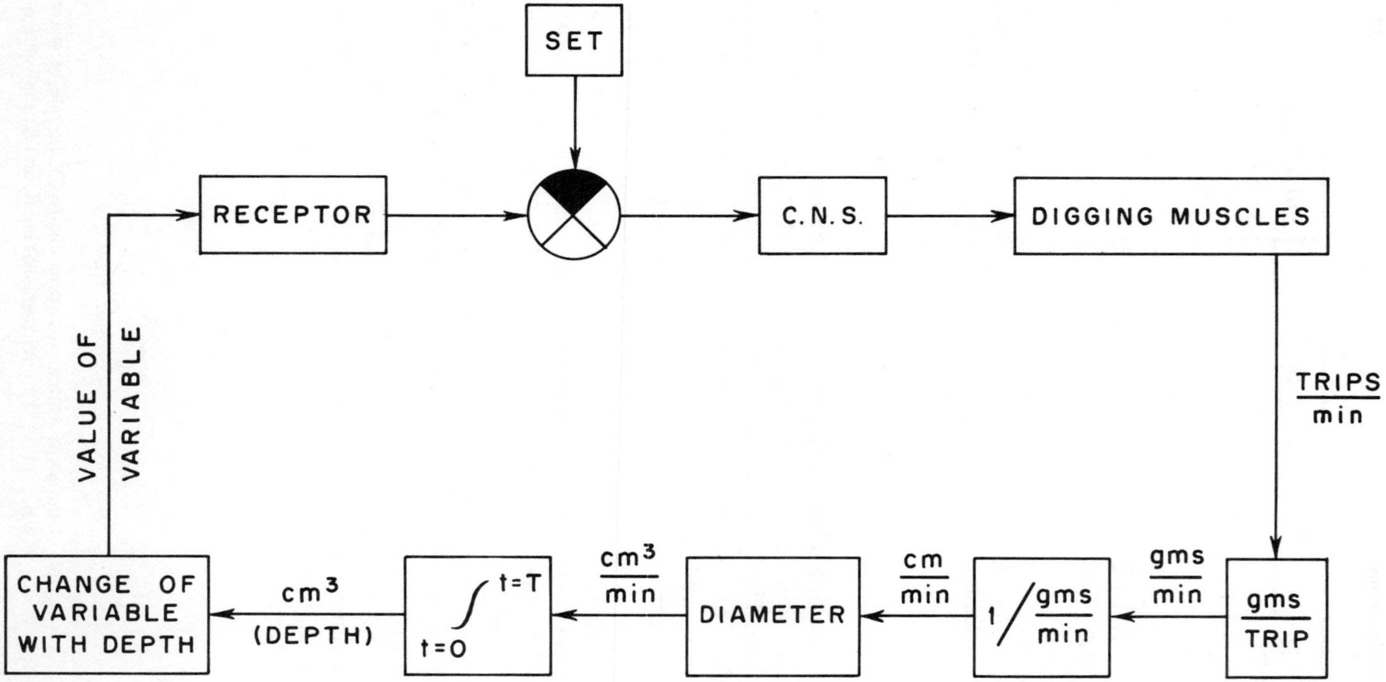

Figure 27. Block diagram of the control system for digging in a predatory wasp (*Sphex ichneumoneus*). The wasp measures some soil variable that varies with depth, such as temperature; the output of the sensory receptor is compared with a reference set point generated within the wasp's own nervous system. The summing point is shown by a circle containing an X; shading of a quadrant of the circle indicates subtraction. (After Brockman 1975.)

Table 1. Types of Systems Diagrams and their Uses

Type	Function	Box (node)	Arrow (branch)	Restrictions
State Approach				
1. Kinematic graph	Temporal patterns of behavior	State	Transition	Only in one state at a time; stochastic or deterministic; any measurement scale
2. State-space diagram	Families of behavioral trajectories	State (Single point in state-space)	Transition	Deterministic; ratio or interval scales for state variables
Structural Approach				
3. Information network (Sociogram)	Information flow in a system	Component	Information channel	Nonadditive; need not obey conservation laws
4. Flow network	Matter or energy flow in a system	Compartment	Amount and direction of matter or energy flow	Must obey laws of conservation

Table 1. ***(Continued)***

Type	Function	Box (node)	Arrow (branch)	Restrictions
Relational Approach				
5. Association diagram	Correlations between variables	Variable	Statistical correlation between variables	Nondirectional
6. Diagram of immediate effects	Causal relations between variables	Variable	Functional relation between variables	Can show only positive or negative relations
7. Signal flow diagram	Causal relations between variables	Variable	Transmission operator	Any relation may be shown; analytic techniques available
8. Block diagram	Functional organization of system	Subsystem or parameter (transfer function)	Variable	Heuristic and communicative; if linear, good analytic techniques available

a compartment equals that flowing out (assuming no storage). No such limitation applies to information networks.

The four relational diagrams reflect increasing knowledge about the relations between variables in the system. The association diagram shows only statistical correlations between variables, with ordinations and other similarity-based diagrams being but refinements of the association diagram. The diagram of immediate effects indicates monotonic direct or inverse causal relations between variables (though sometimes a single arrow type is used to indicate unspecified causal relationships). The signal flow graph is more general, allowing any form of functional relationship, and in the case of linear systems permits easy manipulation of the graph through the use of Mason's Rules. The block diagram differs from the signal flow graph in its emphasis on parameters and subsystems rather than the emphasis on variables shown in the signal flow graph. The block diagram has useful heuristic and communicatory functions. Imposing requirements of linearity and using transfer functions and Laplace transform representation allows powerful analytic methods to be used.

The variety of systems diagrams is of value to ethologists in two main respects. First, certain diagrams may permit analytic techniques to gain further knowledge about the system and to make predictions for untested cases. Second, and of more importance, the diagrams allow visualization of the ethological system. This conceptual aid is useful to the researcher in planning future research and in improving the model itself, while also being useful to the reader in understanding the model and the results of experiments that have been performed.

CHAPTER

11

Modeling

PATRICK W. COLGAN
Queen's University at Kingston

A "model" in ethology once meant a ball of feathers or piece of painted wood with which an ethologist studied the effective eliciting properties of naturally occurring stimuli in an animal's Umwelt. The use of such stimulus dummies remains a valuable experimental approach but, as ethology has gained a broader observational base and accompanying theoretical framework, "model" has come to stand more frequently, as it does in the rest of the natural sciences, for the conceptual tools that ethologists develop to organize and interpret their observations, to generate predictions, and thus to act as heuristic guides for subsequent research.

This chapter discusses the role, in ethological research, of modeling, the process in which models are proposed, tested, modified, supported and used to generate predictions, or rejected. Following this introductory section on general aspects of modeling Section 1 examines the source and nature of assumptions made in setting up a model, and Section 2 on Method describes the criteria used in judging models and the interplay between data collection and modeling. Section 3 delineates a number of current models as examples of the diversity of both the kinds of model systems that can be used and of the ethological situations that can be modeled. Section 4 considers the problems associated with the interpretation and limitations of models.

A model of a system consists of two components with specified characteristics. The two components are a description of a mathematical system (e.g. a system of differential equations) or an understood physical system (e.g. an electric or hydraulic system) and a mapping between this system and the one under study. The mapping links the variables in the model system with those in the observed ethological system. The specified characteristics are that the model generates a set of statements different from other models, that none of these statements is false, that some of them are true, and some testably predictive. Models cover aspects of behavior from neurophysiological mechanisms to ecological communities, from the quantification of flush toilets to the imitation of human higher cognitive processes. With the use of models, complex behavioral problems can be analyzed into their components and reassembled into the functioning, interactive whole.

The general role of models and model building in science is dealt with in an enormous (and ever growing) literature. Leatherdale (1974) has considered the linguistic foundations of modeling. A very readable and detailed introduction to the application of mathematical modeling is given by Maki and Thompson (1973). Model building in ethology is paralleled by similar developments in mathematical psychology (Estes 1975), integrative physiology (Milsum 1966), and population genetics

(Kojima 1970) and ecology (Goel and Richter-Dyn 1974; Maynard Smith 1974b; Patten 1971). This chapter focuses on ethological models only and does not examine these parallel developments or general considerations to any great depth.

In considering nonmodeling approaches to the study of behavior, it should be noted that the nervous systems of those animals whose behavior is of most interest to us are enormously complex. A case in point is *Caenorhabditis elegans,* a lowly nematode with a nervous system of only about 260 neurons in total, over which Gould (1974) has lamented that "it is striking how complicated the wiring seems to be in such a simple animal." The nervous system of sticklebacks and herring gulls contain cells numbering orders of magnitude greater than 260. As the apparently insuperable problems of executing a comprehensive analysis of behavior using a physiological approach become more obvious, one realizes that such physiological reductionism is a principle, not a research strategy, and that holistic models are the only viable options for the full investigation of ethological phenomena.

Modeling is the vital activity through which researchers embody their general theoretical principles and specific hypotheses concerning the behavioral phenomena under study as explicit forms that provide bridges between data and theory. (In an epistemologically fundamental sense, all intellectual activity involves the relating of sensory impressions and a model universe.) Models are built in conjunction with the techniques discussed earlier in this book, especially the systems diagrams outlined in the previous chapter. Different types of models are used for different purposes in the various theoretical (i.e. nondata collecting) aspects of research, which include experimental design, data analysis and interpretation, prediction, and explanation. Simple physical systems appropriately mapped can be very useful heuristically. The two dominant forms of mathematical models are statistical models and simulation models. As is clear from the many statistical procedures outlined in earlier chapters of this book, inferential statistical models provide methods for estimating parameters, testing hypotheses, and making predictions.

Simulation modeling is undoubtedly one of the most active areas in the application of mathematics to ethology over the recent past. In a simulation model a researcher attempts to mimic explicity the process that produces the observed output of the system. His assumptions about the dynamics of the system form a strong hypothesis. Simulation modelers are dissatisfied with less mimetic approaches because they feel that these approaches, like a bikini, reveal what is interesting but hide what is crucial. The fundamental contrast here is between prediction, which can often be achieved satisfactorily using relatively superficial statistical models, and explanation, which requires a deeper analysis. It is this contrast that moved Nelson (1973) to say of Markov process analysis "I do not suppose that description in terms of a Markov process model will in itself lead to very great understanding of the behavioral organization." Similar sentiments have been expressed by workers in other disciplines. For instance, MacArthur and Wilson (1967) state with respect to biogeographical theory that "while multiple regression analysis, or some equivalent quantitative analysis, is indispensible for sorting out components that are not relevant, it is not a very fruitful way of generating new hypotheses."

The necessary basis for comprehensive models capable of providing an explanation of behavior will be very broad and cut across traditional disciplines. Sources of such synthetic theory include: from physiology, neural transmission and network theory (Fienberg 1974); from genetics, behavioral gene flow and frequency change theory, and theory of kinship (Hamilton 1964) and interdemic (Wilson 1973a) selection; from psychology, signal detection theory (Swets 1973) and more general perceptual theory (Hoffman 1970), remnants of neo-Hullian ideas, and even Skinnerian approaches as developed by Herrnstein (1974) for relative response rates and Staddon (1972) for the study of causal linkage using superstitious behavior; from ethology, motivational theory (Heiligenberg 1974; McFarland 1971, 1974), communication theory as developed in Altmann's (1965) work, general social theory as applied to play behavior (Bekoff 1975; Fagen 1977), parent-offspring conflict (Trivers 1974), and selfish herding (Hamilton 1971); from ecology, theory of competitive (MacArthur and Levins 1964), predatory-prey (Holling 1966), and host-parasite interactions, general sociobiological theory (Wilson 1975), and niche theory (Cody 1974, Vandermeer 1972); and from evolutionary biology, theory of adaptive strategies (Pulliam 1975; Schoener 1971). Thus the task is enormous, but no less broad an approach can be hoped to succeed or is therefore worthwhile.

1. ASSUMPTIONS

It is both the great strength and danger of the modeling approach that a modeler may assume any set of postulates that provides him with a sufficient and noncontradictory basis upon which to build. Models may be deterministic or stochastic, and often the latter type, which can include the former as special cases, seem especially appropriate either because the behavioral process of interest appears to be intrinsically probabilistic or because we are limited in our ability to minimize the noise in our measurement procedure. The major mathematics used to date are traditional calculus, probability and statistical theory, differential equations, optimization theory, and control theory, all aided by such techniques as flow graphs and graphical analysis (see McFarland 1971). Beyond these mathematics, it appears certain that, as before in the history of science, new mathematics will have to be developed as the quantification of ethology proceeds and novel probelms are encountered. Systems theory, used in conjunction with the diagrams discussed in Chapter 10, will likely be very important in this regard. Often an ethologist will be well advised to consult an experienced modeler in the form of an applied mathematician. Appropriate initial choice of assumptions, and subsequent later testing were possible, will help to avoid the "garbage in—garbage out" pitfall in which arrival at any set of conclusions can be achieved by beginning with assiduously chosen (but possibly invalid) assumptions.

Two specific points should be made with respect to assumptions for models. First, most behavioral systems involve nonlinear functions of the parameters and nonstationarities (i.e. the values of the parameters vary over time). It is therefore extremely important to note that models can be made to take into account both nonlinearities and nonstationarities, rather than in Procrustian fashion falsely assuming linearity and stationarity as most current analytic techniques do. The Monte Carlo method described below is very applicable for these models.

A second point centers around a principle often invoked under the aegis of the Darwinian theory of evolution, that of optimality. The argument is made that those animals that are most nearly optimal in some sense will be selected for (indeed this may be a tautology) and that therefore an optimality principle can be invoked appropriately to provide a means for modeling the pattern of operation of some behavioral system under study. There is no logical connection, however, between the optimality of performance of some system and that of any of its subsystems. A well known ethological example is found in the study by Tinbergen and others (1962) of egg shell removal by black-headed gulls, in which this behavior is concluded to be suboptimal because of a countervailing force. Egg shell removal helps to protect the brood from heterospecific predators but such removal is delayed because of predation by conspecifics on wet chicks. The fact that realized performance is a compromise of many conflicting demands must be kept in mind when optimal solution techniques are used.

The examples that follow illustrate the variety of assumptions that an experimenter may believe appropriate to make under given circumstances.

2. METHOD

The manner in which a researcher assembles the foundation of assumptions upon which he bases his model is largely a private matter. He draws on both his knowledge of existing findings and his personal hunches. By its very nature, this process is not amenable to objective description. By contrast, the resulting model can be judged through the use of a variety of criteria (Holling 1963, 1964; Levins 1966). The assumptions of a model can be examined for their internal consistency (lack of mutual contradictions), necessity (i.e. no assumption can be deleted without altering the set of resulting theorems), and compatibility with the assumptions made in other areas of research. A prediction of the model can be assessed as to its accuracy and precision (i.e., respectively, is the predicted value correct on average and what is the average deviation about the true value?). The overall model can be scrutinized for its generality (breadth of applicability), simplicity (reflecting the paucity of assumptions made), realism (validity of the mapping between the model system and the system under study), and, according to some, such more aesthetic qualities as elegance. Confidence in a model is particularly enhanced when the model correctly makes predictions that are counter-intuitive (i.e. which would not result from some simpler model). Not all of these qualities of a model can be simultaneously optimized, and compromises must be struck among the possible trade-offs.

Modeling is a ritualized activity in the usual ethological sense in having a very specific function, the emission of complex signals termed models. Such models are emancipated from the vague world of intuitive leaps and superfluous meanings to which many traditional theoretical concepts are bound. In addition, models are under strong selection pressures for appropriate reception by receivers, and are perhaps generated by different neurological loci. The resulting stereotypy and exaggeration make modeling to science what ballet is to art, with similar problems with respect to prima donnas.

Perhaps the most important feature of modeling methodology is the interplay between data collection on the one hand and model construction on the other (see Holling 1963; Platt 1964). Preliminary data enable the construction of a provisional model which in turn generates predictions that are tested in subsequent experiments which provide new observations that permit a revision of the model, and so on through the cycle. Excellent examples of this interplay are: from behavioral physiology, the work on the optomotor responses of the beetle *Chlorophanus* by Hassenstein, Reichardt, and co-workers (see Thorson 1966) and, from behavioral ecology, that on predator-prey systems by Holling (e.g. 1966). In particular, this feedback enables an experimenter to choose between rival models (and there is an indefinite number of models possible for any data set) by performing crucial experiments for which the models make different predictions. This linking of experimental and theoretical work avoids a major problem of contemporary science neatly expressed by Platt (1964): "We speak piously of taking measurements and making small studies that will 'add another brick to the temple of science.' Most such bricks just lie aroung the brickyard."

Monte Carlo simulation is a method of considerable importance to ethological modelers dealing with analytically intractable systems. In probability theory, when unknown parameters, such as moments, of a particular probability distribution cannot be analytically calculated, they can be estimated by repeated sampling in accordance with the sampling procedure and tallying the observed values of the desired parameters. The name Monte Carlo is derived from the similarity of the repeated sampling to the successive spins of a roulette wheel. Nowadays such sampling is generally done on a computer, and a large sample is taken to gain high precision. Since ethologists are often faced with stochastic systems that are not solvable analytically because of such complexities as nonlinearities and nonstationarities, they should consider the method of Monte Carlo simulation. An example using Monte Carlo simulation is included in the following section.

3. EXAMPLES

The best introduction to ethological modeling is an examination of some of the representative types of models employed by various researchers. The earliest synthetic model, involving both ethological and ecological considerations, appears to be an optimization using partial differential equations outlined by Lotka (1925). Behavioral properties associated with aspects of the animal's ecology are varied such that the relative growth rate of the species r is maximized. Indeed, Lotka showed that if the partial derivatives, with respect to pertinent external variables, of r and of the animal's total pleasure (for which we could now read "reinforcement") are proportional, maximizing r is equivalent to maximizing the hedonic value of a dynamic psychology. This is an intriguing examination of the linkage between proximate and ultimate causes in motivation.

Examples of deterministic models applied to motivational situations have been formulated by Heiligenberg and co-workers in the study of aggression in the cichlid fish *Haplochromis burtoni*. One report (Heiligenberg and Kramer 1972) showed that internal changes in attack readiness (AR), which declines in the absence of social stimulation, could be modeled by an exponential decay process with a half-life of 7 days. With regard to the effects of external stimuli, Leong (1969) found that a stimulus dummy with a black eye-bar increased AR by 2.79 bites per minute, a dummy with an orange patch above the pectoral fin decreased AR by 1.77 bites per minute, and a dummy with both features increased AR by the algebraic sum of the changes caused by each feature: 1.08 (≃ 2.79 − 1.77) bites per minute. Subsequent work (Heiligenberg 1976) has revealed that the effect of the black eye-bar consists of two effects, one due to the orientation of the body and the other due to the orientation of the eye-bar thus providing a quantification of the Law of Heterogeneous Summation for this behavior system.

Examples of stochastic models applied to social situations are Cohen's (1971) linear one-step transition (LOST) models of group formation in primates. The deterministic portions of the equations are differential expressions that state that the number of groups contining i individuals increases either because of the arrival of individuals at groups of size $i - 1$ or because of the departure of individuals from groups of size $i + 1$. Arrivals and departures are taken to be dependent on both the numbers of groups and the number of individuals in these groups. The fundamental equation for the rate of change of the number of groups of size $i\,(= 2, 3, \ldots)$ is

$$\frac{dn_i(t)}{dt} = an_{i-1}(t) + b(i-1)n_{i-1}(t) - an_i(t) - bin_i(t)$$
$$- cn_i(t) - din_i(t) + cn_{i+1}(t) + d(i+1)n_{i+1}(t)$$

The first two terms deal with arrivals of individuals at groups of size $i - 1$, the first term reflecting arrivals proportional to the number of such groups an_{i-1} and the second reflecting arrivals proportional to the number of individuals in such groups $[b(i-1)n_{i-1}]$. Similarly the last two terms deal with departures of individuals from groups of size $i + 1$. The middle four terms reflect these same arrival and departure processes operating on groups of size i. (Another equation deals with groups of single individuals.) Certain parameter values yield equilibrium solutions of these equations leading to such distributions as the zero-truncated Poisson and negative binomial, which describe certain data. Other solutions fit other data not accounted for by any simple probability distribution. Thus a set of assumptions about the processes underlying the formation of social groups can be formalized mathematically and used to explain a wide variety of data on group size (see also Cohen 1972, 1975).

An example of modeling using Monte Carlo simulation is the work of Saila and Shappy (1963) on the orienting behavior of migrating oceanic salmon. The problem of interest is whether the observed return rates of tagged Pacific salmon can be predicted by random movements alone or whether oriented behavior is involved. Taking reasonable parameter values from published data, Saila and Shappy describe a two-stage model. In the first stage an individual begins in the ocean 1200 nautical miles west of the mouth of the home stream and makes a succession of linear steps whose direction is selected according to a distribution that includes a parameter reflecting the strength of the directional sense of the migrating fish. These steps continue until either the search time elapses or the fish reaches the coast. The second stage of the model deals with individuals that reach the coast by providing a random walk along the coast, with successful fish reaching their home streams. Repeated runs through the model provide an estimate of the proportion of fish returning under specified conditions. This Monte Carlo procedure shows that the data on the returns of tagged fish are not consistent with a hypothesis of random orientation by salmon, but that a directed component of movement is necessary. However, this directed component need be only weak, and in particular it is not necessary to postulate full navigational ability. The model thus provides insight into possible behavioral capacities of salmon that would lead to the observed return rates.

The topic of migrating salmon also affords an example of rival models and crucial data. Stimulated by Saila and Shappy's model, Patten (1964) proposed an alternative model based on statistical decision theory. The models differ *inter alia* in their predictions on the occurrence of mortality, with Saila and Shappy's model predicting 40 percent loss of the population at sea and 67 percent loss of those fish that make contact with the coast, and Patten's model predicting virtually all mortality occurring at sea. Thus the models are clearly distinguished by the appropriate crucial (but as yet apparently unavailable) data on mortality.

Models can be useful not only for demonstrating the types of mechanisms that may be operative in behavioral systems but also for ruling out the types that cannot be operative. A good example is the work by Chase (1974) on the formation of linear or near linear dominance hierarchies. Chase shows that two commonly held conceptions of how such hierarchies form, when expressed and examined mathematically, cannot be reasonably regarded as valid. The first conception is that a dominance hierarchy arises as the result of a round robin tournament involving a series of pairwise contests. Chase uses a hierarchy measure that ranges from 0 to 1. If the size of the group n is odd, the measure takes on the value 0 when each animal dominates $(n - 1)/2$ other members of the group; if even, the measure takes on a value close to 0 when each animal dominates $(n - 1)/2$ or $(n + 1)/2$ other members. The measure takes on the value of 1 when one animal dominates all others, the next animal dominates all but this one, and so forth. He then shows that for a reasonably linear dominance hierarchy to form (i.e. one whose hierarchy measure is at least .9) the odds of one individual defeating another must be very extreme (on the order of 99:1) especially as the number of individuals involved increases. Available data for chickens reveal odds much less extreme than this, thus casting doubt that the model process is operative. The second conception, that there is a high correlation between some trait that predicts dominance and realized position in a hierarchy, can be similarly refuted. Thus a modeling approach has shown that two popular hypotheses or hierarchy formations require unreasonably stringent conditions.

An intriguing new development in behavioral modeling is the application of game theory, principally by Maynard Smith (e.g. Maynard Smith and Parker 1976). The behavior of animals in conflict situations is modeled by considering an "evolutionarily stable strategy" (ESS) whose mathematical defintion can be interpreted as a strategy whose payoffs are not exceeded by those of any "mutant" strategy. Let $E_J(I)$ be the expected payoff of strategy I against any strategy J. Then I is an ESS if $E_I(I) > E_I(J)$ or, if $E_I(I) = E_I(J)$, if $E_J(I) > E_J(J)$. The latter result can be seen by considering a population, proportion p of which adopt strategy I and $1 - p$ strategy J. Then the overall payoff of straregy I is

$$pE_I(I) + (l - p)E_J(I)$$

while that of strategy J is

$$pE_I(J) + (1 - p)E_J(J)$$

For the overall payoff of strategy I to be greater than that of J, given $E_I(I) = E_I(J)$, it follows that $E_J(I) > E_J(J)$. Strategies may be pure, in which case the animal always makes the same response given his opponent's action, or mixed, in which case the animal may make one of several responses according to a probability distribution.

In particular Maynard Smith considers the questions of how long an animal engaged in display contests (i.e. contests in which there is no physical contact) should persist, and what strategies are ESS when physical contact, and possibly damage, occurs. It can be shown that in contests symmetric with respect to such factors as payoffs and weapons, an ESS is a mixed strategy in which the duration of persistence is selected according to the exponential probability density function with mean equal to the gain from a victory. This may be shown as follows. Let $p(x)\,dx$ be the probability of selecting a duration between x and $x + dx$. Then the expected payoff of a pure strategy J with fixed duration m against this mixed strategy I is a weighted result of its wins minus losses

$$E_I(J) = \int_0^m (v - x)p(x)\,dx - \int_m^\infty xp(x)\,dx$$

This follows, assuming a gain v from a victory and losses proportional to time spent displaying, since J wins $v - x$ when I chooses a duration x less than m, and loses otherwise. We wish to find $p(x)$ such that I has the same payoff regardless of m (i.e. $dE/dm = 0$). Using Leibniz' rule that if

$$E(m) = \int_{\alpha(m)}^{\beta(m)} f(x, m)\,dx$$

then

$$\frac{dE}{dm} = \int_{\alpha(m)}^{\beta(m)} \frac{\partial f(x, m)}{\partial m}\,dx + f(\beta, m)\frac{d\beta}{dm} - f(\alpha, m)\frac{d\alpha}{dm}$$

and remembering that

$$\int_0^\infty p(x)\,dx = 1$$

we find that

$$p(x) = \frac{1}{v}e^{-x/v}$$

The mixed strategy can be achieved by variability in individual behavior or by a genetic polymorphism of pure strategists in the population. In asymmetric contests, the asymmetry will usually enable settlement of a conflict without escalation.

Data supporting the results of such strategic analyses of conflict behavior include those on the lengths of time that male dung flies remain at cowpats where females oviposit. Behavioral models based on game theory have two particularly interesting features. First, to the familiar proximate forms of motivation such as drives, rhythms, and engrams, these models add the ultimate concept of strategy as a means of treating motivation as the result of optimizing selection. Second, the models are synthetic by involving strong assumptions about the nature of the behavior of the animal, genetic considerations, a motivational construct, selection pressures, and social communication and behavior.

Two promising approaches on the vanguard of ethological modeling, both deterministic, are catastrophe theory and the motivational modeling of McFarland and co-workers. Catastrophe theory deals with the behavior of systems that undergo apparently discontinuous changes ("catastrophes"). The discontinuities occur in a behavior space, functions within which are controlled by continuously varying parameters in a control space. Catastrophe theory has been developed by the French mathematician, Thom, who has shown in a long and difficult proof that for systems with not more than four controlling variables there are only seven basic types of catastrophes. The possible applications of this theory appear very diverse (see Zeeman 1976) and include those to ethological phenomena involving attention switches in which control is transferred from one behavioral system to another.

The applications of catastrophe theory have become notoriously controversial. The mathematics of the theory is based on differential topology, a field understood by few. By contrast, enthusiasts for applications have been criticized for using the theory in a very fast and loose manner (see Sussman and Zahler 1977). A major problem is the task of selecting and mapping appropriate behavioral measures to the variables of the model system. It is clear that a careful examination of proposed applications is necessary. Nevertheless it should be noted that catastrophe theory has a useful role in ethology. For instance, Colgan and Nowell (in preparation) have collected field data to test certain specific hypotheses of a catastrophe model of aggressive behavior in nesting male pumpkinseed sunfish. These data prove to be economically described by a cusp catastrophe.

McFarland's work (e.g. Houston and McFarland 1976) is an ambitious attempt to formalize in abstract language the problems and methods of motivational research. Behavior is viewed as being controlled by functions in a "causal factor space" spanned by axes representing internal ane external factors (such as hormone levels and conspecifics, respectively) and involving isoclines relating motivationally equivalent combinations of these factors (equivalent in the sense of leading to the same "behavioral final common path"). A central feature of the approach is the search for concurrent variables. "A variable is said to be concurrent if the ordering of the responses that arise from changing the value of the variable is the same whatever the value of the other variables may be (the values of the other variables are held constant during each determination of an ordering)." If variables are concurrent, then further properties, including eventually the additive and multiplicative interactions studied by Heiligenberg, can be sought.

The approach has both theoretical and experimental implications. On the theoretical side, the approach avoids what is held to be an artificial dissection of the motivational machinery of an animal into "functional systems." On the experimental side, the approach has received support from work within McFarland's group on the drinking and feeding behavior of Barbary doves and from other such work as the "titration method" of Baerends and Kruijt (1973) which establishes the strength of a stimulus through comparison with a series of standard stimuli. It may be that the motivational framework pursued by McFarland will prove to be elusive as such structuralist concepts as deep grammars, but the attempt to solve basic ethological problems without reliance on arbitrary scales imposed by experimenter preconceptions is to be encouraged.

4. INTERPRETATION AND LIMITATIONS

Two characteristics of a model are worth emphasizing: namely its ability to generate some true statements and some predictions. With respect to true statements, it should be clear that models are useful only to the extent that they relate theory and data. Conceptual schemata are cheap (a moderately imaginative mind can generate three before breakfast every day) but those for which no, or only an imperfect, mapping to existing facts is provided serve no purpose. The appearance of such purposeless schemata, and the general erosion of the term "model" in the published literature to stand for any loose conjecture is most unfortunate since it provides a basis for suspicion by sceptics of the modeling technique who are unaware of the essential value of model building as a procedure for coordinating data, enabling generalizations, and generating predictions.

With respect to the production of predictions as a characteristic of models, I have included predictability as a part of the definition of a model because a second deep and justifiable suspicion held by many middle-ground researchers, who are themselves neither quantitative modelers nor biased against any and all mathematical approaches, is "What have I gained, in terms of understanding and therefore predictive ability, as a result of modeling?" This is indeed the pragmatic acid test, and the creations of modelers must pass it. It is essential to note that the better work in the discipline does pass.

The chief pitfalls of modeling are found in the traditionally problematic area of explanation. One problem is that for any set of data an indefinite number of explanations, in the sense of simulation models, can be produced. To choose among these models the criteria described above are helpful although not always sufficient. A second problem, of particular importance to ethologists, is the ease with which the content of models can be blurred with excess meaning and unclear mappings. The classic case is that of energy models of motivation (Hinde 1960). Despite these possible pitfalls, a modeler can achieve, in conjunction with the data-analytic techniques discussed in the previous chapters, the necessary interaction of experimental and theoretical ethology.

EPILOGUE

It is now appropriate to honor the promissory note in the Prologue of an evaluation of the impact of the foregoing techniques on the evolution of ethological theory. In a young science like ethology, analytical techniques have a particularly important propaedeutic function. The preceding chapters provide an ethologist with a wide variety of powerful and general tools for sophisticated quantitative analysis of the diversity of problems that he is likely to encounter. None of these techniques was originated for ethological use, but rather all have been borrowed from other disciplines: systems diagrams from engineering, response clustering from numerical taxonomy, and response repertory estimation from ecology. This lack of indigenous ethomathematics must surely end, and it will be fascinating to witness, on the one hand, the development of these unique procedures and, on the other, the gradual coordination of such disciplines as quantitative ethology, mathematical psychology, and theoretical ecology—disciplines distinct more by historical accident than subject matter—as appreciation of the many formal similarities shared by these areas grows. This coordination of several levels of behavioral study, far from producing the cannibalization of ethology predicted by Wilson (1975), will continue the process of disciplinary interdependence that Julian Huxley (1942) has termed the modern synthesis.

Contemporary ethologists cannot but help feel akin to the proverbial blind men, each grasping and describing one part of an elephant, resulting in widespread contradiction and conflict. Existing quantitative techniques equip benighted researchers with but crude ocelli. For instance, Markov analysis provides a first approximation of sequential dependencies, cluster analysis a glimpse of hierarchically organized systems, and factor analysis a test for a unitary motivating mechanism (Heiligenberg 1973). The development of more powerful tools will be hindered by issues intrinsic to the subject matter such as, in the study of motivational systems, the problem of the observability of behavior as a function of the realizability of the behavioral commands involved (McFarland and Sibly 1972). But the advent of such tools will eventually illuminate the entire elephant.

Inevitably, the hypotheses of ethologists, tested and supported through collection and analysis of relevant data, will come to be incorporated in theoretical models. The advantages of modeling (Chapter 11) are its capacity to generate specific, quantitative, testable predictions and to solve the traditional reductionism-holism problem by presenting on the one hand the components to which a system can be reduced, and on the other the interactions of these components resulting in "emergent properties."

Particularly for students of motivation, the problems of the nature and interactions of motivated responses and the underlying causal processes, heretofore conceptualized within the context of the chiefly speculative, albeit intelligently, hierarchical scheme of Tinbergen (1951), can be approached through both simulation models and multivariate techniques that will provide an initial description of motivational systems and circumscribe the set of possible simulation models (cf. Slater 1973). Old concepts founded on entrenched myths with visceral appeal will be replaced with empirically based and logically consistent ones. A start has already been made in these areas by the reformulation of the concepts of "attention" and "energy" and the introduction of new terms such as "state space" and "time-sharing" by McFarland (e.g. 1971, 1974c). And Frey and Miller (1972) state that their "concept of a dominance vector is proposed to provide a probabilistic construct (greater or lesser likelihood of winning) as an alternate to the mentalistic concepts of 'set' or 'mood'." (Recall the conclusion of Lotka (1925) that the ideal definition is the quantitative definition.) A similar conceptual re-evaluation is taking place in social ethology with such terms as "territoriality" (Myrberg and Thresher 1974). These improvements corroborate Russell's (1927) assertion that "vagueness, no doubt, is omnipresent and unavoidable, but is in only in proportion as we overcome it that exact science becomes possible."

The diversity of current quantitative approaches to ethology discussed in the preceding chapters reflect a diversity of methodological attitudes that can be arranged on a positivist–rationalist continuum. Information theory and cluster analysis tend to be strongly data bound; multivariate procedures used inferentially rely on more abstract conceptualization; and the models discussed in Chapters 4 and 11 incorporate great theorization. The limits at each end of the continuum are underlaid by opposing pressures. On the one hand, there is an essential need for strong hypotheses and more general theories to make explicit the critical assumptions that an investigator makes about the operation of the system under study. On the other hand, the complexity of theoretical structures, in ethology as in the rest of science, must remain limited by our ability to gather data to test specific models appropriately. It is within this framework that theoretical ethology will come of age.

As regards to the future, ethological theory will be based equally on the early analytical work such as McFarland's and the synthetic attempts such as Tinbergen's. The etholgoist studying behavior in the year 2000, glimpsed through an optimistic telescope, will have at his disposal both powerful data-acquisition apparatus including on-line computers, multimodality sensors, and monitors of neurophysiological and ecological correlates of behavior for the collection, storage, retrieval, rearrangement, and analysis of detailed information, and robust, sensitive, and appropriate theoretical tools including systems modeling techniques and their metatheory, such as theory of model types, complex stochastic distributions, means of testing specific hypotheses, and procedures to deal with nonlinearity and nonstationarity. It is the need for the development of these latter theoretical tools, for general and applicable constructs beyond the current ones based on electrical or hydrodynamic analogies, which places ethological modeling at its present point of inflection. The conceptual framework in operation will be generated directly from this apparatus and these theoretical tools, and the ethologist will then be able to do honor to Tinbergen by completing the quantification of the entire study of behavior.

REFERENCES

Aitchison, J., & J. A. C. Brown. 1963. *The lognormal distribution.* Cambridge: Cambridge University Press.

Altman, N. D. E. 1975. *Contingency table analysis in geographic research: a bibliographic essay.* Technical Report, Geography Department, Queen's University at Kingston.

Altmann, J. 1974. Observational study of behaviour: Sampling methods. *Behaviour* **49**: 227–267.

Altmann, S. A. 1965. Sociobiology of rhesus monkeys. II. Stochastics of social communication. *J. Theor. Biol.* **8**: 490–522.

Altmann, S. A. 1974. Baboons, space, time and energy. *Am. Zool.* **14**: 221–248.

Altmann, S. A., & S. S. Wagner. 1970. Estimating rates of behavior from Hansen frequencies. *Primates* **11**: 181–183.

Anderberg, M. R. 1973. *Cluster analysis for applications.* New York: Academic Press.

Anderson, T. W. 1971. *The statistical analysis of time series.* New York: Wiley.

Andrews, H. C. 1972. *Introduction to mathematical techniques in pattern recognition.* New York: Wiley.

Arabie, P. 1973. Concerning Monte Carlo evaluations of nonmetric multi-dimensional scaling algorithms. *Psychometrika* **38**: 607–608.

Arbib, M. A. 1972. *The metaphorical brain: An introduction to cybernetics as artificial intelligence and brain theory.* New York: Wiley-Interscience.

Archer, J. 1973a. Effects of testosterone on immobility responses in the young male chick. *Behav. Biol.* **8**: 551–556.

Archer, J. 1973b. Tests for emotionality in rats and mice. A review. *Anim. Behav.* **21**: 205–235.

Ashby, W. R. 1956. *An introduction to cybernetics.* London: Chapman and Hall.

Aspey, W. 1974. *Wolf spider sociobiology: An ethological and informational theory analysis of agonistic behavior in* Schizocosa crassipes. Ph.D. thesis, Ohio University.

Aspey, W., & J. Blankenship. 1976. *Aplysia* behavioral biology: I. A multivariate analysis of burrowing in *A. brasiliana. Behav. Biol.* **17**: 279–299.

Attneave, F. 1959. *Applications of information theory to Psychology.* New York: Holt, Rinehart, & Winston.

Baerends, G. P. 1941. Fortpflanzungsverhalten und Orientierung der Grabwespe *Ammophila campestris*. *J. Tijdschr. Entomol.* **84**: 68–275.

Baerends, G. P. 1970. A model of the functional organization of incubation behaviour. *Behav. Suppl.* **17**: 261–312.

Baerends, G. P., R. H. Drent, P. Glas, & H. Groenewold. 1970. An ethological study of incubation behaviour in the herring gull. *Behav. Suppl.* **17**: 135–235.

Baerends, G. P., & J. P. Kruijt. 1973. Stimulus selection. Chap. 2 in R. A. Hinde & J. Stevenson-Hinde (Eds.) *Constraints on learning*. London: Academic Press. Pp 23–50.

Baker, M. C. 1973. Stochastic properties of the foraging behavior of six species of migratory shorebirds. *Behaviour* **45**: 242–270.

Bailey, N. T. J. 1967. *The mathematical approach to biology and medicine*. New York: Wiley.

Ballantyne, P. K., & P. W. Colgan. In press. Sound production during agonistic and reproductive behaviour in the pumpkinseed (*Lepomis gibbosus*), the bluegill (*L. macrochirus*) and their hybrid sunfish. I. Context. *Biology of Behaviour*.

Bard, Y. 1974. *Nonlinear parameter estimation*. New York: Academic Press.

Barlow, G. W. 1968. Ethological units of behavior. Chap. 1 in D. Ingle (Ed.) *Central nervous system and fish behavior*. Chicago: University of Chicago Press. Pp. 217–232.

Barlow, R. E., D. J. Bartholomew, J. M. Bremner, & H. D. Brunk. 1972. *Statistical inference under order restrictions: The theory and application of isotonic regression*. New York: Wiley.

Barlow, R. E., A. W. Marshall, & F. Proschan. 1963. Properties of probability distributions with monotone hazard rate. *Ann. Math. Stat.* **34**: 375–389.

Baylis, J. R. 1975. *A quantitative, comparative study of courtship in two sympatric species of the genus* Cichlasoma *(Teleostei, Cichlidae)*. Ph.D. thesis, University of California at Berkeley.

Bekoff, M. 1975. Animal play: Problems and perspectives. Chap. 4 in P. P. G. Bateson & P. H. Klopfer (Eds.) *Perspectives in ethology*. Vol. 2. New York: Plenum Press. Pp. 165–188.

Bekoff, M., H. Hill, & J. Mitton. 1975. Behavioral taxonomy in canids by discriminant function analyses. *Science* **190**: 1223–1225.

Bellman, R., K. L. Cooke, & J. A. Lockett. 1970. *Algorithms, graphs, and computers*. New York: Academic Press.

Beukema, J. 1968. Predation by the three-spined stickleback (*Gasterosteus aculeatus* L.): The influence of hunger and experience. *Behaviour* **31**: 1–126.

Billingsley, P. 1961. Statistical methods in Markov chains. *Ann. Math. Stat.* **32**: 12–40.

Bindra, D. 1959. *Motivation: A systematic reinterpretation*. New York: Ronald Press.

Binjnen, E. J. 1974. *Cluster analysis*. Portland, Or.: International Scholarly Book Service.

Binkley, S. 1973. Rhythm analysis of clipped data: examples using circadian data. *J. Comp. Physiol.* **85**: 141–146.

Bishop, Y. M. M., S. E. Fienberg, & P. W. Holland. 1975. *Discrete multivariate analysis: Theory and practice*. Cambridge, Mass.: M.I.T. Press.

Blachman, N. M. 1968. The amount of information that *y* gives about *x*. *IEEE Trans. Info. Theory* **14**: 27–31.

Blackith, R. E. 1965. Morphometrics. Ch. 9 in T. H. Waterman & H. J. Morowitz (Eds.) *Theoretical and mathematical biology*. New York: Ginn and Co.

Blackith, R. E., & R. A. Reyment. 1971. *Multivariate morphometrics*. London: Academic Press.

Bloxom, B. 1968. Individual differences in multidimensional scaling. *Research Bulletin* **68-45**. Princeton, N. J.: E. T. S.

Blumenthal, L. M. 1953. *Theory and application of distance geometry*. Oxford: Clarendon Press.

Blurton Jones, N. G. 1968. Observations and experiments on causation of threat displays of the great tilt *(Parus major)*. *Anim. Behav. Monogr.* **1**: 74–158.

Bock, R. D. 1975. *Multivariate statistical methods in behavioral research*. McGraw Hill: New York.

Borko, H. (Ed.) 1962. *Computer applications in the behavioral sciences*. Englewood Cliffs: Prentice-Hall.

Box, G. E. P., & G. M. Jenkins. 1970. *Time series analysis: forecasting and control*. San Francisco: Holden-Day.

Brainerd, B. 1972. On the relation between types and tokens in literary text. *J. Appl. Probab.* **9**: 507–518.

Brillinger, D. R. 1975. *Time series: data analysis and theory*. New York: Holt, Rinehart and Winston.

Brockmann, H. J. 1975. Systems diagrams workshop. *Anim. Behav. Soc. Paper*.

Brown, J. L. 1975. *The evalution of behaviour*. New York: Norton.

Brown, R., S. Freeman, & D. McFarland. 1974. Towards a model for the copulatory behaviour of the male rat. Chap. 11 in D. J. McFarland (Ed.) *Motivational control systems analysis*. London: Academic Press. Pp. 461–510.

Bulmer, M. G. 1974. On fitting the Poisson lognormal distribution to species-abundance data. *Biometrics* **30**: 101–110.

Busacker, R. G., & T. L. Saaty. 1965. *Finite graphs and networks: An introduction with applications*. New York: McGraw-Hill.

Cane, V. 1959. Behaviour sequences as semi-Markov chains. *J. R. Stat. Soc. B* **21**: 36–58.

Cane, V. 1961. Some ways of describing behaviour. Chap. 14 in W. H. Thorpe & O. L. Zangwill (Eds.) *Current problems in animal behaviour*. Cambridge: Cambridge University Press. Pp. 361–388.

Carroll, J. B., P. Davies, & B. Richman. 1971. *The American Heritage word frequency book*. Boston, Mass.: Houghton Mifflin.

Carroll, J. D., & J. J. Chang. 1970. Analysis of individual differences in multidimensional scaling via an N-way generalization of "Eckart-Young" decomposition. *Psychometrika* **35**: 288–319.

Carroll, J. D., & M. Wish. 1974. Models and methods for three-way multidimensional scaling. In D. H. Krantz, R. C. Atkinson, R. D. Luce, & P. Suppes (Eds.) *Contemporary developments in mathematical psychology*. Vol. 2. San Francisco: W. H. Freeman.

Chance, M. R. A., & W. M. S. Russell. 1959. Protean displays: a form of allaesthetic behavior. *Proc. Zool. Soc. Lond.* **132**: 65–70.

Chase, I. D. 1974. Models of hierarchy formation in animal societies. *Behav. Sci.* **19**: 374–382.

Chatfield, C., & R. E. Lemon. 1970. Analyzing sequences of behavioural events. *J. Theor. Biol.* **29**: 427–445.

Chatfield, C., & R. E. Lemon. 1971. Organization of song in cardinals. *Anim. Behav.* **19**: 1–17.

Chow, I. A. 1975. Estimation of characteristics of animal behavior by stochastic methods. *Math. Biosci.* **24**: 281–288.

Cliff, N. 1966. Orthogonal rotation to congruence. *Psychometrika* **31**: 33–42.

Cochran, W. G. 1954. Some methods for strengthening the common χ^2 tests. *Biometrics* **10**: 417–451.

Cochran, W. G., & G. M. Cox. 1957. *Experimental design*. 2nd ed. New York: Wiley.

Cody, M. L. 1974. Optimization in ecology. *Science* **183**: 1156–1164.

Cohen, J. E. 1971. *Casual groups of monkeys and men*. Cambridge, Mass.: Harvard University Press.

Cohen, J. E. 1972. Markov population processes as models of primate social and population dynamics. *Theor. Pop. Biol.* **3**: 119–134.

Cohen, J. E. 1975. The size and demographic composition of social groups of wild orang-utans. *Anim. Behav.* **23**: 543–550.

Cole, A. J. (Ed.) 1969. *Numerical taxonomy*. New York: Academic Press.

Colgan, P. 1973. Motivational analysis of fish feeding. *Behaviour* **45**: 38–66.

Colgan, P. 1975. Self-selection of photoperiod as a technique for studying endogenous rhythms in fish. *J. Interdis. Cycle Res.* **6**: 203–211.

Colgan, P. W., & M. R. Gross. 1977. Dynamics of agression in male pumpkinseed sunfish (*Lepomis gibbosus*) over the reproductive phase. *Zeit. Tierpsychol.* **43**: 139–151.

Cooley, W. W., & P. R. Lohnes. 1971. *Multivariate data analysis*. New York: Wiley.

Coombs, C. H. 1964. *Theory of Data*. New York: Wiley.

Cooper, L. G. 1972. A new solution to the additive constant problem in metric multidimensional scaling. *Psychometrika* **37**: 311–322.

Coulon, J. 1971. Influence de l'isolement social sur le compartement du cobaye. *Behaviour* **38**: 93–120.

Cox, D. R. 1972. The analysis of multivariate binary data. *Appl. Stat.* **21**: 113–120.

Cox, D. R., & P. A. W. Lewis. 1966. *The statistical analysis of series of events*. London: Methuen.

Crook, J. 1970. Social organization and the environment: Aspects of contemporary social ethology. *Anim. Behav.* **18**: 197–209.

Crow, J. F., & M. Kimura. 1970. *An introduction to population genetics theory*. New York: Harper & Row.

Dane, B., & W. G. van der Kloot. 1964. Analysis of the display of the goldeneye duck (*Bucephala clangula* (L.)). *Behaviour* **22**: 282–328.

Dane, B., C. Walcott, & W. H. Drury. 1959. The form and duration of the display actions of the goldeneye (*Bucephala clangula*). *Behaviour* **14**: 265–281.

Darroch, J. N., & D. Ratcliff. 1972. Generalized iterative scaling for loglinear models. *Ann. Math. Stat.* **43**: 1470–1480.

David, H. T., W. H. Kruskal, L. Augenstein, & H. Quastler. 1956. Approximate distribution of sample information for use in estimating true information by confidence intervals. *Control Systems Laboratory, University of Illinois, Report R-76:* 1–37.

Dawkins, M., & R. Dawkins. 1974. Some descriptive and explanatory stochastic models of decision-making. Chap. 3 in D. J. McFarland (Ed.) *Motivational control systems analysis*. London: Academic Press. Pp. 119–168.

Dawkins, R. 1971. A cheap method of recording behavioural events, for direct computer access. *Behaviour* **40**: 162–173.

Dawkins, R. 1976. Hierarchical organisation: a candidate principle for ethology. Chap. 1 in P. P. G. Bateson, & R. A. Hinde (Eds.) *Growing points in ethology*. Cambridge: Cambridge University Press. Pp. 7–54.

Dawkins, R., & M. Dawkins. 1973. Decisions and the uncertainty of behaviour. *Behaviour* **45**: 83–103.

DeFries, J. 1969. Pleiotropic effects of albinism on open field behavior in mice. *Nature* **221**: 65–66.

DeFries, J., J. Hegmann, & M. Weir. 1966. Open-field behavior in mice: Evidence for a major gene effect mediated by the visual system. *Science* **154**: 1577–1579.

Degerman, R. L. 1972. The geometric representation of some simple structures. In R. N. Shepard, A. K. Romney, & S. B. Nerlove (Eds.) *Multidimensional scaling: Theory and applications in the behavioral sciences. Vol. 1.* New York: Seminar Press.

De Ghett, V. J. 1972. *The behavioral and morphological development of the Mongolian gerbil (*Meriones unguiculatus*) from birth until thirty days of age.* Ph.D. thesis, Bowling Green University.

De Leeuw, J., & S. Pruzansky. 1975. A new computational method to fit the weighted Euclidean distance model. *Bell Laboratories Technical Memorandum TM-75-1229-12*.

Delius, J. D. 1969. A stochastic analysis of the maintenance behaviour of skylarks. *Behaviour* **33**: 137–178.

Deming, W. E., & F. F. Stephan. 1940. On a least squares adjustment of a sample frequency table when the expected marginal totals are known. *Ann. Math. Stat.* **11**: 427–444.

Dingle, H. 1969. A statistical and information analysis of aggressive communication in the mantis shrimp *Gonodactylus bredini* Manning (Crustacea: Stomatopoda). *Anim. Behav.* **17**: 567–581.

Dingle, H. 1972. Aggressive behavior in stomatopods and the use of information theory in the analysis of animal communication. In H. E. Winn & B. L. Olla (Eds.) *Behavior of marine animals*. Vol. 1. New York: Plenum Press. Pp. 126–156.

Dixon, W. 1970a. *BMD-Biomedical computer programs*. 2nd ed. Berkeley: University of California Press.

Dixon, W. 1970b. *BMD-Biomedical computer programs X-series supplement*. Berkeley: University of California Press.

Dixon, W. J. 1973. *BMD-Biomedical Computer Programs*. 3rd ed. Berkeley: Universtiy of California Press.

Dowling, J. E. 1970. Organization of vertebrate retinas. *Invest. Ophthalmol.* **9**: 655–680.

Drake, A. W. 1967. *Fundamentals of applied probability theory*. New York: McGraw-Hill.

Duncan, I. J. H., A. R. Horne, B. O. Hughes, & D. G. M. Wood-Gush. 1970. The pattern of food intake in female Brown Leghorn fowls as recorded in a Skinner box. *Anim. Behav.* **18**: 245–255.

Duran, B. J., & P. L. Odell. 1974. *Cluster analysis: A survey*. New York: Springer-Verlag.

Edwards, A. W. F., & L. L. Cavelli-Sforza. 1964. Reconstruction of evolutionary trees. In

V. H. Heywood & J. McNeill (Eds.) *Phenetic and phylogenetic classification.* London: Systematics Association Publication No. 6. Pp. 67–76.

Edwards, A. W. F., & L. L. Cavalli-Sforza. 1965. A method for cluster analysis. *Biometrics* **21**: 362–375.

Engen, S. 1974. On species frequency models. *Biometrika* **61**: 263–270.

Engen, S. 1975. Coverage of a random sample from a biological community. *Biometrics* **31**: 201–208.

Estes, W. K. 1975. Some targets for mathematical psychology. *J. Math. Psychol.* **12**: 263–282.

Everitt, B. 1974. *Cluster analysis.* New York: Halsted.

Ewing, A. W. 1969. The genetic basis of sound production in *Drosophila pseudoobscura* and *D. persimilis. Anim. Behav.* **17**: 555–560.

Eysenck, H. 1953. The logical basis of factor analysis. *Am. Psychol.* **8**: 105–114.

Fagen, R. M. 1977. Selection for optimal age-dependent schedules of play behavior. *Am. Nat.* **111**: 395–414.

Fagen, R. M. MS. Behavioral transitions: Cell-by-cell tests for statistical significance.

Fagen, R. M. MS. Information measures: statistical confidence limits and inference.

Fagen, R. M., & R. N. Goldman. 1977. Behavioural catalogue analysis methods. *Anim. Behav.* **25**: 261–274.

Fager, E. W. 1972. Diversity: a sampling study. *Am. Nat.* **106**: 293–310.

Fentress, J. C. 1973. Specific and nonspecific factors in the causation of behavior. Chap. 6 in P. P. G. Bateson, & P. H. Klopfer (Eds.) *Perspectives in Ethology. Vol. 1.* New York: Plenum Press. Pp. 155–224.

Fentress, J. C. 1976. Dynamic boundaries of patterned behaviour: Interaction and self-organisation. Chap. 4 in P. P. G. Bateson, & R. A. Hinde (Eds.) *Growing points in ethology.* Cambridge: Cambridge University Press. Pp. 135–169.

Fienberg, S. E. 1970. The analysis of multidimensional contingency tables. *Ecology* **51**: 419–433.

Fienberg, S. E. 1971. A statistical technique for historians: Standardizing tables of counts. *J. Interdis. Hist.* **1**: 305–315.

Fienberg, S. E. 1972a. On the use of Hansen frequencies for estimating rates of behavior. *Primates* **13**: 323–326.

Fienberg, S. E. 1972b. The analysis of incomplete multidimensional contingency tables. *Biometrics* **28**: 177–202.

Fienberg, S. E. 1974. Stochastic models for single neuron firing trains: A survey. *Biometrics* **30**: 300–428.

File, S., & S. Day. 1972. Effects of time of day and food deprivation on exploratory activity in the rat. *Anim. Behav.* **20**: 758–762.

Fisher, R. A. 1936. The use of multiple measurements in taxonomic problems. *Ann. Eugen.* **7**: 179–188.

Fleiss, J. L., & J. Zubin. 1969. On the methods and theory of clustering. *Multiv. Behav. Res.* **4**: 235–250.

Frey, D. F., & R. J. Miller. 1972. The establishment of dominance relationships in the blue gourami, *Trichogaster trichopterus* (Pallas). *Behaviour* **42**: 8–62.

Fruchtgott, E., & E. Cureton. 1964. Factor analysis of emotionality and conditioning in mice. *Psychol. Rep.* **15**: 787–794.

Gani, J. (Ed.) 1976. *Perspectives in probability and statistics.* Applied Probability Trust, Sheffield, England. (U.S. distributor, Academic Press, New York.)

Garner, W. R., & W. J. McGill. 1956. Relation between uncertainty, variance, and correlation analyses. *Psychometrika* **21**: 219–228.

Garnett, J. 1919. On certain independent factors in mental measurement. *Proc. R. Soc. Lond.* **46**: 91–111.

Gasking, D. 1960. Clusters. *Australas. J. Phil.* **38**: 1–36.

Gates, M. A., E. E. Powelson, & J. Berger. 1974. Syngenic ascertainment in *Paramecium aurelia. Syst. Zool.* **23**: 482–489.

Geertsema, S., & H. Reddingius. 1974. Preliminary considerations in the simulation of behaviour. Chap. 8 in D. J. McFarland (Ed.) *Motivational control systems analysis.* London: Academic Press. Pp. 355–405.

Geisser, S. 1964. Posterior odds for multivariate normal classifications. *J. R. Stat. Soc.* **26**: 69–76.

Gleason, T. C. 1967. A general model for nonmetric multidimensional scaling. *Report No. MMPP 67-3.* Ann Arbor: University of Michigan.

Goel, N. S., & N. Richter-Dyn. 1974. Stochastic models in biology. New York: Academic Press.

Golani, I. 1973. Non-metric analysis of behavioral interaction sequences in captive jackals *(Canis aureus* L.). *Behaviour* **44**: 89–112.

Golani, I. 1976. Homeostatic motor processes in mammalian interactions: A choreography of display. Chap. 2 in P. P. G. Bateson & P. H. Klopfer (Eds.) *Perspectives in ethology.* Vol. 2. New York: Plenum Press. Pp. 69–134.

Gold, H. 1976. *Correlates of support for separation amongst French Canadians in Quebec.* M.A. thesis, Queen's University at Kingston.

Goldman, R. N. 1973. *The species sampling problem.* Ph.D. thesis, Harvard University.

Good, I. J. 1953. The population frequencies of species and the estimation of population parameters. *Biometrika* **40**: 237–264.

Good, I. J., & G. H. Toulmin. 1956. The number of new species, and the increase in population coverage when a sample is increased. *Biometrika* **43**: 45–63.

Goodman, L. A. 1968. The analysis of cross-classified data: Independence, quasi-independence and interactions in contingency tables with or without missing entries. *J. Am. Stat. Ass.* **63**: 1091–1131.

Goodman, L. A. 1970. The multivariate analysis of qualitative data: Interactions among multiple classifications. *J. Amer. Stat. Ass.* **65**: 226–256.

Goodman, L. A. 1971. The analysis of multidimensional contingency tables: Stepwise procedures and direct estimation methods for building models for multiple classifications. *Technometrics* **13**: 33–61.

Goodman, L. A. 1972. A general model for the analysis of surveys. *Am. J. Sociol.* **77**: 1035–1086.

Gould, J. L. 1974. Genetics and molecular ethology. *Zeit. Tierpsychol.* **36**: 267–292.

Gower, J. C. 1966. Some distance properties of latent root and vector methods used in multivariate analysis. *Biometrika* **53**: 325–338.

Grant, P. 1970. Experimental studies of competitive interaction in a two-species system. II. The behaviour of *Microtus, Peromyscus,* and *Clethrionomys* species. *Anim. Behav.* **18**: 411–426.

Gray, L. 1976. An analysis of behavioral diversity in resource utilization strategies. *Anim. Behav. Soc. Paper.*

Griffith, J. S. 1971. *Mathematical neurobiology: An introduction to the mathematics of the nervous system.* New York: Academic Press.

Grizzle, J. E., C. F. Starmer, & G. G. Koch. 1969. Analysis of categorical data by linear models. *Biometrics* **25**: 489–504.

Guttman, L. 1954. Some necessary conditions for common-factor analysis. *Psychometrika* **19**: 149–161.

Guttman, L. 1968. A general nonmetric technique for finding the smallest coordinate space for a configuration of points. *Psychometrika* **33**: 469–506.

Haberman, S. J. 1973. *CTAB: Analysis of multidimensional contingency tables by log-linear models (Users' Guide).* Ann Arbor: National Educational Resources.

Halliday, T. R. 1975. An observational and experimental study of sexual behaviour in the smooth newt, *Triturus vulgaris* (Amphibia: Salamandridae). *Anim. Behav.* **23**: 291–322.

Hamerton, J. L., L. Dallaire, J. R. Miller, L. Siminovitch, & N. E. Simpson. 1977. *Diagnosis of genetic disease by amniocentesis during the second trimester of pregnancy.* Report No. 5, Medical Research Council: Ottawa.

Hamilton, W. D. 1964. The genetical evolution of social behaviour. I & II. *J. Theor. Biol.* **7**: 1–16, 17–52.

Hamilton, W. D. 1971. Geometry for the selfish herd. *J. Theor. Biol.* **31**: 295–311.

Harman, H. 1967. *Modern factor analysis.* Chicago: University of Chicago Press.

Harman, H. 1968. Factor analysis. In D. Whitla (Ed.) *Handbook of measurement and assessment in behavioral sciences.* Menlo Park: Addison-Wesley.

Harris, B. 1959. Determining bounds on integrals with applications to cataloguing problems. *Ann. Math. Stat.* **30**: 521–548.

Harris, R. 1975. *A primer of multivariate statistics.* New York: Academic Press.

Harshman, R. A. 1970. Foundations of the PARAFAC procedure: Models and conditions for an "explanatory" multimodal factor analysis. *U.C.L.A. Working Papers in Phonetics* **16**: 1–86.

Harshman, R. A. 1972. Determination and proof of minimum uniqueness conditions for PARAFAC1. *U.C.L.A. Working Papers in Phonetics* **22**: 111–117.

Hartigan, J. A. 1967. Representation of similarity matrices by trees. *J. Am. Stat. Ass.* **62**: 1140–1158.

Hartigan, J. A. 1975. *Clustering algorithms.* New York: Wiley.

Hazlett, B. A., & W. H. Bossert. 1965. A statistical analysis of the aggressive communication systems of some hermit crabs. *Anim. Behav.* **13**: 358–373.

Hazlett, B. A., & G. F. Estabrook. 1974a. Examination of agonistic behavior by character analysis. I. The spider crab *Microphrys bicornutus. Behaviour* **48**: 131–144.

Hazlett, B. A., & G. F. Estabrook. 1974b. Examination of agonistic behavior by character analysis. II. Hermit crabs. *Behaviour* **49**: 88–110.

Heiligenberg, W. 1973. Random processes describing the occurrence of behavioural patterns in a cichlid fish. *Anim. Behav.* **21**: 169–182.

Heiligenberg, W. 1974a. A stochastic analysis of fish behaviour. Chap. 2 in D. J. McFarland (Ed.) *Motivational control systems analysis*. London: Academic Press. Pp. 87–118.

Heiligenberg, W. 1974b. Processes governing behavioral states of readiness. In D. S. Lehrman et al. (Eds.) *Advances in the study of behaviour*. Vol. 5. New York: Academic Press. Pp. 173–200.

Heiligenberg, W. 1976. The interaction of stimulus patterns controlling aggressiveness in the cichlid fish *Haplochromis burtoni*. *Anim. Behav.* **24**: 452–458.

Heiligenberg, W., & U. Kramer. 1972. Aggressiveness as a function of external stimulation. *J. Comp. Physiol.* **77**: 332–340.

Heinroth, O. 1911. Beitrage zur Biologie, namentlich Ethologie und Psychologie der Anatiden. *Verh. 5th Int. Orn. Kong.*: 589–702.

Herrnstein, R. J. 1974. Formal properties of the matching law. *J. Exp. Anal. Behav.* **21**: 159–164.

Hinde, R. A. 1953. Appetitive behaviour, consumatory act and the hierarchical organization of behaviour: With special reference to the great tit *(Parus major)*. *Behaviour* **5**: 189–224.

Hinde, R. A. 1960. Energy models of motivation.*Symp. Soc. Exp. Biol.* **14**: 199–213.

Hinde, R. A. 1970. *Animal behaviour: A synthesis of ethology and comparative psychology*. 2nd ed. New York: McGraw-Hill.

Hinde, R. A. 1960. Energy models of motivation. *Symp. Soc. Exp. Biol.* **14**: 199–213.

Hinz, P., & J. Gurland. 1970. A test of fit for the negative binomial and other distributions. *J. Am. Stat. Ass.* **65**: 887–903.

Hodges, J. L. Jr., & E. L. Lehmann. 1970. *Basic concepts of probability and statistics*. 2nd ed. San Francisco: Holden Day.

Hoffman, W. C. 1970. Higher visual perception as prolongation of the basic Lie transformation group. *Math. Biosci.* **6**: 437–471.

Hogg, R. V., & A. T. Craig. 1965. *Introduction to mathematical statistics*. 2nd ed. New York: Macmillan.

Holling, C. S. 1959. Some characteristics of simple types of predation. *Can. Entomol.* **91**: 385–398.

Holling, C. S. 1963. An experimental component analysis of population processes. *Mem. Entomol. Soc. Can.* **32**: 22–32.

Holling, C. S. 1964. The analysis of complex population processes. *Can. Entomol.* **96**: 335–347.

Holling, C. S. 1966. The functional response of invertebrate predators to prey density. *Mem. Entomol. Soc. Can.* **48**: 1–86.

Holloway, J. D., & N. Jardine. 1968. Two approaches to zoogeography: A study based on the distribution of butterflies, birds, and bats in the Indo-Australian area. *Proc. Linn. Soc. Lond.* **179**: 153–188.

Hope, K. 1969. *Methods of multivariate analysis with handbook of multivariate methods programmed in atlas autocode*. New York: Gordon and Breach, Science Publishers, Inc.

Horan, C. B. 1969. Multidimensional scaling: Combining observations when individuals have different perceptual structures. *Psychometrika* **34**: 139–165.

Houston, A., & D. J. McFarland. 1976. On the measurement of motivational variables. *Anim. Behav.* **24**: 459–475.

Hubbell, S. P. 1973. Populations and simple food webs as energy filters. II. Two-species systems. *Am. Nat.* **107**: 122–151.

Huggins, W. H., & D. R. Entwistle. 1968. *Introductory systems and design.* Waltham, Mass.: Blaisdell.

Hull, D. L. 1970. Contemporary systematic philosophies. *Ann. Rev. Ecol. Syst.* **1**: 19–54.

Hutchinson, G. E. 1953. The concept of pattern in ecology. *Proc. Acad. Nat. Sci. Phila.* **105**: 1–12.

Hutchinson, G. E. 1957. Concluding remarks. *Cold Spring Harbor Symp. Quant. Biol.* **22**: 415–427.

Hutt, S. J., & C. Hutt. 1970. *Direct observation and measurement of behavior.* Springfield, Ill.: C. C. Thomas.

Huxley, J. S. 1942. *Evolution, the modern synthesis.* New York: Harper & Row.

Isaac, D., & P. Marler. 1963. Ordering of sequences of singing behaviour of Mistle Thrushes in relation to timing. *Anim. Behav.* **11**: 179–188.

Isaac, P. D., & D. D. S. Poor. 1974. On the determination of appropriate dimensionality in data with error. *Psychometrika* **39**: 91–109.

Ito, K., & W. J. Schull. 1964. On the robustness of the T_0^2 test in multivariate analysis of variance when variance-covariance matrices are not equal. *Biometrika* **51**: 71–82.

Jardine, N., & R. Sibson. 1971. *Mathematical taxonomy.* New York: Wiley.

Jenkins, G. M., & D. G. Watts. 1975. *Spectral analysis and its applications.* 2nd ed. San Francisco: Holden-Day.

Johnson, S. C. 1967. Hierarchical clustering schemes. *Psychometrika* **32**: 241–254.

Jolicoeur, P. 1959. Multivariate geographical variation in the wolf, *Canis lupus* L. *Evolution* **13**: 283–299.

Kaufman, I. C., & L. A. Rosenblum. 1966. A behavioral taxonomy for *Macaca nemestrina* and *Macaca radiata,* based on longitudinal observation of family groups in the laboratory. *Primates* **7**: 205–258.

Kerr, G. E. 1976. Uncertainty (information) analysis of the behaviour of the Carolina locust: Relationships within individuals. *Anim. Behav. Soc. Paper.*

Kim, J. 1975. Factor analysis. In N. Nie et al. (Ed.) *SPSS statistical package for the social sciences.* 2nd ed. New York: McGraw-Hill.

Klahr, D. 1969. A Monte Carlo investigation of the statistical significance of Kruskal's nonmetric scaling procedure. *Psychometrika* **34**: 319–330.

Klecka, W. 1975. Discriminant analysis. In N. Nie et al. (Ed.) *SPSS statistical package for the social sciences.* 2nd ed. New York: McGraw-Hill.

Kleiman, D. G. 1972. Maternal behaviour of the green acouchi *(Myoprocta pratti* Pocock), a South American caviomorph rodent. *Behaviour* **43**: 48–84.

Kloot, W. van der, & M. J. Morse. 1975. A stochastic analysis of the display behavior of the Red-breasted merganser *(Mergus serrator). Behaviour* **54**: 181–216.

Kojima, K. (Ed.) 1970. *Mathematical topics in population genetics.* New York: Springer-Verlag.

Koopowitz, H. 1970. Feeding behaviour and the role of the brain in the polyclad flatworm, *Planocera gilchristi. Anim. Behav.* **18**: 31–35.

Kortlandt, A. 1955. Aspects and the prospects of instinct. *Arch. Neerl. Zool.* **11**: 155–284.

Krane, W. R. 1976. Least squares estimation of individual differences in multidimensional scaling. *York University Psychology Department Report No. 43.*

Krebs, J. R. 1974. Colonial nesting and social feeding as strategies for exploiting food resources in the Great blue heron *(Ardea herodias). Behaviour* **51**: 99–134.

Krebs, J. R. 1976. Why don't animals cheat? *Nature* **260**: 481.

Krebs, J. R., J. C. Ryan, & E. L. Charnov. 1974. Hunting by expectation or optimal foraging? A study of patch use by chickadees. *Anim. Behav.* **22**: 953–964.

Kruskal, J. B. 1964a. Multidimensional scaling by optimizing goodness of fit to a nonmetric hypothesis. *Psychometrika* **29**: 1–27.

Kruskal, J. B. 1964b. Nonmetric multidimensional scaling: A numerical method. *Psychometrika* **29**: 115–129.

Kruskal, J. B. 1976. More factors than subjects, tests and treatments: An indeterminacy theorem for canonical decomposition and individual differences scaling. *Psychometrika* **41**: 281–300.

Kruskal, J. B., & M. Wish. In Press. *Multidimensional scaling.* Beverley Hills: Sage Publications.

Kruskal, J. B., F. W. Young, & J. B. Seery. 1973. How to use KYST, a very flexible program to do multidimensional scaling and unfolding. Unpublished manuscript, Bell Laboratories.

Ku, H. H., & S. Kullback. 1974. Loglinear models in contingency table analysis. *Am. Stat.* **28**: 115–122.

Kuczura, A. 1973. Piecewise Markov processes. *SIAM J. Appl. Math.* **24**: 169–181.

Kullback, S. 1959. *Information Theory and Statistics.* New York: Wiley. Reprinted 1968. New York: Dover.

Kummer, H. 1971. *Primate societies: Group techniques of ecological adaptation.* Chicago: Aldine-Atherton.

Lance, G. N., & W. T. Williams. 1967. A general theory of classificatory sorting strategies. I. Hierarchical systems. *Comput. J.* **9**: 373–380.

Leatherdale, W. H. 1974. *The role of analogy, model and metaphor in science.* New York: American Elsevier.

Lefebvre, L., & R. Joly. 1976. The organization of comfort activities in the American Kestrel *(Falco sparverius). Anim. Behav. Soc. Paper.*

Lemon, R. E., & C. Chatfield. 1971. Organization of song in Cardinals. *Anim. Behav.* **19**: 1–17.

Leong, C. Y. 1969. The quantitative effect of releasers on the attack readiness of the fish *Haplochromis burtoni* (Cichlidae, Pisces). *Zeit. Vgl. Physiol.* **65**: 29–50.

Lerwill, C. J., & P. Makings. 1971. The agonistic behaviour of the golden hamster *Mesocricetus auratus* (Waterhouse). *Anim. Behav.* **13**: 478–492.

Levelt, W. J. M., J. P. Van de Geer, & R. Plomp. 1966. Triadic comparisons of musical intervals. *Br. J. Math. Stat. Psychol.* **19**: 163–179.

Levins, R. 1966. The strategy of model building in population biology. *Am. Sci.* **54**: 421–431.

Levins, R. 1974. The qualitative analysis of partially specified systems. *Ann. N. Y. Acad. Sci.* **231**: 123–138.

Levitsky, D. A. 1970. Feeding patterns in rats in response to fasts and changes in environmental conditions. *Physiol. Behav.* **5**: 291–300.

Lewis, D. 1960. *Quantitative methods in psychology*. New York: McGraw-Hill.

Leyhausen, P. 1973. Verhaltensstudien an Katzen. *Zeit. Tierpsychol. Suppl. 2*. 3rd ed. Berlin: Paul Parey.

Lingoes, J. C., & E. E. Roskam. 1973. A mathematical and empirical analysis of two multidimensional scaling algorithms. *Psychometric Monogr. 19*.

Lloyd, M., J. H. Zar, & J. R. Karr. 1968. On the calculation of information-theoretical measures of diversity (ecosystem). *Am. Nat.* **79**: 257–272.

Lorenz, K. Z. 1950. The comparative method in studying innate behaviour patterns. *Symp. Soc. Exp. Biol.* **4**: 221–268.

Lorenz, K., & N. Tinbergen. 1939. Taxis und instinkthandlung in der Eirollbewegung der Graugans: I. *Zeit. Tierpsychol.* **2**: 1–20.

Losey, G. S. 1972. The ecological importance of cleaning symbiosis. *Copeia 1972*: 820–833.

Losey, G. S. 1975. Information theory. *Anim. Behav. Soc. Paper.*

Lotka, A. J. 1925. *Elements of physical biology*. Baltimore: Williams & Wilkins. Reprinted 1956. New York: Dover.

MacArthur, R. A. 1958. Population ecology of some warblers of northeastern coniferous forests. *Ecology* **39**: 599–619.

MacArthur, R. A. 1960. On the relative abundance of species. *Am. Nat.* **94**: 25–36.

MacArthur, R. A., & R. Levins. 1964. Competition, habitat selection, and character displacement in a patchy environment. *Proc. Nat. Acad. Sci.* **51**: 1207–1210.

MacArthur, R. H., & E. O. Wilson. 1967. *The theory of island biogeography*. Princeton, N. J.: Princeton University Press.

Machlis, L. E. 1977. An analysis of the temporal patterning of pecking in chicks. *Behaviour* **63**: 1–70.

Magnen, J. K., & S. Tallon. 1966. Le periodicité spontanée de la prise d'aliments ad libituen du rat blanc. *J. Physiol. Paris* **58**: 323–349.

Maki, D. P., & M. Thompson. 1973. *Mathematical models and applications*. Englewood Cliffs: Prentice-Hall.

Maurus, M., & H. Pruscha. 1973. Classification of social signals in squirrel monkeys by means of cluster analysis. *Behaviour* **47**: 106–128.

May, R. M. 1973. *Stability and complexity in model ecosystems*. Princeton, N. J.: Princeton University Press.

May, R. M. 1975. Patterns of species abundance and diversity. Chap. 4 in M. L. Cody & J. Diamond (Eds.) *Ecology and evolution of communities*. Cambridge, Mass.: Harvard University Press. Pp. 81–120.

Maynard Smith, J. 1972. *On evolution*. Edinburgh: University Press.

Maynard Smith, J. 1974a. The theory of games and the evolution of animal conflicts. *J. Theor. Biol.* **47**: 209–221.

Maynard Smith, J. 1974b. *Models in ecology*. Cambridge: Cambridge University Press.

Maynard Smith, J. 1976. A comment on the Red Queen. *Am. Nat.* **110**: 325–330.

Maynard Smith, J., & G. A. Parker. 1976. The logic of asymmetric contests. *Anim. Behav.* **24**: 159–175.

McFarland, D. J. 1965. Control theory applied to the control of drinking in the Barbary dove. *Anim. Behav.* **13**: 478–492.

McFarland, D. J. 1971. *Feedback mechanisms in animal behaviour.* London: Academic Press.

McFarland, D. J. (Ed.) 1974a. *Motivational control systems analysis.* London: Academic Press.

McFarland, D. J. 1974b. Experimental investigation of motivational state. Chap. 6 in D. J. McFarland (Ed.) *Motivational control systems analysis.* London: Academic Press. Pp. 251–282.

McFarland, D. J. 1974c. Time-sharing as a behavioural phenomenon. In D. S. Lehrman et al. (Eds.) *Advances in the study of behaviour.* Vol. 5. New York: Academic Press. Pp. 201–225.

McFarland, D. J., & R. M. Sibly. 1972. "Unitary drives" revisited. *Anim. Behav.* **20**: 548–563.

McFarland, D. J., & R. M. Sibly. 1975. The behavioural final common path. *Phil. Trans. R. Soc. B* **270**: 265–293.

McGee, V. E. 1966. The multidimensional analysis of "elastic" distances. *Br. J. Math. Stat. Psychol.* **19**: 181–196.

McGill, W. J. 1954. Multivariate information transmission. *Psychometrika* **19**: 97–116.

McNeil, D. R. 1973. Estimating an author's vocabulary. *J. Am. Stat. Ass.* **68**: 92–99.

Meisel, W. S. 1972. *Computer-oriented approaches to pattern recognition.* New York: Academic Press.

Meredith, W. M., C. H. Frederiksen, & D. H. McLaughlin. 1974. Statistics and data analysis. *Ann. Rev. Psychol.* **25**: 453–505.

Messick, S. J., & R. P. Abelson. 1956. The additive constant problem in multidimensional scaling. *Psychometrika* **21**: 1–16.

Metz, H. 1974. Stochastic models for the temporal fine structure of behaviour sequences. Chap. 1 in D. J. McFarland (Ed.) *Motivational control systems analysis.* London: Academic Press. Pp. 5–86.

Michener, C. 1974. *The social behavior of the bees.* Cambridge, Mass.: Harvard University Press.

Miller, E. H. 1975. Walrus ethology. I. The social role of tusks and applications of multidimensional scaling. *Can. J. Zool.* **53**: 590–613.

Miller, G. A., & F. C. Frick. 1949. Statistical behavioristics and sequences of responses. *Psychol. Rev.* **56**: 311–324.

Miller, G. A., & W. G. Madow. 1954. On the maximum likelihood estimate of the Shannon-Wiener measure of information. *Air Force Cambridge Research Center Technical Report AFCRC-TR-54-75*: 1–22.

Miller, R. G. 1974. The jackknife—a review. *Biometrika* **61**: 1–15.

Milsum, J. H. 1966. *Biological control systems analysis.* New York: McGraw-Hill.

Mock, D. W. 1976. A comparison of great blue heron and great egret communication strategies. *Anim. Behav. Soc. Paper.*

Moehring, J. L. 1972. *Communication systems of a goby-shrimp symbiosis.* Ph.D. thesis, University of Hawaii.

Morgan, B. J. T., M. J. A. Simpson, J. P. Hanby, & J. Hall-Craggs. 1976. Visualizing interactions and sequential data in animal behaviour: Theory and application of cluster-analysis methods. *Behaviour* **56**: 1–43.

Morris, D. 1957. "Typical intensity" and its relation to the problem of ritualisation. *Behaviour* **11**: 1–12.

Mosteller, C. F., & J. W. Tukey. 1968. Data analysis, including statistics. Chap. 10 in G. Lindzey & E. Aronson (Eds.) *Handbook of social psychology*. Vol. 2. 2nd ed. Reading: Addison-Wesley. Pp. 80–203.

Moynihan, M. H. 1970. Control, suppression, decay, disappearance and replacement of displays. *J. Theor. Biol.* **29**: 85–112.

Murchison, C. 1935. The experimental measurement of a social hierarchy in *Gallus domesticus*: IV: Loss of body weight under conditions of mild starvation as a function of social dominance. *J. Gen. Psychol.* **12**: 296–312.

Myrberg, A. A. Jr., & R. E. Thresher. 1974. Interspecific aggression and its relevance to the concept of territoriality in reef fishes. *Am. Zool.* **14**: 81–96.

Nelson, K. 1964. The temporal patterning of courtship behavior in the glandulocaudine fishes (Ostariophysi, Characidae). *Behaviour* **24**: 90–146.

Nelson, K. 1973. Does the holistic study of behavior have a future? Chap. 8 in P. P. G. Bateson & P. H. Klopfer (Eds.) *Perspectives in ethology*. Vol. 1. New York: Plenum Press. Pp. 281–328.

Nerlove, M., & S. J. Press. 1973. *Univariate and multivariate log-linear and logistic models*. Rand Corporation Technical Report R-1306-EDA/NIH, Santa Monica, Calif.

Oatley, K. 1974. Circadian rhythms and representations of the environment in motivational systems. Chap. 10 in D. J. McFarland (Ed.) *Motivational control systems analysis*. London: Academic Press. Pp. 427–459.

Odum, H. T., & J. Ruiz-Reyes. 1970. Holes in leaves and the grazing control mechanism. In H. T. Odum (Ed.) *A tropical rain forest: A study of irradiation and ecology at El Verde, Puerto Rico*. Division of Technical Information, U. S. Atomic Energy Commission. Pp. I-69-I-80.

Ollason, J. C., & P. J. B. Slater. 1973. Changes in the behaviours of the male zebra finch during a twelve hour day. *Anim. Behav.* **21**: 191–196.

Oortmerssen, G. A. van. 1971. Biological significance, genetics and evolutionary origin of variability in behaviour within and between inbred strains of mice *(Mus musculus)*: A behaviour genetic study. *Behaviour* **38**: 1–91.

Orloci, L. 1969. Information theory models for hierarchic and nonhierarchic classifications. In A. J. Cole (Ed.) *Numerical taxonomy*. New York: Academic Press. Pp. 148–164.

Overall, J. E., & C. J. Klett. 1972. *Applied multivariate analysis*. New York: McGraw-Hill.

Pahl, P. J. 1969. On testing for goodness-of-fit of the negative binomial distribution when expectations are small. *Biometrics* **25**: 143–152.

Painter, R. J., & R. P. Yantis. 1971. *Elementary matrix algebra with linear programming*. Boston: Prindle, Weber & Schmidt.

Parker, G. A. 1970. The reproductive behaviour and the nature of sexual selection in *Scatophaga stercoria* L. (Diptera: Scatophagidae). V. The female's behaviour at the oviposition site. *Behaviour* **37**: 140–168.

Patil, G. P., & S. W. Joshi. 1968. *A dictionary and bibliography of discrete distributions*. Edinburgh and London: Oliver & Boyd.

Patrick, R., M. G. Hohn, & J. H. Wallace. 1954. A new method for determining the pattern of the diatom flora. *Notulae Naturalae Acad. Nat. Sci. Phila. No. 259.*

Patten, B. C. 1964. The rational decision process in salmon migration. *J. Cons. Perm. Int. Explor. Mer* **28**: 410–417.

Patten, B. C. (Ed.) 1971–1975. *Systems analysis and simulation in ecology.* 4 Vols. New York: Academic Press.

Peirce, C. S. 1873. *Theory of errors of observations.* Report of Superintendent of U.S. Coast Survey (for the year ending Nov. 1, 1870). Washington, D. C.: Government Printing Office. Appendix No. 21, pp. 200–224 and Plate No. 27. (Cited in Mosteller and Tukey 1968; original not seen.)

Phillips, R. E., & O. M. Youngren. 1971. Brain stimulation and species-typical behaviour: activities evoked by electrical stimulation of the brains of chickens *(Gallus gallus). Anim. Behav.* **19**: 757–779.

Pielou, E. C. 1966. Shannon's formula as a measure of specific diversity: its use and misuse. *Am. Nat.* **100**: 463–465.

Pielou, E. C. 1969. *An introduction to mathematical ecology.* New York: Wiley-Interscience.

Platt, J. R. 1964. Strong inference. *Science* **146**: 347–353.

Poley, W., & J. Royce. 1973. Behavior genetic analysis of mouse emotionality: II. Stability of factors across genotypes. *Anim. Learn. Behav.* **1**: 116–120.

Poole, R. W. 1974. *An introduction to quantitative ecology.* McGraw-Hill Series in Population Biology.

Preston, F. W. 1948. The commonness, and rarity, of species. *Ecology* **29**: 254–283.

Preston, F. W. 1962. The canonical distribution of commonness and rarity. *Ecology* **43**: 185–215, 410–432.

Pulliam, H. R. 1975. The principle of optimal behavior and the theory of communities. Chap. 9 in P. P. G. Bateson & P. H. Klopter (Eds.) *Perspectives in ethology* Vol. 2. New York: Plenum Press. Pp. 311–332.

Räber, H. 1948. Analyse des Balzverhaltens eines domestizierten Truthahns *(Meleagris). Behaviour* **1**: 237–266.

Ramsay, J. O. 1975. Solving implicit equations in psychometric data analysis. *Psychometrika* **40**: 337–360.

Ramsay, J. O. 1977. Maximum likelihood estimation in multidimensional scaling. *Psychometrika* **42**:241–266.

Rand, W. M., & A. S. Rand. 1976. Agonistic behavior in nesting iguanas: a stochastic analysis of dispute settlement dominated by the minimization of energy cost. *Zeit. Tierpsychol.* **40**: 279–299.

Richardson, M. W. 1938. Multidimensional psychophysics. *Psychol. Bull.* **35**: 659–660.

Riechert, S. 1976. Web-site selection in the desert spider *Agelenopsis aperta. Oikos* **27**: 311-315.

Riggs, D. S. 1963. *The mathematical approach to physiological problems.* Cambridge, Mass.: M.I.T. Press.

Riggs, D. S. 1970. *Control theory and physiological feedback mechanisms.* Baltimore: Williams and Wilkins.

Robertson, C. M. MS. Individual differences in the perception of sex by Siamese fighting fish *(Betta splendens* Regan): An INDSCAL analysis.

Rohlf, F. J. 1970. Adaptive hierarchical clustering schemes. *Syst. Zool.* **19**: 58–82.

Roper, T. J. 1973. Nesting material as a reinforcer for female mice. *Anim. Behav.* **21**: 733–740.

Rosenberg, S., C. Nelson, & P. S. Vivehanathan. 1968. A multidimensional approach to the structure of personality impressions. *J. Pers. Soc. Psychol.* **9**: 283–294.

Roskam, E. E. 1969. *A comparison of principles for algorithm construction in nonmetric scaling*. Report No. MMPP 69-2. Ann Arbor: University of Michigan.

Rowell, C. H. F. 1961. Displacement grooming in the chaffinch. *Anim. Behav.* **9**: 38–63.

Royce, J., A. Carran, & E. Howerth. 1970. Factor analysis of emotionality in ten inbred strains of mice. *Multiv. Behav. Res.* **5**: 19–48.

Rubenstein, D. I., & B. A. Hazlett. 1974. Examination of the agonistic behaviour of the crayfish *Oronectes virilis* by character analysis. *Behaviour* **50**: 193–216.

Russell, B. 1927. *The analysis of matter*. London: Allen & Unwin. Reprinted 1954. New York: Dover.

Saila, S. B., & R. A. Shappy. 1963. Random movement and orientation in salmon migration. *J. Cons. Perm. Int. Explor. Mer* **28**: 153–166.

Schleidt, W. M. 1964a. Über das Wirkungsgefuge von Balzbewegungen des Truthahnes. *Naturwissenschaften.* **51**: 445–446.

Schleidt, W. M. 1964b. Über die Spontaneität von Erbkoordinationen. *Zeit. Tierpsychol.* **21**: 235–256.

Schleidt, W. M. 1965. Gaussian interval distributions in spontaneously occurring innate behaviour. *Nature* **206**: 1061–1062.

Schleidt, W. M. 1973. Tonic communication: continual effects of discrete signs in animal communication systems. *J. Theor. Biol.* **42**: 359–386.

Schleidt, W. M. 1974. How "fixed" is the Fixed Action Pattern? *Zeit. Tierpsychol.* **36**: 184–211.

Schmidt, R. S. 1971. A model of the central mechanisms of male anuran acoustic behavior. *Behaviour* **39**: 288–317.

Schoener, T. W. 1971. Theory of feeding strategies. *Ann. Rev. Ecol. Syst.* **2**: 369–404.

Schönemann, P. H. 1970. Fitting a simplex symmetrically. *Psychometrika* **35**: 1–21.

Schönemann, P. H. 1972. An algebraic solution for a class of subjective metrics. *Psychometrika* **37**: 441–451.

Schönemann, P. H., & R. M. Carroll. 1970. Fitting one matrix to another under choice of a central dilation and a rigid motion. *Psychometrika* **35**: 245–256.

Seal, H. 1964. *Multivariate statistical analysis for biologists*. London: Methuen.

Shannon, C. E., & W. Weaver. 1949. *The mathematical theory of communication.* Urbana: University of Illinois Press.

Shepard, R. N. 1958. Stimulus and response generalization: Tests of a model relating generalization to distance in psychological space. *J. Exp. Psychol.* **55**: 509–523.

Shepard, R. N. 1962. The analysis of proximities: Multidimensional scaling with an unknown distance function. I & II. *Psychometrika* **27**: 125–140, 219–246.

Shepard, R. N. 1974. Representation of structure in similarity data: Problems and prospects. *Psychometrika* **39**: 373–421.

Sibly, R. M., & D. J. McFarland. 1974. A state-space approach to motivation. Chap. 5 in D. J. McFarland (Ed.) *Motivational control systems analysis*. London: Academic Press. Pp. 213–250.

Siegel, S. 1956. *Nonparametric statistics for the behavioral sciences*. New York: McGraw-Hill.

Siemens, N., C. H. Marting, & F. Greenwood. 1973. *Operations research*. New York: The Free Press.

Simpson, M. J. A., & A. E. Simpson. 1977. One-zero and scan methods for sampling behaviour. *Anim. Behav.* **25**: 726–731.

Singh, J. 1968. *Great ideas of operations research*. New York: Dover.

Slater, P. J. B. 1973. Describing sequences of behavior. Chap. 5 in P. P. G. Bateson & P. H. Klopfer (Eds.) *Perspectives in ethology*. Vol. 1. Pp. 131–153.

Slater, P. J. B. 1974a. Bouts and gaps in the behaviour of zebra finches, with special reference to preening. *Rev. Comp. Anim.* **8**: 47–61.

Slater, P. J. B. 1974b. The temporal pattern of feeding in the zebra finch. *Anim. Behav.* **22**: 506–515.

Slater, P. J. B., & J. C. Ollason. 1972. The temporal patterns of behaviour in isolated male zebra finches: transition analysis. *Behaviour* **42**: 248–269.

Slatkin, M. 1975. A report on the feeding behavior of two East African baboon species. *Contemp. Primatol.* **5**: 418–422.

Slocomb, J., B. Stauffer, & K. L. Dickson. 1977. On fitting the truneated lognormal distribution to species-abundance data using maximum likelihood estimation. *Ecology* **58**: 693–696.

Smith, W. J. 1969. Messages of vertebrate communication. *Science* **165**: 145–150.

Sneath, P. H. A. 1969. Evaluation of clustering methods. In J. A. Cole (Ed.) *Numerical taxonomy*. New York: Academic Press. Pp. 257–271.

Sneath, P. H. A., & R. R. Sokal. 1973. *Numerical taxonomy*. San Francisco: W. H. Freeman.

Snedecor, G., & W. Cochran. 1967. *Statistical methods*. Ames: Iowa State University Press.

Sokal, R. R. 1966. Numerical taxonomy. *Sci. Am.* **215(6)**: 106–116.

Sokal, R. R. 1974. Classification: Purposes, principles, progress, prospects. *Science* **185**: 1115–1123.

Sokal, R. R., & F. J. Rohlf. 1969. *Biometry*. San Francisco: W. H. Freeman.

Sokal, R. R., & P. H. A. Sneath. 1963. *Principles of numerical taxonomy*. San Francisco: W. H. Freeman.

Soucek, B., & F. Vencl. 1975. Bird communication study using digital computer. *J. Theor. Biol.* **49**: 147–172.

Spearman, C. 1904. General intelligence, objectively determined and measured. *Am. J. Psychol.* **15**: 201–293.

Spence, I. 1970. *Multidimensional scaling: An empirical and theoretical investigation*. Ph.D. thesis, University of Toronto.

Spence, I. 1972. A Monte Carlo evaluation of three nonmetric multidimensional scaling algorithms. *Psychometrika* **37**: 461–486.

Spence, I. 1974. On random rankings studies in nonmetric scaling. *Psychometrika* **39**: 267–268.

Spence, I. In press. Incomplete experimental design for multidimensional scaling. In R. B. Gooledge & J. N. Rayner (Eds.) *Multidimensional analysis of large data sets*. Columbus: Ohio State University Press.

Spence, I., & D. W. Domoney. 1974. Single subject incomplete designs for nonmetric multidimensional scaling. *Psychometrika* **39**: 469–490.

Spence, I., & J. Graef. 1974. The determination of the underlying dimensionality of an empirically obtained matrix of proximities. *Multiv. Behav. Res.* **9**: 331–341.

Spence, I., & J. C. Ogilvie. 1973. A table of expected stress values for random rankings in nonmetric multidimensional scaling. *Multiv. Behav. Res.* **8**: 511–517.

Spurway, H., & J. B. S. Haldane. 1953. The comparative ethology of vertebrate breathing. I. Breathing in newts, with a general survey; Appendix: The biometry of ethology. *Behaviour* **6**: 8–34.

Staddon, J. E. R. 1972. A note on the analysis of behavioural sequences in *Columba livia*. *Anim. Behav.* **20**: 284–292.

Stamps, J. A., & G. W. Barlow. 1973. Variation and stereotypy in the displays of *Anolis geneus* (Sauria:Iguanidae). *Behaviour* **47**: 67–94.

Steinberg, J. B., & R. C. Conant. 1974. An informational analysis of the intermale behaviour of the grasshopper *(Chortophaga viridifasciata)*. *Anim. Behav.* **22**: 617–627.

Stenson, H. H., & R. L. Knoll. 1969. Goodness of fit for random rankings in Kruskal's nonmetric scaling procedure. *Psychol. Bull.* **72**: 122–126.

Stephenson, G. R. 1974. Social structure of mating activity in Japanese macaques. *Symp. 5th Cong. Int. Primat. Soc.*: 63–115.

Struhsaker, T. T. 1967. Social structure among vervet monkeys *(Cercopithecus aethiops)*. *Behaviour* **29**: 83–121.

Sudd, J. H. 1967. *An introduction to the behaviour of ants*. New York: St. Martin's Press.

Sussman, H. J., & R. S. Zahler. In press. Catastrophe theory as applied to the social and biological sciences: a critique. *Synthese*.

Sustare, B. D., & E. H. Burtt, Jr. 1976. Sequential analysis of house fly *(Musca domestica)* grooming behavior. *Anim. Behav. Soc. Paper*.

Swets, J. W. 1973. The relative operating characteristic in psychology. *Science* **182**: 990–1000.

Takane, Y., F. W. Young, & J. De Leeuw. 1977. Nonmetric individual differences multidimensional scaling: An alternating least squares method with optimal scaling features. *Psychometrika* **42**: 7–67.

Theberge, J. B., & J. B. Falls. 1967. Howling as a means of communication in timber wolves. *Am. Zool.* **7**: 331–338.

Thomas, D. W., & J. Mayer. 1968. Meal taking and regulation of food intake by normal and hypothalmic hyperphagic rats. *J. Comp. Physiol. Psychol.* **66**: 642–653.

Thomas, G. B. Jr. 1968. *Calculus and analytical geometry*. Reading: Addison-Wesley.

Thomas, M. U., & D. R. Barr. 1977. An approximate test of Markov chain lumpability. *J. Am. Stat. Ass.* **72**: 175–179.

Thorson, J. 1966. Small-signal analysis of a visual reflex in the locust. II. Frequency dependence. *Kybernetik* **3**: 53–66.

Thurstone, L. 1931. Multiple factor analysis. *Psychol. Rev.* **38**: 406–426.

Thurstone, L. 1935. *The vectors of mind*. Chicago: University of Chicago Press.

Tinbergen, N. 1942. An objectivistic study of innate behaviour of animals. *Bibl. Biotheor.* **1**: 39–98.

Tinbergen, N. 1950. The hierarchical organization of nervous mechanisms underlying instinctive behaviour. *Symp. Soc. Exp. Biol.* **4**: 305–312.

Tinbergen, N. 1951. *The study of instinct*. Oxford: Clarendon Press.

Tinbergen, N., G. J. Brockhuysen, F. Feekes, J. C. W. Houghton, H. Kruuk, & E. Szulc. 1962. Egg shell removal by the black-headed gull, *Larus ridibundus* L.; a behaviour component of camouflage. *Behaviour* **19**: 74–117.

Toates, F. M., & K. Oatley. 1970. Computer simulation of thirst and water balance. *Med. Biol. Eng.* **8**: 71–87.

Tobler, W. In press. The relationship between surveying and scaling. In R. G. Golledge & J. N. Rayner (Eds.) *Multidimensional analysis of large data sets*. Columbus: Ohio State University Press.

Torgerson, W. S. 1952. Multidimensional scaling: I. Theory and method. *Psychometrika* **17**: 401–419.

Torgerson, W. S. 1958. *Theory and methods of scaling*. New York: Wiley.

Trivers, R. L. 1974. Parent-offspring conflict. *Am. Zool.* **14**: 249–264.

Tryon, R. C., & D. E. Bailey. 1970. *Cluster analysis*. New York: McGraw-Hill.

Tschudi, F. 1972. *The latent, the manifest and the reconstructed in multivariate data reduction models*. Ph.D. thesis, University of Oslo.

Tucker, L. R. 1972. Relations between multidimensional scaling and three-mode factor analysis. *Psychometrika* **37**: 3–27.

Vandermeer, J. H. 1972. Niche theory. *Ann. Rev. Ecol. Syst.* **3**: 107–132.

Van Sommers, P. 1974. Studies in learned behavioural regulation. Chap. 7 in D. J. McFarland (Ed.) *Motivational control systems analysis*. London: Academic Press. Pp. 283–350.

Van Valen, L. 1974. Molecular evolution as predicted by natural selection. *J. Mol. Evol.* **3**: 89–101.

Van Valen, L. 1976. The Red Queen lives. *Nature* **260**: 575.

Washburn, S. L., & D. A. Hamburg. 1965. The implications of primate research *and* The study of primate behavior. Chaps. 1 & 18 in I. DeVore (Ed.) *Primate behavior*. New York: Holt, Rinehart, & Winston. Pp. 1–13, 607–622.

Weast, R. C., S. M. Selby, & C. D. Hodgman (Eds.) 1964. *Handbook of chemistry and physics*. 45th ed. Cleveland: The Chemical Rubber Co.

Webb, D. J. 1974. The statistics of relative abundance and diversity. *J. Theor. Biol.* **43**: 277–291.

Wecker, S. 1963. The role of early experience in habitat selection by the prairie deer mouse, *Peromyscus maniculatus bairdi*. *Ecol. Monogr.* **33**: 307–325.

Weiss, L., & J. Wolfowitz. 1974. *Maximum probability estimators and related topics*. (Lecture notes in mathematics, vol. 424). Berlin: Springer-Verlag.

Weiss, P. 1941. Self-differentiation of the basic patterns of coordination. *Comp. Psych. Monogr.* **17**: 1–96.

Whimbey, A., & V. Denenberg. 1967. Experimental programming of life histories: the factor structure underlying experimentally created individual differences. *Behaviour* **29**: 296–314.

White, R. E. C. 1971. WRATS: a computer compatible system for automatically recording and transcribing behavioural data. *Behaviour* **40**: 135–161.

Whitman, C. 1919. The behavior of pigeons. *Publ. Carnegie Inst.* **257**: 1–161.

Whittaker, R. H. 1970. *Communities and ecosystems*. New York: Macmillan.

Whittaker, R. H. 1972. Evolution and measurement of species diversity. *Taxon* **21**: 213–251.

Wiener, N. 1948. *Cybernetics*. New York: Wiley.

Wiepkema, P. R. 1961. An ethological analysis of the reproductive behaviour of the bitterling *(Rhodeus amarus* Bloch). *Arch. Neerl. Zool.* **14**: 103–199.

Wiepkema, P. R. 1968. Behaviour changes in CBA mice as a result of one goldthioglucose injection. *Behaviour* **32**: 179–210.

Wiley, R. H. 1973. The strut display of the male sage grouse: a "fixed" action pattern. *Behaviour* **47**: 129–152.

Wiley, R. H. 1975. Multidimensional variation in an avian display. *Science* **190**: 482–483.

Wiley, R. H. 1976. Personal communication.

Williams, W. T. 1971. Principles of clustering. *Ann. Rev. Ecol. Syst.* **2**: 303–326.

Wilson, E. O. 1973a. Group selection and its significance for ecology. *Bioscience* **23**: 631–638.

Wilson, E. O. 1973b. The natural history of lions. *Science* **179**: 466–467.

Wilson, E. O. 1975. *Sociobiology: the new synthesis*. Cambridge, Mass.: Harvard University Press.

Wilson, E. O., & R. Fagen. 1974. On the estimation of total behavior repertories in ants. *J. N. Y. Entomol. Soc.* **82**: 106–112.

Wirtz, J. H. 1967. Social dominance in the golden-mantled ground squirrel, *Citellus lateralis chrysodeirus* (Merriam). *Zeit. Tierpsychol.* **24**: 342–350.

Yadava, R. P. S., & M. V. Smith. 1971. Aggressive behaviour of *Apis mellifera* L. workers towards introduced queens. I. Behavioural mechanisms involved in the release of worker aggression. *Behaviour* **39**: 212–236.

Young, F. W. 1968. A FORTRAN IV program for nonmetric multidimensional scaling. *Report 56*. Chapel Hill, N. C.: Thurstone Psychometric Laboratory.

Young, F. W. 1970. Nonmetric multidimensional scaling: Recovery of metric information. *Psychometrika* **35**:455–473.

Young, F. W. 1972. A model for polynomial conjoint analysis algorithms. In R. N. Shepard, A. K. Romney, & S. B. Nerlove (Eds.) *Multidimensional scaling: Theory and applications in the behavioral sciences*. New York: Seminar Press.

Young, G., & Householder, A. S. 1938. Discussion of a set of points in terms of their mutual distances. *Psychometrika* **3**: 19–22.

Zahn, C. T. 1971. Graph-theoretical methods for detecting and describing Gestalt clusters. *IEEE Trans. Comput. C-20*: 68–86.

Zeeman, E. C. 1976. Catastrophe theory. *Sci. Am.* **234(4)**: 65–83.

Author Index

Subject Index